Fundamentals of Soil Science

Designed as a text book, but equally useful as a reference source for scholars and others, this book offers all the necessary and desired information about soils and their culture. Beginning with classification of soils and their physical and chemical properties, it deals systematically with all such topics as soil acidity, soil moisture, soil organisms, accumulation of organic matter in soils, effect of manures and fertilizers on soil, soil fertility maintenance and development and management of alkali soils. Soil requirements for specific fruit crops have also been discussed.

On the whole the book introduces the reader to soil as natural entities and their inherent characteristics; explains the basic relationship between soils and plants; and gives a clear understanding about the fundamental principles involved in the use of soil management practices.

An exhaustive subject index for easy reference hunting and a detailed glossary of terms are other attractions of the book.

Fundamentals of SOIL SCIENCE

by

C.E. Millar

&

L.M. Turk

BIOTECH

2024

Biotech Books

ISBN13 978-81-7622-061-3

Published by : **BIOTECH BOOKS**
4762-63/23, Ansari Road, Darya Ganj,
New Delhi - 110002
Phones : 011 4300 3222
Email : biotechbooks@yahoo.co.in

PRINTED IN INDIA

ACKNOWLEDGMENT

THE AUTHORS take pleasure in expressing appreciation to *Dr. C. H. Spurway, Professor J. O. Veatch, Dr. G. J. Bouyoucos,* and *Mr. A. H. Mick,* of the Soil Science Department, of Michigan State College, for much valuable assistance in the preparation of this book.

PREFACE

Fundamentals of Soil Science was written for use as a college textbook, as a reference book for farmers and owners of farm lands, and as an aid to anyone desiring information on soils and their culture.

The purpose of the book is fourfold: first, to give the reader the opportunity to become familiar with soils as natural units or entities and with their inherent characteristics; second, to develop in the student an understanding of the significance of fundamental soil properties; third, to set forth basic relationships between soils and plants; and last, to give the reader an understanding of the principles involved in the use of proved soil-management practices.

General principles of soil science are emphasized and explained in simple terms, omitting, so far as possible, technical discussions, particularly of debatable points. Details of practice have been omitted except where their inclusion is necessary for clarity.

It has been necessary to go somewhat into the fields of chemistry, botany, geology, physics, microbiology, crop production, and agricultural engineering because soil science is closely related to them all.

C. E. Millar
L. M. Turk

East Lansing, Michigan
January, 1943

CONTENTS

CHAPTER I

SOIL DEVELOPMENT

Soils are developed; they are not merely an accumulation of debris resulting from decay of rock and organic materials. Soil formation is, accordingly, a constructive as well as a destructive process. Destructive forces predominate in the breaking down chemically and physically of minerals and plant and animal structures, with the resultant partial loss of the more soluble and volatile products. Particles of the original materials in various stages of decomposition also remain, together with newly developed compounds. The constructive forces develop new chemical compounds, both mineral and organic, and provide a new distribution or association of the materials so that the resultant soil has a characteristic structural and textural as well as chemical composition which influences plant growth. In other words, a soil is an entity—an object in nature which has characteristics that distinguish it from all other objects in nature.

There are two major divisions of soils: (1) those composed largely of mineral material with comparatively small quantities of organic matter; and (2) those in which there is such a large proportion of organic matter that the characteristics of this constituent dominate the properties of the soil. The latter are designated as organic soils or muck and peat. In discussing the subject of soil development it is helpful to keep in mind certain definite objectives.

Objectives:

A. Sources of materials from which soils are developed.
B. Characteristics of rocks and minerals which decay during soil formation.
C. Chemical and physical processes active in soil formation.
D. Biological agencies which aid in producing soils.
E. Products and results of mineral-decomposing processes.
F. Constructive process of soil development.
G. The soil profile.

SOURCES OF MATERIAL FROM WHICH SOILS ARE DEVELOPED

Before the forces of soil development can be operative, it is necessary that raw materials on which they can work be present. Let us first study, therefore, the sources of the raw or parent materials from which soils are made.

Mineral soils are often classified according to the origin and nature of the materials in which they have developed. This is not a good method of *soil classification* because the true characteristics of the soil itself, such as texture, color, and structure, are either partially or wholly ignored. A better classification (p. 45) is founded only on inherent soil properties. Nevertheless, a classification of parent materials is useful in defining technical and practical terms that are universally employed. Many soil characteristics are, moreover, either directly or indirectly influenced by the parent material; so there is some justification for the use of terms such as "glacial soils," "sedentary soils," or "alluvial soils"; but when using these terms their true meaning, that is, "soils developed in ice-transported materials," "soils developed in residual materials," or "soils developed in water transported materials" must be kept in mind.

In considering the sources of soil material, a number of questions suggest themselves.

Questions:

1. What is meant by "residual" and "transported" soils?
2. What differences in the rocks of "residual" areas influenced the nature of soil developed from them?
3. What agencies were active in transporting soil material?
4. In what forms were glacial materials deposited?
5. What differences were there in materials deposited by ice?
6. Rivers, lakes, and oceans played what part in accumulating parent materials?
7. What kinds of parent material were deposited by wind?

Residual and Transported Materials. Soils have developed in materials that resulted from the decay of hard rocks or of unconsolidated sediments raised above ocean level and thus exposed to soil-forming processes. Where the products of decay have not been moved from the site of accumulation but remain in their original location, the soils produced in them are often loosely termed *residual soils;* they are more awkwardly, but nevertheless more correctly, designated as "soils developed in residual materials." On the other hand, many soils have developed from rocks, sand, silt, and clay, which had been moved from their original sites. Such soils are commonly designated as *transported soils;* more correctly speaking, they are soils that have developed in transported parent materials, since it is impossible for a soil to be naturally transported in toto. Soils developed in residual materials are sometimes described as "old" because a very long time

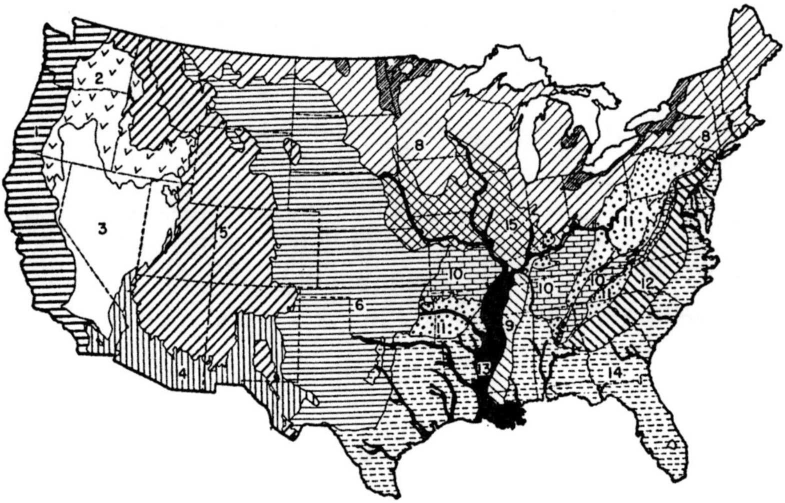

FIG. 1. A generalized map showing the distribution of areas of soil material classified on a geological basis. [Drawn from a map prepared by the Division of Soil Survey and presented in Bul. 96, U.S.D.A. For a more detailed map showing the distribution of these materials and giving more information concerning their nature refer to Plate 4, Atlas of American Agriculture, Part III.]

LEGEND OF AREAS

1. Pacific coast region. 2. Northwest intermountain region. 3. Great Basin region. 4. Southwest arid region. 5. Rocky Mountain region. 6. Great Plains region. 7. Glacial lake and river terraces. 8. Glacial region. 9. Loessial deposits. 10. Limestone valleys and uplands. 11. Appalachian Mountains and Plateaus. 12. Piedmont Plateaus. 13. River flood plains. 14. Atlantic and Gulf coastal plains. 15. Loessial deposits over glacial material.

is required for the weathering forces of nature to convert rocks into soil material and for soils to develop in this material. A much shorter time is required for soil-forming forces to produce soils in transported sediments, and hence transported soils are often designated as "young." The general distribution of residual and transported soil material in the United States is shown in Fig. 1.

Differences in Residual Parent Materials. Since they have not suffered the mixing that accompanies transportation by ice and water there are many variations in the characteristics, both physical and chemical, of sedentary materials. Indeed, there are nearly as many variations in these residual materials as there were variations in the original rocks from which they were derived. Igneous, sedimentary, and metamorphic rocks all served as sources; and even within these broad groups there are, of course, many differences. Sedimentary rocks

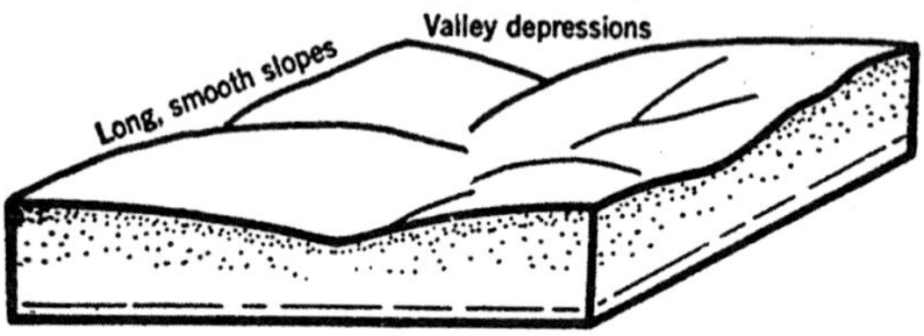

FIG. 2. A diagrammatic presentation of the mature topography of a residual soil area. The long smooth slopes and moderate valley depressions are characteristic. (Compare with Fig. 3.)

(rocks formed from sediments deposited in bodies of water and more or less solidified) exhibit a great variation in composition. Sandstones usually are high in content of quartz and other minerals rich in silica. The cementing materials frequently are compounds of iron with a calcium carbonate content varying from zero to a high percentage. Clay sediments have been converted into shale and slate, which may contain either no calcium carbonate or a considerable

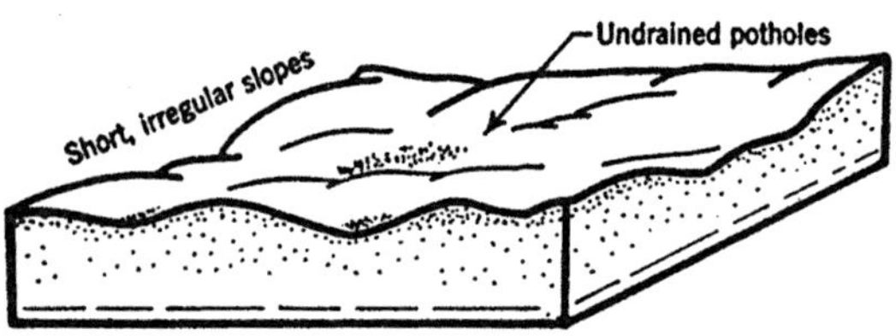

FIG. 3. Glaciated areas are geologically young and present a landscape with short, irregular slopes and small, poorly drained depressions known as "pot holes." (Compare with Fig. 2.)

amount. Some shales also contain appreciable quantities of potassium compounds and other salts.

Deposits of calcium carbonate and magnesium carbonate may be converted into soft, unconsolidated, or slightly consolidated materials, termed marl, or into limestone of varying degrees of hardness containing varying percentages of impurities like iron oxide, sand, or clay. Some limestones contain practically no magnesium carbonate while others may contain up to approximately 45 per cent and are termed dolomite. Some limestone also contains appreciable quantities of phosphorus compounds, as in the Bluegrass region of Kentucky, in parts of Tennessee, and in Florida.

Igneous rocks, those which have been molten, likewise are found in a great variety of physical and chemical forms. Granites, felsites, gabbros, basalt, and many others are combinations of silicon with varying proportions of aluminum, calcium, potassium, magnesium, and sodium. Iron, too, is an important element in some, whereas it is practically absent in others. The decay of igneous rocks may produce residues either high in calcium, magnesium, and other basic elements or low in such constituents and high in acidic siliceous compounds. Some igneous rocks undergo rapid physical disintegration to produce sandy or gravelly residues, whereas other kinds decay into silt and clay, thus giving rise to fine-textured soils.

There are then many wide differences in soil parent materials of the residual group. These differences influence to a greater or lesser extent the characteristics of the soils that develop in them. Great differences and extreme variations are, therefore, common in soils developed from these residual materials.

Transported materials are separated into four groups on the basis of the transporting agency involved, namely, ice, water, wind, and gravity. Hence there are four kinds of transported parent material: glacial, alluvial, aeolian, and colluvial.

Glacial Materials. Ice was the transporting agency for most of the surface mantle in northeastern United States and in the upper Mississippi valley. The Ohio and Missouri rivers form a general southern boundary for ice-carried material. As the great continental ice-sheets moved southward from their accumulation centers in the Canadian plateaus, they first followed and filled the great drainage valleys and then gradually spread out over the intervening upland and divides. As the ponderous ice-mass moved forward it pushed before it and gathered within itself a large part of the unconsolidated surface layer. It also scooped up great rock fragments which scraped at the rock floor over which they passed. Sharp corners and edges of even the hardest rocks were ground smooth by this abrasive action to form the rounded rocks and boulders that are characteristics of glaciated landscapes. Large quantities of weathered and unweathered rocks, varying in size from fine rock powder to massive boulders, were thus incorporated into the ice and carried along in the glacier.

The movements of this continental ice-sheet depended upon the climatic conditions prevailing at that time. During mild periods the ice melted rapidly. In cold seasons melting ceased and the ice front would creep southward. Sometimes during extremely mild periods the ice would melt faster than it was pushed forward. This would lead to a rapid recession of the ice front, and all debris carried in the ice was, of course, dropped. Generally, after this type of recession, the land surface appeared as a rolling plain, called a *till plain* or *ground moraine*. At certain times climatic conditions allowed the glacier to melt back just as fast as its rate of advance, which resulted in the front of the ice remaining at one place for some time. All debris

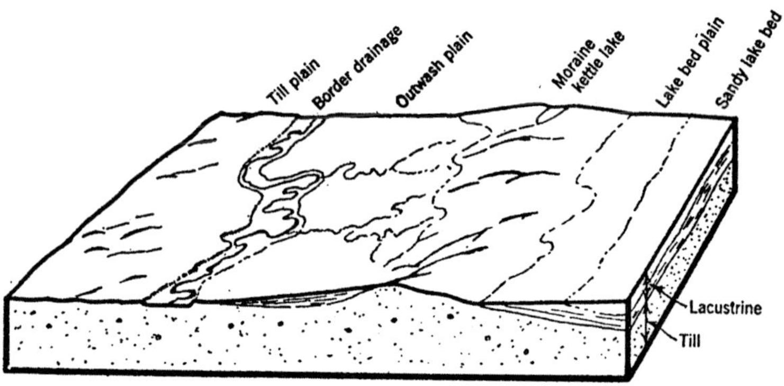

FIG. 4. A generalized view of physiographic features in a glaciated area. In the left is a till plain separated from the moraine by an outwash plain and a border-drainage way developed while the ice front was at the moraine line and melting as rapidly as it advanced to deposit the morainic material. To the right a glacial lake has receded to leave a plain of heavy sediment partially covered with lacustrine sands (deposit nearer the water's edge).

carried by the ice was brought to the line of the stationary ice-front and there dumped as melting proceeded. This process resulted in the formation of ridges or a series of hills, called *terminal* and *recessional moraines*. Moraines may be formed from an unassorted, heterogeneous mass of boulders, rocks, sand, silt, and clay, briefly called *till*, but in places a water-sorting also occurred. For the greater part, the glacial drift is a mixture of non-uniform materials that range from the finest particles to large boulders. Not only does the material vary in texture but the topographical features are also variable.

As the ice melted, giving rise to moraines, great volumes of water rushed away from the glacier front. These streams carried quantities of sediment which were dropped as the current diminished to form coarse-textured *out-wash plains* which are generally associated with moraines. *Kames* and *eskers* were also formed but they are only of local significance. Most of the finer silt and clay was carried into lake basins where it settled in quiet water to form lake-bed or *lacustrine plains.*

Minor readvances of the ice-sheet resulted in the modification of these glacial formations in many places. Sometimes till was relaid over a lake plain or sandy and gravelly out-wash was pushed up to form gravelly moraines. Major readvances "streamlined" glacial drift in some parts of the country to form *drumlins.* In general, the effect of glaciation was to scrape off and smooth the tops of high land forms while filling valleys and depressions. Thus continental glaciation decreased the local relief intervals and flattened topography.

Differences in the Composition of Glacial Drift. The chemical composition of the debris carried by ice-sheets depends on the nature of the rocks over which they passed. Glaciers that originated near Hudson Bay gathered grayish-colored till as they moved southward. Glaciers that came from north and west of Lake Superior gathered boulder clay that was high in iron-bearing minerals and red in color. Glacial formations, such as moraines, till plains, and lake deposits, produced by these two different glaciers therefore show many differences in chemical composition. In some places these two kinds of till, even though from distant sources, are closely associated; red materials, for example may be directly on top of the gray drift of an earlier ice invasion.

The chemical composition of glacial drift often was modified by being mixed with an earlier, more weathered material. While some portions were covered by only one ice-sheet, most of the glaciated area was invaded at least twice and in places as many as five times by major continental glaciers. Between invasions there elapsed long periods of mild weather, when vegetation flourished. During these periods soils developed in the glacial till. The next advance of the glacier scraped off these soils and redeposited them with other fresh debris. The old material, crushed and ground anew and mixed with rock powder of more recent origin, supplied a large amount of fresh, undecayed mineral surface, potentially rich in plant nutrients. The resulting till, however, differed from the old drift as well as from the new unmixed boulder clay, thus introducing another variation in the chemical differences of glacial materials in general.

Within the area covered by drift from a single source there is, moreover, considerable variation in physical as well as chemical properties. Indeed, chemical and physical differences are closely associated, the chemical differences generally originating in the mechanical separation by sedimentation of the various mineral components of boulder clay because of their different physical characteristics. For example, rocks containing a high proportion of quartz are hard and do not break down as readily as minerals containing more calcium, magnesium, potassium, sodium, or iron. It follows that sands and gravels, being the coarse components of drift, must also be the siliceous component. Clays, on the other hand, are the least siliceous part of boulder clay. When glacial streams, then, segregate the sand and gravel in out-wash plains and the clays in lacustrine plains, they are at the same time separating the siliceous minerals from the basic minerals. But even within out-wash plains or other formations there is variation in the content of limestone and soft minerals.

Glacial materials vary, then, not only in texture but also in chemical composition. This variation is due in part to the nature of the sources of the material. Other factors furthering this difference are the method of transportation, the many readvances of the continental glaciers, and finally the reworking and sorting by glacial streams. A common characteristic of all glacial parent materials is the extreme variability which naturally leads to differences in the soils which develop from them.

Alluvial Materials. Alluvial parent materials are scattered in narrow, irregular strips bordering streams and rivers throughout glacial

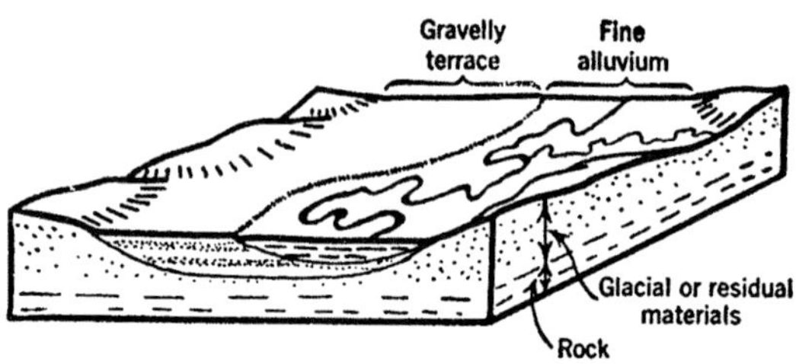

FIG. 5. A meandering stream is reworking previously accumulated soil material and has deposited fine alluvium in the flood plain. When the stream was younger and was cutting its channel deeper, it deposited coarse gravelly material during flood periods; this material now appears as a terrace.

as well as residual areas. The largest uniform body of alluvium is located in the southern Mississippi valley (Fig. 6). Alluvial material that has been transported by water ranges in texture from fine clay to large rocks and even to boulders. The fine sediments are washed into the drainage valley from the watershed surface. When the current is swift the carrying capacity of the stream is high, but as the rate of flow decreases first the coarse and then the fine materials settle out. A common characteristic of all alluvial material is stratification, layers of different-sized particles overlying each other.

Most alluvium is carried and deposited during floods because it is at this period that erosion is most active and the carrying capacity of streams is at a maximum. When a flooding stream overflows its banks, its carrying power is suddenly reduced as the flow area increases and

FIG. 6. Soils developed from alluvial materials of the Mississippi delta are naturally fertile but require good management to maintain their supply of active organic matter. This is a winter cover crop of burr clover in a peach orchard. [Courtesy of Delta Experiment Farm, Stoneville, Mississippi.]

velocity decreases. This causes the coarse sands and gravels to settle along the bank where they sometimes form conspicuous ridges called *natural levees.* As the water reaches the *flood plains* of the valley, the rate of flow is slow enough to permit the silt to settle. Finally the water is left in quiet pools, where it seeps away or evaporates leaving the fine clay. Levees are characterized by good internal drainage during periods of low water, whereas flood plains exhibit poor internal drainage.

Large mature valleys like the Mississippi are characterized by rivers with many meanders which develop in a well-recognized cycle. As a stream swings back and forth across its flood plain, cutting new channels in flood periods, sections of the old bed, known as ox-bows, are left as lakes and swamps which fill with sediments during subsequent overflows.

Terraces are developed from flood plains as streams cut deeper channels, owing to lowered outlets. Several terraces may be found along a stream or a lake which has undergone repeated changes in level. Terraces usually are quite well drained and may be droughty. They exhibit stratification.

Streams flowing from hills or mountains into dry valleys or basins drop their sediments in a fanlike deposit as the water spreads out at their debouchures. These *alluvial fans* are usually coarse textured, being composed of sands and gravels, and are well or excessively drained.

Sediments not deposited as flood plains are carried to the lake, gulf, or other body of water into which a stream empties. The decrease in velocity at the stream's mouth together with the coagulating effect of the salt content of the receiving water body results in the deposition of much of the suspended material thus producing a delta. These deposits are poorly drained but, where drainage is provided, they constitute important crop-producing areas, as is evidenced by the deltas of the Nile, Po, Tigris, Euphrates, and the Mississippi.

Flood plains as well as deltas are in general rich in plant nutrients and comparatively high in organic matter content. Terraces and alluvial fans, on the other hand, are more likely to be less fertile. Special crops such as vegetables and fruits frequently are grown on the latter formations because the soil warms up quickly and their good drainage and coarse texture permit free root development.

Lake-Bed Deposits. Mineral sediments were deposited in lakes by streams and by glaciers, while organic materials have accumulated through the growth and partial decay of plants. Changes in lake outlets and a diminished supply of water have lowered water levels and in many places have resulted in the disappearance of entire lakes, thus exposing the lake-bottom sediments. In Michigan there is evidence of as many lakes which have become extinct as there are lakes in existence, whereas along the Great Lakes extensive areas have been exposed through the disappearance of ice barriers, lowering of the outlet at Niagara Falls, and tilting of the land surface. The disappearance of Lake Agassiz has laid bare a great land surface in Minnesota and the Dakotas, and the basin of old Lake Bonneville occupies an immense area in Utah.

Soils in lake-bed areas include lacustrine sediments which have been deposited from suspension in water, glacial formations which have been leveled off by water action, sand bars, deltas, and alluvial fans of glacial streams, terraces, and sandy ridges marking the shore lines formed as the lakes receded by stages. In general, lake-bed soils in humid regions are fine textured, comparatively level, high in organic matter content and poorly drained, both internally and externally. The climatic conditions of the Lake Bonneville basin, however, have resulted in the formation of arid soils, and hence they vary greatly from those of the lake-bed areas in the eastern part of the United States.

Soil Material of Ocean and Gulf. Sediments carried by streams to oceans and gulfs are deposited through decreased current velocity combined with chemical coagulation. Extensive areas of sands are interspersed with beds of silt and clay deposited in estuaries, other sheltered bodies of water, or further out in the ocean depths. When raised above sea level, these deposits are subjected to soil-forming processes and become in time a part of our cultivated land area.

The source of these materials as well as the decomposing and erosive processes to which they were submitted before deposition influences their chemical and physical properties. The solvent action of ocean water, coupled with the ceaseless grinding by waves, pulverized soft minerals and extracted soluble salts. Finally the forces of decay and disintegration acted on the deposits after exposure above water level. Large areas of these lands, particularly the more sandy, are composed of particles low in plant nutrient content and high in hard minerals like quartz. Such soils require generous fertilization and careful management for the production of crops. Large acreages are nevertheless successfully used for the production of vegetables, fruits, and other special crops. The high rainfall of the Atlantic coastal region contributes to their usefulness.

Considerable acreages of loams, silt loams, and clay loams also occur, especially in the Gulf coastal plains of Texas and Louisiana. General farming as well as the production of large quantities of special crops is extensive on these soils.

Wind-Transported Soil Material. There are three distinct classes of wind-moved soil material: (1) sand of variable fineness which may be collected into low swells or steep ridges, as is true of the dunes occurring on the leeward side of large bodies of water and on sandy deserts; (2) volcanic ash, deposits of which are found in Kansas, Nebraska, and Montana; and (3) siltlike material, called *loess*, which occupies large areas in the United States, Europe, and Asia.

Dune sand is of little agricultural value although crops are produced on it to a limited extent in humid regions. At times dunes are a hazard to agriculture, for in their movement they sometimes gradually cover good land.

Soils derived from *volcanic ash* are of little agricultural value in the United States. They are very porous and light in weight.

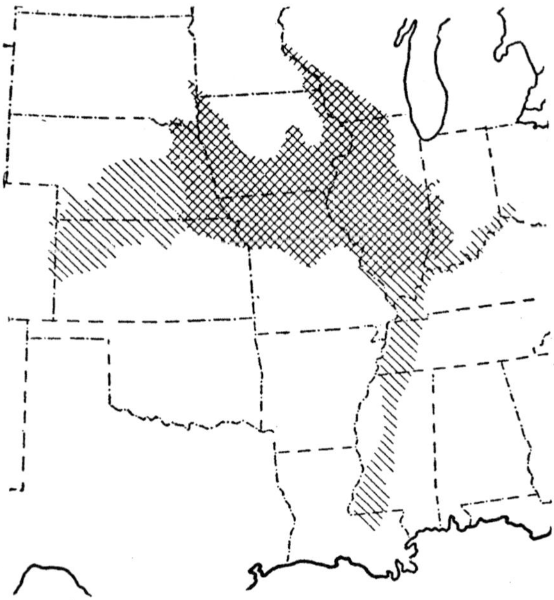

FIG. 7. Map showing the general distribution of loessial material in central United States. Another area of appreciable size lies in southeastern Washington and adjoining parts of Oregon and Idaho. The area in which loess overlies glacial deposits is depicted by crosshatching.

Loess was deposited in central United States after the recession of the ice-sheet. This material was derived in part from sediments deposited by huge rivers which were fed by the melting continental glaciers in a broad belt, even beyond the southern limits of glaciation. A period of aridity after the recession of the glaciers with strong westerly winds set the stage for the transportation of this wind-blown material to its present resting place, as shown in Fig. 7. The great thickness of the deposits of loess on the east and northeast banks of the Mississippi and Missouri rivers is one of the facts which has led to this explanation of the accumulation of the material. Glaciers from the Rockies probably supplied the sediments making up the western part of the loessial deposits.

Extensive deposits of loess are also found along the Rhine and its tributaries and over a large part of the immense valley of the Hwangho. Other deposits occur in southern Russia, several of the Balkan countries, and northern France, Belgium, and Poland.

FIG. 8. Loessial material has the characteristic of standing in almost vertical cliffs, as is shown by this eroded plain. Level or gently sloping areas are used for crop production (foreground). [Sketched from a photograph in Thorpe's "Soils of China." The National Geological Survey of China, 1936.]

Loess is composed largely of silt having a grayish-yellow or buff color. This wind-blown material has the property of standing in almost vertical walls so that gullies and streams cutting through it have very steep banks. Its content of mineral plant nutrients was originally high, as was the quantity of calcium compounds, and the soils derived from it are generally productive. In fact, they constitute some of the most fertile land of this and other countries.

CHARACTERISTICS OF ROCKS AND MINERALS FROM WHICH SOILS ARE DERIVED

Soils are composed very largely of fragments of rocks and minerals and of the products of their decomposition. A student should know at least some of the elementary facts concerning the composition and properties of the rocks and minerals found most abundantly in soils. Three questions serve as helpful guides in the study of soil minerals.

Questions:

1. Minerals represent what kinds of chemical compounds?
2. Into what groups are soil minerals divided?
3. What are the chief characteristics of the mineral groups?

Chemical Composition and Characteristics of Rocks and Minerals. Combinations of oxygen with silicon and iron and in some cases with water also give rise to such compounds as SiO_2 and $Fe_2O_3 \cdot 3H_2O$ and

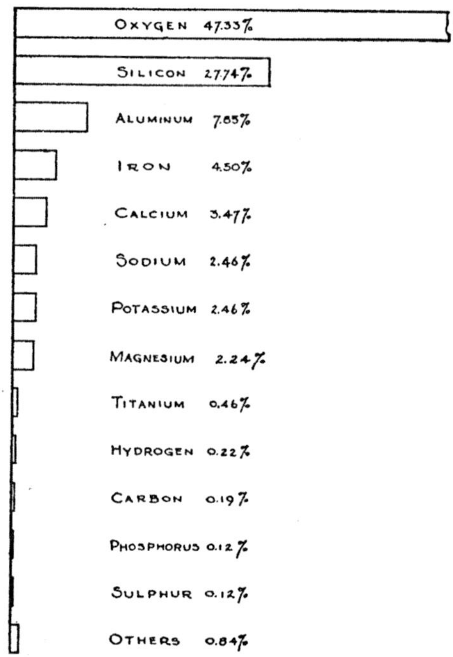

FIG. 9. Composition of the solid crust of the earth (lithosphere) to a depth of 10 miles. Note relatively small percentages of several elements essential for plant growth. [From Bul. 695, U.S.G.S., by F. W. Clarke.]

constitute one group of minerals called oxides. Basic compounds like $Ca(OH)_2$ and $Mg(OH)_2$ combine with CO_2 to form another mineral group very important in soil formation, known as carbonates. By far the largest number of minerals belong to the silicate group, in which Ca, Mg, Na, K, Fe, and Al combine with the various silicic acids like orthosilicic (H_4SiO_4), metasilicic (H_2SiO_3), and polysilicic

($H_4Si_3O_8$). At times the Al is a part of the negative radical forming aluminosilicates. Many of these compounds and the products formed from their decomposition take on water, that is, they become hydrated.

The name, chemical composition, and some information as to occurrence, properties, and decomposition products of some of the most important and widely distributed minerals and rocks are listed below:

Oxide Group		
Quartz	SiO_2	Very resistant to change; highly insoluble; very hard.
Hematite	Fe_2O_3	Red; hydrates with increase in volume to form $2Fe_2O_3 \cdot 3H_2O$.
Magnetite	Fe_3O_4	Dark brown; magnetic, resistant to change; not of wide occurrence.
Limonite	$2Fe_2O_3 \cdot 3H_2O$	Yellow; soft; resistant to change except in degree of hydration; widely distributed.
Other hydrated oxides of Fe	$2Fe_2O_3 \cdot H_2O$ $Fe_2O_3 \cdot H_2O$ $Fe_2O_3 \cdot 2H_2O$	Color varies from yellowish red to yellow with increase in hydration. Are supposed to account for red and yellow colors in soils. Are resistant to change other than in degree of hydration. Are of wide occurrence.
Carbonate Group		
Calcite	$CaCO_3$, quite pure.	Well crystallized. Goes into solution readily as $Ca(HCO_3)_2$.
Limestone	$CaCO_3$ varies greatly in purity. Main impurities are Fe_2O_3, SiO_2, $MgCO_3$, and clay.	Is easily crushed. Goes into solution readily as $Ca(HCO_3)_2$, leaving impurities. Of wide occurrence and great importance in soil formation.
Dolomite	$CaCO_3 \cdot MgCO_3$, proportion of two varies within limits. Impurities as in limestone.	Harder than limestone and decomposes with greater difficulty. Goes into solution slowly as $Ca(HCO_3)_2$ and $Mg(HCO_3)_2$, leaving impurities. Rather widely distributed.
Magnesite	$MgCO_3$, purity varies.	Not of wide occurrence.
Marl	$CaCO_3$, impurities are mud, organic matter, sand, $MgCO_3$. Purity varies widely.	An unconsolidated to somewhat solidified form of limestone.

Carbonate Group		
Calcareous clays, shales, and sandstones	Content of $CaCO_3$ ranges from a trace to several per cent.	Important in soil formation in some glaciated regions and areas composed of sedimentary rocks.
Feldspar Group		
Microcline and orthoclase	$KAlSi_3O_8$	Orthoclase occurs in large quantities as a constituent of granite and gneiss.
Labradorite Albite Anorthite	$NaCaAl_2Si_6O_{16}$ $NaAlSi_3O_8$ $CaAl_2Si_2O_8$	The sodium and calcium feldspars form a part of many crystalline rocks like basalt, diabase, gabbro, diorite, and many lavas. The feldspars decompose fairly readily in the presence of H_2O and CO_2. The Ca goes to the carbonate or more soluble bicarbonate and the Na and K to carbonates. Some SiO_2 is split off and may remain as insoluble SiO_2 or form silicic acids and soluble or colloidal hydrated silicates. Most of the Al and much Si go into the formation of hydrated aluminosilicates, of which kaolinite ($H_2Al_2Si_2O_8 \cdot H_2O$) is an example. The exact hydrated aluminosilicate formed depends on the conditions prevailing during decomposition. Compounds of this nature are important constituents of clay and play prominent roles in determining the properties of soils and the reactions taking place in them.
Micas		
Muscovite (white) Biotite (dark)	$H_2KAl_3Si_3O_{12}$ $(HK)_2(MgFe)_2$-$(AlFe)_2Si_3O_{12}$	Break up into fine glittering scales which are sometimes mistaken for gold when yellow. When much iron is present in biotite, it decomposes readily; otherwise the micas break up physically quite easily but decompose chemically very slowly. They are abundant ingredients of many rocks.

Other Silicates		
Amphiboles (hornblende)	$Ca(MgFe)_2Si_4O_{12}$ with $Na_2Al_2Si_4O_{12}$ and $(MgFe)_2(AlFe)_2$-Si_2O_{12}	The minerals are dark green in thin sections but appear black as they occur. Hornblende decomposes fairly readily because of its easy cleavage and high content of ferrous iron. On decomposition it produces a rust-colored clay which is a constituent of many red soils. Augite does not cleave as easily and often occurs as "black gravel" in soils.
Pyroxenes (augite)	$CaMgSi_2O_6$ with $(MgFe)(AlFe)_2Si_2O_6$ There is considerable variation in the proportions of the different elements especially of Ca, Mg, Al, Fe.	
Olivine	$MgFeSiO_4$	Decomposes in presence of moisture to form serpentine ($H_4Mg_3Si_2O_9$), liberating SiO_2 and FeO.
Leucite	$KAlSi_3O_{12}$	A constituent of many volcanic rocks but otherwise not an abundant mineral. It is of interest because of its high K content.
Talc (soapstone) and Serpentine	$3MgO \cdot 4SiO_2 \cdot H_2O$ $3MgO \cdot 2SiO_2 \cdot 2H_2O$	These two hydrated silicates of magnesium are found extensively in some regions. Both are soft, but serpentine in the absence of definite structure disintegrates slowly unless considerable ferrous iron is present. Talc cleaves readily and is very soft; so disintegration is accelerated. Both give rise to rather infertile soils.
Sulphides		
Pyrite and Marcasite	FeS	These minerals are yellow, often mistaken for gold when small fragments are mixed with sand or gravel. They oxidize under suitable conditions, forming H_2SO_4. They are common constituents of rocks in some areas.
Phosphates		
Apatite	$Ca_3(PO_4)_2 \cdot Fl \cdot Cl$	A dark-brown mineral widely distributed but in small quantities. It is the original source of the phosphorus found in other combinations in the soil.

Igneous Rocks

Igneous rocks represent a molten mass which was submitted to more or less rapid cooling conditions. With slow cooling, deep in the earth, large crystals of the mineral components developed, giving rise to the intrusive or plutonic rock types. Rapid cooling near or on the earth's surface resulted in small crystals or else in a glassy mass. They constitute the extrusive or eruptive types. The mineral compounds have had no opportunity to be acted upon by weathering agencies and so become somewhat adjusted to their new environment. As a result, many of the chemical combinations are very unstable and are easily broken down into more change-resistant compounds.

Granites (intrusive) Rhyolites (extrusive)	Chief mineral constituents are quartz, orthoclase, and occasionally plagioclase. One or more minerals of the mica, hornblende, or pyroxene group are usually present with minute particles of several other minerals.	Members of this group are the most acid, that is, they contain the highest percentages of silica of all igneous rocks. The mineral crystals may be very small or several inches in length in different kinds of granites. In the rhyolites they are small. Color varies from gray to shades of green, yellow, pink, and red. These are among the most common of igneous rocks.
Syenites (intrusive) Trachytes (extrusive)	Differ from granites essentially in the absence of quartz.	Syenites and trachytes are intermediate in acidity between the granites and basalts.
Diorites (intrusive) Andesites (extrusive)	Differ from previous groups in the absence of potash minerals like orthoclase. Principal constituents are plagioclase and hornblende.	Diorites and andesites are of wide distribution. Color varies from green to almost black.
Gabbros (intrusive) Basalts (extrusive)	Gabbros and basalts contain primarily a Na·Ca feldspar, one or more pyroxenes, and sometimes olivine. Apatite, mica, and hornblende are common constituents.	These rocks contain less SiO_2 and more Na, Ca, and Mg than members of preceding groups. They are known as basic rocks. Because of their low silica content and high content of basic elements, the products of decomposition are basic. They are dark colored, running from gray or greenish to black, with browns and reds. Basalts are widely distributed among the younger volcanic rocks. In the United States they are found mainly west of the Mississippi.

Igneous Rocks		
Glasses	These rocks cooled so rapidly that there was no opportunity for crystallization, and hence they are amorphous.	Similar to granite in chemical composition.

CHEMICAL AND PHYSICAL PROCESSES ACTIVE IN SOIL DEVELOPMENT

Having studied briefly the minerals which furnish the raw material from which soils are developed, one wonders as to the forces which are active in converting this material into soil. It will be helpful to keep the following questions in mind while studying this section.

Questions:

1. What are the agencies which break down minerals?
2. How do the forces of disintegration function?
3. What chemical actions decompose rocks and minerals?

Agencies of Mineral Breakdown. Minerals and rocks may change in size without appreciable alteration in chemical composition. As smaller-sized particles are produced, this change is called disintegration. Not only are the larger pieces of rocks and minerals altered in this manner, but even very small soil particles are broken down, sometimes much more readily than larger fragments. The forces which may cause mineral breakdown without alteration in chemical composition are designated as physical forces and include wind, water, ice, temperature change, gravity, mechanical actions of animals, and mechanical forces of growing plants.

The chemical decomposition of minerals is almost invariably accompanied by a change in size of particle. Agencies bringing about chemical breakdown of minerals include oxygen, water, carbon dioxide, organic and inorganic acids, bases, and salts.

The process of physical and chemical breakdown of rocks and minerals of all sizes, from very large to minute particles, is given the general name of "weathering," and the agencies which bring about the changes are designated "forces of weathering."

Disintegration Processes. An illustration of gravity as a force of weathering is the disintegration of rocks and mineral material as they fall or slide down mountain slopes and precipices. Gravity is thus the force involved in grinding colluvial material.

Another grinding force involved in the weathering of soil materials is the abrasive action of wind-borne sand and silt which cannot be appreciated by a person who is unfamiliar with sand and dust storms. Windshields of cars driven continuously in desert regions soon become so etched that replacement is necessary. There are many instances of rock walls and pillars which have been cut into unusual shapes by wind-driven mineral particles. The transportation of loess must have been accompanied by much grinding of the material itself.

FIG. 10. Exfoliation or scaling off of the surface of rocks in the Mohave Desert due to changes in temperature.

Stream-carried sediments are effective cutting agents, as is demonstrated by innumerable canyons, gulches, and other channels through rocky formations. Rocks and gravel, rolled along the bottom, as well as suspended material are ground finer as a result of stream flow. Water in itself has considerable abrasive power, but this is rather small compared to the cutting power of water-borne mineral particles.

In the movement of glaciers there is much grinding of the floor over which the ice moves and some crushing and pulverization of the transported material. The heavy load of rock flour carried by streams flowing from present-day mountain glaciers bears testimony to this grinding action and suggests, at least, the enormous forces that were involved in continental glaciations.

The penetration of roots, particularly small ones, through crevices in rock fragments, enlarges the cracks and gradually splits the rocks as growth continues. Burrowing animals also play a minor part in breaking up rock and mineral constituents of soil materials.

Changes of temperature disrupt rocks by causing repeated expansion and contraction of the constituent minerals. This action is especially effective in arid climates and high altitudes where changes of temperature between night and day are very great. Rocks composed of decidedly unlike minerals, like the granites, are especially susceptible to disintegration since the various minerals have unequal coefficients of expansion, and hence the rock is literally pulled to pieces by repeated heating and cooling. In some arid regions disintegrated granite is used for the same purposes as gravel in more humid climates.

Rocks are disrupted also through changes in temperature by having the outer shell expand more rapidly than the interior when warmed by the morning sun. During the day the rock becomes more or less warmed throughout and hence expanded. At night the outer portion contracts rapidly, owing to cooling after sunset, and hence the shell becomes too small for the interior. Repetition of these changes results in the cracking or scaling off of the rock exterior, as is shown in Fig. 10.

The expansive power of water upon crystallization is another force in disrupting rocks. As small cracks occur and crevices develop through solution or disintegration of compounds of calcium, iron, magnesium, and other basic elements, they are filled with moisture which alternately freezes and thaws, thus gradually enlarging the opening. Some rocks are so porous that they absorb sufficient quantities of water to shatter them upon repeated freezing.

Chemical Processes of Mineral Decay. In humid climates where moisture is present in the soil for long periods of time, chemical forces as well as physical forces are important agents in rock decay. In order for chemical forces to be effective, water, of course, must be present. Combination of mineral constituents, especially iron, with oxygen is an ever-present possibility. The transformation of ferrous salts to ferric compounds continues as fresh surfaces of iron containing minerals are exposed. A simple example of this kind of *oxidation* is as follows: $2FeO + O = Fe_2O_3$.

The union of water with soil minerals may be illustrated by the *hydration* of iron oxide thus: $2Fe_2O_3 + 3H_2O = 2Fe_2O_3 \cdot 3H_2O$. Hydration is a common process in rock decay. Not only does the iron in minerals oxidize and become hydrated, thus increasing in size and softness, and so lead to further mineral breakdown, but also many other compounds like SiO_2 and Al_2O_3 and aluminosilicates as well as organic substances become hydrated.

Minerals containing strongly basic elements are inclined to exchange a part of the base for hydrogen from water, thus:

$$CaSiO_3 + 2HOH = H_2SiO_3 + Ca(OH)_2$$

or

$$KAlSi_3O_8 + HOH = HAlSi_3O_8 + KOH$$

This process is known as *hydrolysis* and is a very effective reaction in changing the chemical composition of many minerals.

The decomposition of organic matter in the soil and the respiration of plant roots liberate large quantities of CO_2 into the soil air. This gas readily combines with bases, producing carbonates or bicarbonates. For example, the $Ca(OH)_2$ and KOH liberated by the hydrolysis reactions shown above readily combine with CO_2, as follows:

$$Ca(OH)_2 + 2CO_2 = Ca(HCO_3)_2$$

and

$$2KOH + CO_2 = K_2CO_3 + H_2O$$

This process, known as *carbonation*, is one of the most effective in decomposing soil minerals. As the bicarbonates of calcium and magnesium are fairly soluble, carbonation is the primary process in the accumulation of residual soil material through decomposition of limestone and in the decomposition of minerals with a high calcium content.

The soil air contains approximately ten times as much CO_2 as atmospheric air. Furthermore, the film of moisture lying immediately next to the surface of the fine soil particles is highly charged with this gas. With this intimate association of oxygen, water, and carbon dioxide in contact with mineral particles, it cannot be expected that each will act independently on the soil minerals. On the other hand, oxidation, hydration, and carbonation usually proceed jointly or in such close association that the individual actions cannot be readily discerned. An example of the combined action of carbonation and hydrolysis is illustrated by the equations

$$\underset{\text{orthoclase}}{2KAlSi_3O_8} + CO_2 + 2H_2O = \underset{\text{kaolinite}}{H_4Al_2Si_2O_9} + 4SiO_2 + K_2CO_3$$

$$\underset{\text{tricalcium phosphate}}{Ca_3(PO_4)_2} + 2CO_2 + 2H_2O = \underset{\text{dicalcium phosphate}}{Ca_2H_2(PO_4)_2} + Ca(HCO_3)_2$$

The process of *solution* is also a chemical force of mineral decomposition, but as pure water is a very weak solvent for most soil minerals, solution comes about largely through the joint action of CO_2 and H_2O, or by water containing traces of mineral or organic acids, or by salt solutions. The principal mineral acids developed in the soil through chemical and biological agencies are HNO_2, HNO_3, H_2SO_4, H_3PO_4, and H_2SO_3. Seldom do these acids occur in more than traces, but their continued formation ultimately results in considerable decomposition of minerals. Activities of soil organisms, such as bacteria and fungi, produce CO_2 and complex organic compounds of acid nature. They also aid in the process of solution. Solutions of salts have a greater solvent action for many minerals than has pure water, and hence in soil areas where soluble salts are abundant the process of solution may be accelerated.

BIOLOGICAL AGENCIES WHICH AID IN SOIL FORMATION

The relative rate of activity of various soil-developing processes is greatly modified by the type of vegetation growing on the soil and by the animal life inhabiting it. At the same time the type of vegetation occupying a soil area and the animal life within it are influenced by the properties of the soil. For example, earthworms are scarce in certain soils of high acidity and low organic matter content. The work of fungi is said to predominate over bacterial activity in the coniferous forests of the cool, humid North. Certain species of plants grow poorly on acid soils, while others require high soil acidity for abundant growth. The interrelation of plant and animal life and soil properties is thus seen to be both intricate and interesting. A few of the most evident relationships between plants and animals and soil-forming processes are suggested in the following questions.

Questions:

1. What forms of plants inhabited the earth's surface in the early stages of soil development?
2. Are bacteria, fungi, and similar life forms still active in soil formation?
3. How does the growth of higher plants affect soil properties?
4. Is animal life concerned in soil development?

Activities of Elementary Plants. In the very early stages of mineral decomposition and soil formation the lower forms of plant life, such as lichens, fungi, and bacteria, undoubtedly played a major role. The ability of lichens to grow on boulders and other rock surfaces exhibiting practically no indication of decomposition demonstrates their ability to extract mineral nutrients from unweathered minerals, to subsist with little water, and to obtain nitrogen from an undetermined source, possibly the air. Under clusters of lichens, even on very hard rocks, is found a thin accumulation of soil material, evidently the result of the decomposing action of the plant on the rock.

The ability of certain bacterial forms to extract nitrogen from the air for their own body needs, when supplied with mineral nutrients and carbonaceous food, and without associative growth of other plants, unquestionably played a considerable part in the building up of nitrogenous organic material in the soil in the early stages of development, thus paving the way for the growth of higher plants.

The work of these elemental plants has continued all during the process of mineral decomposition and soil development, and they remain active throughout the existence of the soil. It is through the action of such organisms that plant and animal remains are decayed, thus liberating the nutrients contained for the use of the next plant generation and simultaneously producing CO_2 and complex organic compounds which aid in mineral decomposition.

Influence of Higher Plants. Higher plants, such as grass and trees, exert a great influence on the characteristics of the soil developed under them. Their influence on erosion and moisture relationships within the soil are contributing factors. The type of root development and the return of vegetable matter to the surface of the ground by both annuals and perennials play an active role in the nutrient cycle. The absorption of soluble nutrients from the soil and their return in plant parts, even the variation in chemical composition of leaves of different species, are important in their effects on soil development. Some of these points are of sufficient significance to warrant a fuller discussion in later chapters.

Work of Animals. Worms, insects, and other small animal forms play a greater part in modifying chemical and physical properties of soils than do larger animals. The continuous burrowing of worms and insects establishes drainage channels in the soil and otherwise changes soil structure. Earthworms pass large quantities of soil through their bodies, thus altering it both chemically and physically. Worms and ants bring up a surprising amount of soil and deposit it on the surface. Many insects consume plant material both on the surface and within the soil. Worms are active in dragging fragments of leaves and grass into the soil to become a part of the soil organic matter. The abundant formation of "middens" in some timbered soils testifies to this activity of worms, whereas the disappearance from the soil surface of leaves of certain species indicates a strong choice on the part of worms as to the kinds of vegetation they consume.

PRODUCTS AND RESULTS OF MINERAL-DECOMPOSING PROCESSES

The relative rate and predominance of the various processes of mineral disintegration and decomposition vary under different climatic conditions and as the actions proceed. In the early periods of the earth's history, when the surface was more or less continuous rock with little if any animal or plant life, the processes must have been largely physical. As mineral fragments became smaller, thus exposing greater and greater surface to the action of water and oxygen and other gases which may have been present, chemical decay played a greater part. As animal and plant life developed, giving rise to CO_2 and other decomposition products, chemical decomposition became increasingly active. In arid climates, however, where moisture is limited and where plant and animal life is not abundant, and in high altitudes where life is also limited, physical processes continue to predominate in mineral breakdown.

As soil material becomes more completely decomposed, all processes of physical decay become slower while chemical decomposition becomes more and more dominant. How far these processes will continue is not determined, for no soil has yet been found which is not undergoing chemical and physical change.

It follows, then, that a logical concept of the nature of soils must necessarily take into consideration the dynamic characteristics of the materials from which, and in which, the soils are forming. A discussion of the results and products of the weathering processes described above serves as a helpful summary of these dynamic properties. In the discussion below the important points can be more easily distinguished if the following questions are kept in mind.

Questions:

1. What are the chief chemical compounds remaining as a result of mineral decomposition?
2. Which soil properties are affected by products of weathering?
3. What are the generalized changes in soil material as a result of weathering processes?

Products of Mineral Decay. As pointed out in the discussion of minerals, the members of the carbonate group are converted into bicarbonates, such as $Ca(HCO_3)_2$ and $Mg(HCO_3)_2$, which are relatively soluble. These compounds may be removed from the soil if moisture and drainage conditions permit or else the cations may pass into other combinations and be retained. Fragments of the original minerals may remain for long periods and make up an appreciable part of the soils subsequently formed. In other places the soils contain none of these fragments but are developed from the impurities that were in the minerals, such as sand, clay, and hydrated oxides of iron and aluminum. Thus, several hundred feet of limestone may be decomposed before enough residues are accumulated to make a soil a few feet in depth. The resulting soil may even be deficient in lime and hence acid in reaction.

The oxide minerals remain largely in the soil as quartz and hydrated oxides of iron.

The micas and talcs remain as particles of various sizes, usually quite fine, and without great decomposition except when iron is a constituent.

The feldspars and other silicates decompose with the production of the carbonates and bicarbonates of calcium, magnesium, sodium, potassium, and other bases which vary in solubility. Some of the silica becomes soluble, and some remains as insoluble SiO_2. Much of the Si and Al remain as claylike and sometimes jelly-like hydrated aluminosilicates as $H_2Al_2Si_2O_8 \cdot H_2O$. Hydrated oxides of iron and of aluminum are formed. With magnesium-bearing silicates, products, including hydrated magnesium silicates like $H_2Mg_2Si_2O_8 \cdot H_2O$, containing less Ca, Fe, and SiO_2 than the original mineral, are formed.

Soil Properties as Affected by Products of Mineral Decay. The processes of mineral decomposition result in the removal of the basic elements like Ca, K, Na, and Mg from the minerals and in the formation of more or less soluble compounds of them. A large part of the negative radicals of the minerals composed of silicon or aluminum and silicon remain as hydrated aluminosilicates, together with SiO_2, Al_2O_3, and Fe_2O_3, in various stages of hydration. They are mixed with remaining particles of the minerals in different stages of decomposition. The physical effects of the process are to accumulate material in a much finer state of division, with part of it sticky or jelly-like because of hydration.

The chemical effects vary with climatic conditions. In areas of low rainfall the soluble products may accumulate near the surface, producing a soil rich in soluble salts and alkaline in reaction. In more humid climates the soluble compounds containing the basic elements may leach away, leaving a soil poor in plant nutrients and in need of bases. Such soils are said to be acid in reaction. Under some climatic conditions the finely divided and hydrated oxides of iron and aluminum together with aluminosilicates may move downward, leaving the silica and undecomposed mineral particles in the surface layers. Under other climatic environments the silica and soluble salts may be carried downward, leaving an accumulation of hydrated iron and alumina.

CONSTRUCTIVE PROCESSES OF SOIL DEVELOPMENT

Up to this point the discussion has been devoted to the processes and products of rock and mineral disintegration and decomposition; in other words, to destructive processes. The conspicuous part of soil development is largely the result of such processes, and they have received detailed study for many years. The constructive processes of soil development are not so evident to the eye and, therefore, have received much less study although they may be more important in determining soil characteristics and the utilization of the soil for plant growth. Destructive and constructive processes, however, are intimately related. In fact, distinctions between the two are often slight. For example, a process may be destructive in its action on parent material and yet constructive in its effect on the resulting soil. There is, then, some confusion and inevitable overlapping in the use of these terms. This section emphasizes those processes that contribute most to the construction of soils even though they may have been discussed before as being destructive to parent material.

Four questions are proposed as guides in studying the constructive processes of soil development.

Questions:

1. What physical properties of soils are influenced by constructive soil processes?
2. What chemical compounds are developed in soil and how do they affect soil properties?
3. What kind of organic compounds are built up in the soil?
4. What general changes occur in soil material during soil development?

Soil Structure. The physical structure of soils presents more visible evidence of constructive action in soils than does chemical or mineralogical composition. The formation of zones of higher clay content, at various depths underlying more sandy or silty horizons, is a result of constructive change. The practical effects of such clay-rich horizons on moisture supply and root development are far reaching. In soils of sandy texture the additional moisture retained because of this horizon makes possible greatly increased yields of crops. On the other hand, if the clay layer is very dense, root development is restricted in places and the soil is too wet for the growth of some crops.

Other constructive soil-forming processes lead to the development of organic compounds that move downward and concentrate in definite zones. Hydrated oxides of iron are also constructive in their action.

These substances sometimes cement the soil material into a semirocklike mass which limits root development and water movement.

Through freezing and thawing, the adhesive and cohesive action of moisture films, and the cementing power of sticky claylike substances and salts, the mineral particles and fragments are bound together in aggregates of irregular but definite form and structure. Evidence of this structural arrangement is shown by the fact that earth removed from a hole will not completely fill the same hole when it is returned if plenty of water and tamping are employed. The development of a characteristic soil structure is a process requiring many years. From a practical standpoint structure is important because it influences moisture movements in both the plow layer and subsoil. The hard massive layers and the clay horizons already mentioned are a part of structure. So are the channels formed by insects, worms, and plant roots. Not only do these and other structures control the speed of moisture movement, but they also control aeration. A farmer spends a large proportion of his time and labor in developing by means of tillage a soil structure favorable to root growth and rapid seed germination. Soil that loses its natural field structure loses its identity and becomes an entirely different medium for plant growth. The many different kinds of aggregates or structural units found in soil have been named and classified, and this knowledge constitutes a definite part of soil science.

Chemical Compounds Developed in Soils. As rocks and minerals decompose, portions of some of the products are recombined into new chemical compounds. There is a considerable list of *secondary* minerals which pedologists have so designated because they are formed through the decomposition of *primary* minerals. Clay is made up largely of such secondary minerals mixed with fragments of quartz (SiO_2), hydrated oxides of iron, some organic matter, and other impurities.

The very fine clay, generally called colloidal material, is comprised largely of minerals developed in the soil. Three groups of these minerals have been recognized, namely, the montmorillonite group $[(OH)_4Al_4Si_8O_{20} \cdot XH_2O]$, the kaolinite group $[(OH)_8Al_4Si_4O_{10}]$, and the illite group $[(OH)_4K_x(Al_4 \cdot Fe_4 \cdot Mg_4Mg_6)(Si_{8-x} \cdot Al_x)O_{20}]$. In each group there are several members which differ from the formula given primarily in the proportion of silicon and aluminum present. Also, there is considerable variation in composition of the members of the illite and montmorillonite groups due to replacement of aluminum by ferric iron and sometimes magnesium. The crystals of these minerals are composed of plates of SiO_2 and Al_2O_3, with varying degrees of hydration. The montmorillonite and the illite minerals have two plates of SiO_2 and one plate of Al_2O_3, whereas the kaolinite minerals have one plate of each.

These minerals give the soil most of its cohesive and adhesive properties and are also very effective in holding certain plant nutrients in such a state that they are not readily lost by leaching and yet may become available for plant use. These materials are now receiving much intensive study and, as improved methods of procedure are developed and new scientific apparatus is perfected, much more information concerning these compounds, which are the result of constructive soil processes and which control many vital soil processes, will be gathered.

Organic Compounds Formed in Soils. The action of soil organisms in decomposing plant and mineral soil materials results in the development of many complex organic compounds which are active agents in soil processes. The bodies of the organisms themselves contribute appreciably to the organic content of the soil. These compounds not only serve as a means of storing a supply of nitrogen in the soil but also influence the chemical and physical soil properties. They affect soil structure, moisture retention and movement, and soil temperature to some degree. Furthermore, they function in retaining plant nutrients against leaching, and in supplying them to plants.

General Changes Leading to Soil Development. There are then many factors active in the evolution of soils. Some of the processes are simple, many are complex and all are more or less closely interrelated. Some changes proceed rapidly in the early stages of formation; others become dominant as the end stages approach. Most of the processes are overlapping and all proceed simultaneously. The result is, of course, the soil, the main steps in the evolution of which may be generalized as follows:

1. The fragments or particles of parent material become smaller.
2. The basic elements are gradually removed from the surfaces of the particles and combined into compounds of variable solubility. In areas of high rainfall these salts are leached out, leaving the soil increasingly deficient in basic elements. In regions of less rainfall the salts accumulate in the surface soil or at various distances below the surface.
3. Easily decomposed minerals gradually disappear, leaving the more resistant minerals, the products of mineral decay, and secondary compounds.

4. In advanced stages of weathering the fine clay or colloidal particles attain a high hydrogen-ion content and a correspondingly low content of calcium, magnesium, sodium, and potassium ions. The particles then move downward until they are coagulated by different environmental conditions. This process gives rise to a zone or horizon of clay concentration.

5. As weathering attacks their surfaces, a coating of the residues of mineral decay accumulates around the particles. This coating consists of hydrated oxides of iron, aluminum, silicon, and hydrated aluminosilicates and frequently is jelly-like.

6. New minerals are developed which have definite crystal structures, although occurring in very minute particles, and which have a far-reaching influence on the chemical and physical properties of the soil.

7. Organic matter, in different stages of decay, accumulates in widely varying quantities on the surface and within the surface layer of the soil material.

8. A characteristic structure develops.

Climate determines the relative rates of activity of the different forces of weathering and soil formation. It also limits the extent to which soil development proceeds. In arid regions, for example, the lack of water has prevented soil development from progressing much beyond step 3. In these places where the moisture supply is limited, physical changes predominate. With increasing moisture content, chemical changes assume a more important role. Not only the moisture factor of climate but also temperature has an important influence on soil development. The speed of chemical reaction in humid regions varies markedly with the average seasonal temperatures and with the length of the warm season, as would be expected from the effect of temperature on the speed of chemical reactions.

In the early stages of soil development the nature of the parent material dominates the characteristics of the soil. Such soils are said to be young. But when parent material has been subjected to soil-forming forces for long periods of time, it is more completely altered and the soil eventually becomes the product of the climate or environment under which it develops. These soils that reflect the characteristics of the climate are said to be mature.

Since weathering and soil-forming processes start at the earth's surface and proceed downward, different stages of weathering are found at different depths. This variation with depth gives rise to layers called *horizons*, in which widely varying chemical and physical changes proceed. The soil *profile* is a vertical cross-section from the surface through all the horizons into the underlying parent material.

THE SOIL PROFILE

The result of soil-forming forces working for long periods of time is the gradual differentiation of layers or horizons within the parent material. Collectively these horizons are called a *profile*. The following questions emphasize the important points in studying soil profiles.

Questions:

1. How are soil horizons named and described?
2. How does topography influence the profile that develops in a given parent material?
3. How does drainage influence soil-profile development?
4. What is the relation between soil reaction and soil development?
5. How do the A_2 and B horizons vary in different soils?
6. What part does organic matter play in determining profile characteristics?
7. Why are salt accumulations found in some soils?
8. What is a soil series?

Differences in Horizons. There are many kinds of soil profiles because no two horizons are ever exactly alike. Not only does their mineral composition vary but also the percentage of different-sized particles as well as their structure and color differ widely within short distances. The nature of the horizons in a profile and the intensity of their development depend upon many factors, a few of which are the composition of the parent material, length of time it has been exposed to soil-forming processes, climate, nature of the vegetative cover, and site characteristics such as relief and topography. Most profiles, however, are generally characterized by two or more of the following horizons: (1) The surface which consists of or contains an accumulation of plant leaves, stems, and roots; this horizon of plant residues varies from less than an inch to 20 inches or more in thickness. (2) An horizon which because of the action of soil-forming forces has lost part of its original constituents. (3) An horizon of accumulation; this is the layer where material removed from the two overlying horizons is deposited. (4) The horizon of parent material, the upper portion of which may have undergone a small amount of alteration; as a whole, this material is assumed to be similar to that which existed on the surface before soil-developing processes began.

In pedology these horizons are conventionally designated by letters. The surface layer, or A horizon, is subdivided into the A_{00}, A_0, A_1, A_2, and A_3 sections.

The A_{00} refers to the undecomposed litter of twigs and plant leaves; in the A_0 only partially decomposed organic matter is found. That

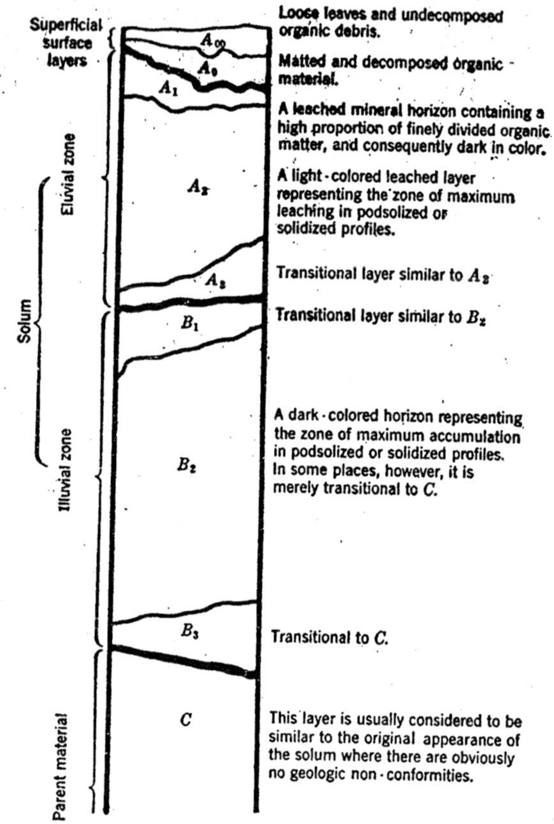

FIG. 11. A generalized profile of Gray-Brown Podsolic soils. The thickness of the various horizons varies as indicated.

layer where mineral particles predominate but the dark color of organic residues persists is designated as the A_1 horizon. The A_2 layer, on the contrary, is comparatively light in color and shows the maximum effects of leaching or *eluviation.*

The horizon of accumulation or *illuviation* is technically designated as the B horizon. It is often subdivided into the B_1, B_2, and B_3 sections, depending on the degree of accumulation in evidence. The parent material is referred to as the C horizon, while underlying strata (such as bed rock) are called the D horizon. These terms and their relationships are presented in Fig. 11. Notice that horizons A and B together constitute the *solum.*

As previously stated, all these horizons are not present in every soil. The central member of Fig. 12 shows a profile in which the horizon of eluviation is not evident. Insufficient rainfall and a grass

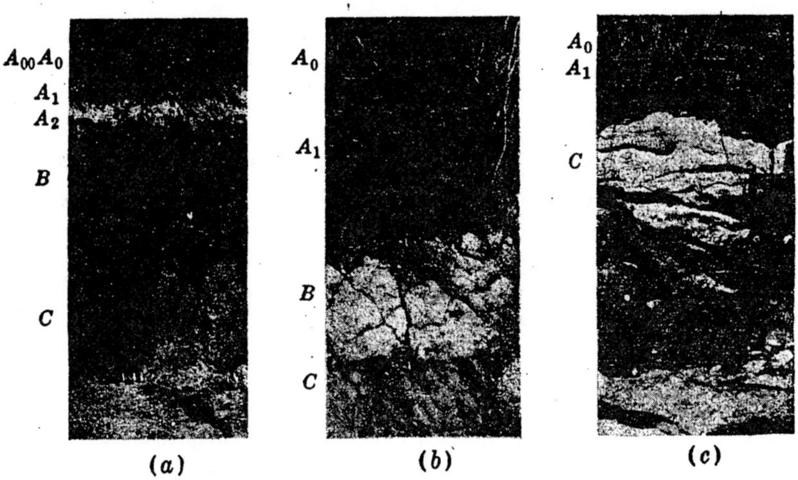

FIG. 12. Compare the thin A_1 and the strongly developed A_2 horizons of the podsol profile on the left with the deep A_1 and the absence of A_2 horizon in the chernozem profile in the center. On the right is the profile of a soil developing from limestone. Note the absence of a *B* horizon. [Center and right profiles by courtesy of Dr. Harper, of the Oklahoma Experiment Station.]

cover are two factors which contribute to the development of this type of profile.

In some cases both the horizons of eluviation and illuviation are not discernible, and a profile similar to one of those in Fig. 12 is produced. A high water table which has limited the activity of weathering agencies, a limited rainfall, and a comparatively short period of activity of soil-developing processes are some of the conditions giving rise to this type of profile.

Topography Influences the Soil Profile. Topography modifies the soil-profile development through two methods: (1) by influencing the quantity of precipitation absorbed and retained in the soil, thus affecting moisture relations; and (2) by influencing the rate of removal of the soil by erosion. As moisture is essential for the action of the chemical and biological processes of weathering and effectively acts in conjunction with some of the physical forces, it is evident that a modification of moisture relationships within a soil will materially influence profile development. It is noteworthy in a glaciated region that, in similar parent material, the soil of steep ridges or hills differs from the soil on gentle slopes or on level to undulating topography.

On steep slopes the continuous removal of surface soil by erosion keeps exposing the lower horizons and so modifies the profile that would otherwise be present. Furthermore, the decreased thickness of the humus-containing soil results in a lessened supply of the products of decay of organic matter which contribute to the decomposition of minerals and to horizon-forming processes.

Topography indirectly plays another part in profile development by influencing the supply of moisture available for plant growth. It has a bearing on the agricultural value of the land because it is related not only to both external and internal drainage conditions but also to the ease of performing tillage operations.

Drainage Conditions Affect Soil Development. As pointed out before, drainage materially influences soil-forming processes. The impeding influence of excessive surface drainage, for example, is shown by the development of more conspicuous and deeper profiles in level than in sloping parent material which, aside from surface topography, is similar. This difference is particularly noticeable in light sands where the internal drainage is also excessive. Free internal drainage, compared with restricted drainage, has a marked influence on the characteristics of the profile in all kinds of parent material.

Poor drainage may have even more effect on soil development than excessive drainage. Colors of the soil material at shallow depths are changed from the yellows, reds, and browns denoting good aeration to the drabs, grays, and mottled yellows resulting from chemical reduction where drainage is poor. The accumulation of organic materials is usually facilitated because they are preserved by water. And because of their low topographic position, poorly drained soils generally receive organic matter and colloidal minerals that have eroded from the adjacent slopes. The horizon of eluviation is also modified or it may be entirely lacking. It is often replaced by a gray or bluish-gray horizon, known as *glei,* which develops at various depths from the reducing action of wet conditions. In this layer iron compounds as well as compounds of calcium, magnesium, and manganese are changed to soluble forms. In some places, however, the underlying layers may be so slightly changed that the surface horizon appears to rest directly on the parent material.

Intermediate drainage conditions give rise to other profiles. A water table, for instance at a depth of 2 to 3 feet below the surface, encourages the development of a conspicuous eluviated horizon underlain by an intensely developed illuviated layer, which is sometimes indurated or cemented into a hard rocklike mass.

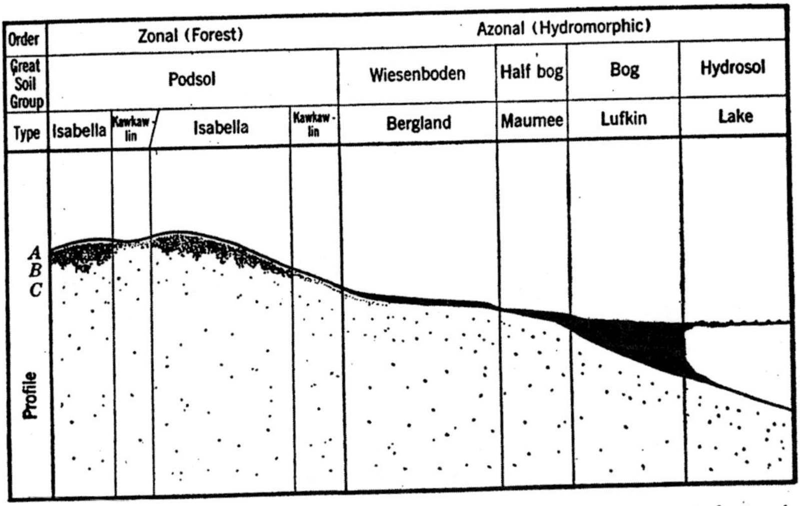

FIG. 13. Topography through its influence on drainage is a potent factor in soil development in humid regions. The moderate humus accumulation, well-defined horizon of eluviation, and strongly developed *B* horizon of the zonal profile are superseded by an increase in humus, diminution in eluviation, and accumulation in the *B* horizon as poorer drainage limits the action of soil-development processes. Finally the bog is encountered and then the hydrosol or body of water itself. The figure illustrates the relationships between the soil types of a *catena* in a humid glaciated area (Michigan).

Soil Reaction as a Characteristic of Soil Profiles. Parent material ranges from that composed largely of acid minerals, through that containing various percentages of limestone, to limestone itself. Compounds containing calcium and magnesium, which give it an alkaline reaction, make up varying proportions of the parent material. As the processes of chemical weathering proceed, these basic elements are transformed into soluble compounds that either accumulate or leach out to a greater or less extent. In time, then, either the soil horizons may become more basic in reaction or they may become acid, depending upon the prevailing climatic environment. Accordingly, within a single region a soil-reaction test may give some indication of the extent to which soil development has progressed.

This is because the reaction of a soil horizon indicates chemical conditions which are conducive either to the flocculation of the finer clay, on the one hand, or to its dispersion with the possibility of being carried downward by water, on the other. The formation of soluble compounds of iron, aluminum, silicon, manganese, and many other

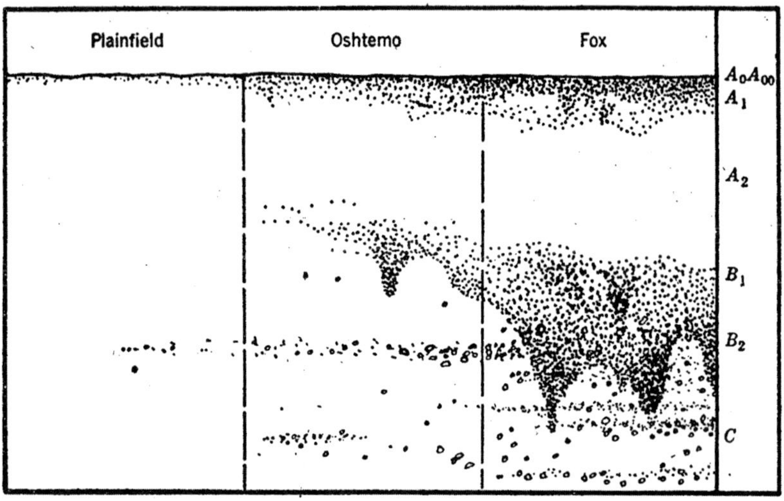

FIG. 14. The chemical composition of coarse material, deposited as an outwash plain, greatly influences the characteristics of the soil subsequently developed. The material on the left was highly siliceous and resulted in a soil having a minimum of clay in the *B* horizon. The Oshtemo soil developed a well-defined *B* horizon containing some clay because of the presence of appreciable quantities of limestone fragments in the parent material. The outwash on the right contains considerable quantities of limestone, and hence the Fox soil has a strongly developed *B* horizon with sufficient clay to bind the sand and gravel into a water-retentive layer which gives the soil a much greater agricultural value than is possessed by the Oshtemo or Plainfield.

elements is also related to soil reaction. Moreover the activity of bacteria and other soil organisms as well as the growth of higher plants is affected by soil reaction. A determination of the reaction of the various soil horizons is therefore useful in studying profile development and soil properties.

Properties of the Lower Soil Horizons Vary. The possibilities of variation in the horizon of illuviation have already been mentioned. In some sandy soils it contains very little more clay than the parent material and differs from adjoining horizons mainly in color. This kind of a B horizon contributes little to the crop-producing capacity of the soil. In other sandy soils, however, the illuviated horizon is higher in clay than the other horizons and hence serves as a reservoir for the storage of moisture for crop use. In some cases the B horizon may be either so firmly cemented or composed of such a dense accumulation of clay that plant root growth is restricted. On the other hand, in many soils derived from heavy clay, the B horizon has a more open structure than the parent material and thus offers the opportunity for a greater root penetration than would otherwise be true. Accumulations of calcium and other nutrients and concentrations of humus sometimes are found in the horizon of illuviation. Of course, these substances add materially to the fertility of the soil.

The thickness of the B horizon is also a contributing factor to the properties of a soil. A thin layer has little influence, favorable or unfavorable as the case may be, on plant growth but a thick layer of the same nature may decidedly affect the crop-producing power of the soil.

The horizon of eluviation is frequently lower in clay content, more acid, and poorer in nutrients than the other horizons of the profile. This condition prevails more often in humid than in arid regions. The thickness of this horizon therefore influences the fertility of soils in a humid climate. In many soils the leached layer is not a good medium for rapid seed germination or for the development of young roots. As the surface decreases in thickness through erosion and more and more of the eluviated horizon is incorporated in the plow layer, it therefore becomes necessary to supply additional quantities of organic matter, manure, and commercial fertilizers to maintain the productivity of the land.

Organic Matter Is a Distinguishing Feature of Soils. The quantity of organic matter that accumulates in soil depends upon many factors, such as the length of growing season, soil-moisture relationships, type of vegetative cover, soil reaction, temperature, rainfall, and nutrient content. Because all these factors affect other soil characteristics, the amount of organic matter is often considered as an index of soil conditions. Organic matter both through its presence and through its decomposition products exerts so great an influence on soil properties that the quantity becomes an important consideration in comparing soil profiles.

Salt Accumulation in Soils. In climatic regions where precipitation percolating through the decaying minerals is not sufficient to carry away the soluble compounds developed, they accumulate as salts in the profile. The depth and extent of these accumulations together with their chemical composition play important parts in determining soil properties and soil utilization. Accumulations of salts in soils occur chiefly, but are not confined to, arid climates; occasionally they are also found in soils of humid regions.

Soil Series. It has been found that in spite of the differences that exist in profiles there are also certain broad similarities. Closely related profiles are grouped together as a *series*. The profiles of a series are similar in origin; that is, they developed under the same climatic conditions through the same forces acting on similar, but not quite the same, parent material for approximately the same length of time. Topographic and drainage conditions associated with the profiles in any particular series are the same. Moreover, corresponding horizons in the different profiles are similar in depth, color, reaction, and chemical characteristics. The organic matter content as well as the nature and supply of humus is also similar.

Variations within the series are confined to difference in texture. Thus there may be loams, silt loams, and clay loams within a series. At present, however, the tendency is to restrict this grouping to closely related textural classes. Clay soils and sandy soils do not, therefore, occur in the same series; few are even broad enough to include sandy loams, loamy sands, and loams.

A soil series is usually given the name of a town, river, or other geographical or political feature near which it was first identified. The Miami series, for example, was first recognized near Miami, Ohio.

CHAPTER II

CLASSIFICATION OF SOILS

The properties and the management of soils may be most effectively studied and remembered if the soils are classified according to a definite system. Several schemes for classifying soils have been suggested at various times but one based on the properties of the soils themselves—rather than on kind of material from which they were derived, or on agency by which the material was accumulated, or on other factors having to do with soil development—appears most logical and advantageous. Such a system was suggested about 1880 by the Russian scientist Dokuchaev and has been further developed by European and American workers. The system is based on the theory that soils have a definite morphology (form and structure) which is related to a particular association of climate, vegetative cover, relief, age, and parent material. The classification of soils on both a textural basis and as related to environmental condition is presented and the utilization of the soil groups by man is discussed under seven general headings.

Objectives:

A. A textural classification of soils.
B. The systematic classification of soils.
C. Soil mapping and the soil survey.
D. The relation between soil groups and climatic conditions.
E. Age, relief, and parent material in relation to soil groups.
F. The relation between soil groups and vegetative cover.
G. Soil groups in relation to population density and the production of food and fiber.

A TEXTURAL CLASSIFICATION OF SOILS

The casual observer is inclined to confine his observations of soil properties to the surface soil or that portion turned by the plow, paying particular attention to color and texture. Texture refers to the sizes of particles in the soil. This property is important because it is directly related to the total surface of the soil particles and therefore to many phenomena in soils that are important in determining productivity.

The steps followed in developing a classification of soils based on particle size are indicated by questions.

Questions:

1. What are soil separates? Soil classes?
2. What are the names and distinguishing characteristics of several common soil classes?
3. Is there a more generalized textural classification of soils?
4. Does the class name convey any information concerning a soil beyond the percentage of separates contained?
5. What is a soil series and a soil type?
6. How does the pedologist define a soil?

Soil Separates. One of the first facts to attract the attention in examining soils is that they are composed of particles of different sizes. Needless to say, the proportions of different-sized particles in a soil have much to do with its physical and chemical properties, and hence with plant relationships.

Soil particles have been divided into groups entirely on the basis of size, that is without regard to their chemical composition, color, weight, or other properties. These groups of particles are termed soil *separates* because any soil composed mainly of mineral matter may be *separated* into these particle groups. The process of dividing a soil into its separates and determining the relative proportion of each

TABLE 1

SOME CHARACTERISTICS OF SOIL SEPARATES

Name	Diameter *	No. of Particles per Gram	Surface Area in One Gram
Fine gravel	2.00–1.00 mm.	90	11.3
Coarse sand	1.00–0.50 mm.	722	22.7
Medium sand	0.50–0.25 mm.	5,777	45.4
Fine sand	0.25–0.10 mm.	46,213	90.7
Very fine sand	0.10–0.05 mm.	722,074	226.9
Silt	0.05–0.002 mm.	5,776,674	453.7
Clay	below 0.002 mm.	90,260,853,860	11,342.5

* As established by the U.S.D.A.

separate present is called *mechanical analysis*. In Table 1 are given the names of the separates, together with their diameters, the number of particles in one gram, assuming them to be spheres with maximum diameter of the group, and the surface area in square centimeters exposed by the particles in one gram of the separate.

Soil Classes. Never is a soil composed of only one separate. Usually, at least small quantities of the majority of the separates are present. The next step in classifying soils, therefore, is to group them on the basis of the proportion of the different separates present. These groups are designated as soil *classes,* which are named according to the separate or separates which contribute most to their characteristics. This does not mean that a class is necessarily named after the separate present in largest quantity. It takes a very large quantity of coarse particles to exert as much influence on soil properties as a comparatively small quantity of the finest particles—clay. Clay is the most potent separate in imparting its properties to a mixture of separates, and hence the adjective clay is found in the class name of many soils which contain a higher percentage of the other separates than they do of clay.

It is customary to designate a soil by the class name which is in accord with the physical make-up of the surface or plow soil with little regard to the texture of the subsoil. But there is usually a rather close relationship between the textural qualities of the surface soil and the underlying material from which it has been derived through soil-forming processes; so the texture of the surface usually is a good clue as to the nature of the subsoil. For example, a sandy surface usually rests on material high in sand content. Likewise, surface layers high in clay or silt usually are underlaid by materials high in these separates.

Sometimes, however, consideration is given to the nature of the layers below plow depth in naming the textural class, particularly if the qualities of this underlying material are quite different from those of the surface, so that its plant relationships are distinctly modified. For example, if a few inches of sandy soil rest on a mixture of clay, silt, and sand, there is an inclination to rank the surface soil somewhat heavier than its actual silt and clay content warrant. Another way to recognize the effect of underlying material on the plant relationships of the surface soil is to designate the textural class of the subsurface layers also, for example, so many inches of sandy loam resting on a given depth of silt loam.

Common Soil Classes. A soil composed of a mixture of the separates such that the properties of no one separate or group of separates dominates its characteristics is called a *loam.* The stickiness of the clay and the finely pulverant (floury) nature of the silt are balanced by the non-sticky and mealy or gritty characteristics of the fine to coarse sands. Water drains through the soil with ample freedom, and yet considerable moisture is retained for plant use. A loam readily assumes and retains a structure which is conducive to rapid germination of seeds and to easy penetration by roots; in other words, it can easily be kept in a state of good tilth.

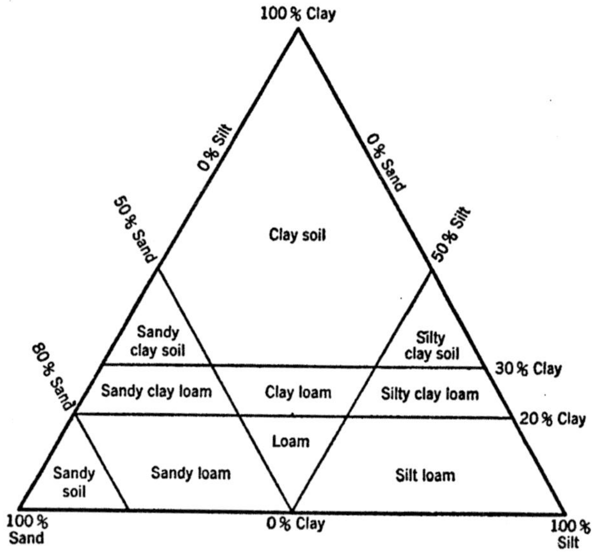

FIG. 15. The percentages of sand, silt, and clay in the textural classes of soil may be determined from this diagram.

Where a soil of loamlike qualities contains sufficient sand separates so that their characteristics predominate over those of silt and clay, the texture is a *sandy loam*. If the quantity of one of the sand separates is adequate to impart the properties of that separate to the soil, the name of the separate is added to make the soil class name. Thus, in a *fine sandy loam* the qualities of fine sand predominate over those of the other sand separates; in a *medium sandy loam* the properties of the medium sand are evident. If a soil contains very little more silt and clay than are necessary to throw it into the sandy loam group, it is frequently called a *light sandy loam*. On the other hand, in a *heavy sandy loam* the sand content approaches the lower limit, and the silt and clay content the upper limit for a sandy loam.

Soils with a silt and clay content just below that required for sandy loam are generally designated *loamy sands*. This name was adopted because such soils produce moderately good yields of crops when rainfall is adequate, and therefore it seemed undesirable to classify them as sands which are decidedly lower in productivity characteristics.

Soils containing appreciably less silt and clay than is required for a sandy loam are classed as *sands*. The humus content, as well as the clay and silt content, of sands is very low, which distinguishes them in part from loamy sands.

A comparatively small percentage of clay imparts to a soil the characteristics of clay to a sufficient extent to warrant naming the soil a *clay loam* (Fig. 15). To persons accustomed to working in sandy loams, a clay loam seems very plastic when wet, and hard when dry. Many farmers and others unfamiliar with the correct textural classification of soils call clay loams "clay soils." As with sandy loams there are *heavy* and *light clay loams* in which the proportion of clay is near the upper and lower limits, respectively.

A soil containing enough clay to classify it in the clay loam group (usually well above the average for clay loams) and, in addition, containing near the maximum quantity of silt is in many cases very dense as well as plastic, and moisture moves through it slowly. Such a soil, texturally classified as a *silty clay loam,* is highly productive when skillfully tilled.

So dominant are the properties of clay that the presence of over 30 per cent results in the designation of the soil as a *clay*. Usually, though not always, a considerable percentage of silt is present and, of course, sand, even coarse sand, may constitute an appreciable part of a clay soil. Clay soils, like silty clay loams, are productive when in good physical condition, but they require special management methods to prevent the puddling or breaking down of the granules.

Silt loams contain more than 50 per cent of silt. Lacking the cohesive properties of clay and the grittiness of sand, silt must be present in large quantities to impart its velvety or flourlike properties to a soil. Although silt loams are water retentive and similar to clay loams in other respects, they lack the cohesiveness of clay loams and are therefore easier to keep in good tilth. Large acreages of high-quality land in the corn belt states are silt loams in texture.

A Generalized Classification of Soils. The various soil classes are not separated from one another by definite lines of division but rather grade into the adjoining classes of lighter and heavier texture. For example, a loam gradually merges into a silt loam or clay loam as the percentages of silt and clay increase and, on the other hand, a sandy loam is produced as the percentage of sand separates replace silt and clay in the loam. It is difficult, therefore, for those not well acquainted with the specific properties which characterize the different soil classes to distinguish accurately between them, and hence a more generalized textural classification of soil is used. For instance, it is customary when speaking in general terms to refer to loams and silt loams as medium-textured soil, to clay loams, silty clay loams, and clays as heavy soils, and to sandy loams, loamy sands, and sands as light soils. An area made up of heavy soils is frequently called *clay land.*

The Class Name Is Descriptive of Soil Properties. It should be remembered that a soil class name is meant to convey not only an idea as to the percentages of the various separates present but also *information concerning the soil properties that are associated with such a mixture.* Thus the ideas concerning tillage practices, moisture-retaining power, aeration, crop adaptation, and so forth, brought to mind by the term sandy loam should be quite different from those conveyed by the class name clay loam.

The Soil Series. A soil series is usually given the name of a town, river, or other geographical or political feature near which it was first identified or mapped. The series name carries with it a knowledge of all the characteristics involved in differentiating the series. Thus to one who is familiar with soils the mention of a series name conveys a mental picture of the soil as to topography, drainage conditions, character of profile, reaction, color and structure of the horizons, approximate humus content, and something concerning the general nutrient content and productivity for crops commonly grown. It is thus seen that a series name is a highly informative term. Note, however, that specific knowledge of the soil texture is not conveyed by the series name. This information is supplied by the textural or soil class name.

Soil Type. A combination of the series and class names is necessary, then, to describe fully both the profile and texture of a soil. The combination of a series and class designation is called the *soil-type name.* Thus the type name Miami (series) loam (class) brings to the mind of one familiar with the timbered soils of the northeast central states a great deal of information concerning a particular soil. In New York Honeoye loam is a very informative name, as is Sassafras sandy loam to people living on the Atlantic coastal plain. Cecil sandy loam carries much meaning to persons familiar with soils of the southeastern states. Barnes silt loam brings to the mind of the soil scientist a picture of a widely known soil type in the central plains, whereas

Mohave sandy loam describes to him a soil of the southwestern desert region. A type name can be considered a shorthand term which briefly and concisely describes one particular soil.

A Pedologic Definition of Soil. The noun *soil* is derived, by way of Old French, from the Latin *solum* which commonly had the same meaning as our modern word. In general, soil refers to the loose surface of the earth as distinguished from solid rock, although many people, when they think of the word soil, have in mind that material which nourishes and supports growing plants. This latter meaning is even more general since it not only includes soil, in the common sense, but also such things as rocks, water, snow, and even air—all of which are capable of supporting plant life. The farmer, of course, has a more practical conception of the meaning of soil—he says that it is the medium in which crops can grow. The civil engineer, on the other hand, looks upon soil as that material which supports foundations, roads, or sewers. In short, the word soil has many meanings. In talking or writing about soils, one must understand that the word can be used in these various ways.

By soil the soil scientist generally means that part of the earth's crust which has been changed as a result of soil-forming processes, that is, the *solum*. Pedology has adopted the old Latin root in order to make clear this definition of the word soil. A little more than just *solum* is intended, however, because the pedologist deals with areas, not just single vertical profiles in a landscape. In discussing pedologic topics like development and classification, soil always means either the *solum* or the profile of a soil type. In dealing with other subjects such as fertility and productivity, soil is often used in the looser sense of something that supports plant life; in a vague manner, this meaning implies a definite profile, but it also implies to many people the concept of productive land. In reading this book, then, it is well to remember the different meanings that the word soil may express. Some students are often confused because they make no attempt to interpret the term in its appropriate sense.

A SYSTEMATIC CLASSIFICATION OF SOILS

The modern system of classification is based upon the recognition of soils as natural entities possessing definite unique characteristics. This scientific classification is often confused with better-known special systems that have evolved or have been devised from time to time to meet specific requirements. Some examples of these special groupings are found in the classification of soils according to their lime requirements, according to their abilities to produce special crops, or according to their moisture-retaining characteristics. The textural classification is an example of a system that has evolved through common usage. As new facts have been collected the pedologic system gradually has grown. The growth of soil science necessarily has paralleled the accumulation of knowledge concerning soil relationships. Here are important questions to consider in a discussion of this subject.

Questions:

1. Is there a similarity between the classification of soils and the classification of plants?
2. What are the differences between the three soil orders?
3. What divisions are there in category V of the soil-classification system?

Soil Classification Compared to Botanical Classification. This fundamental system of classifying soils is similar in many respects to systems used in classifying other natural objects, such as plants and animals. In botany the species name together with the genus name is the common method of technically designating a plant. An analogous binomial method is used in soils, but the above groups are called class and series, respectively. The name Coloma loamy sand is comparable to the botanical name *Quercus alba.* Note, however, that English words are used in soils in contrast to the Latin terms of botany. Plants have both a scientific name and a common name. Soils have only one,

Chart 1. A Comparison between Botanical and Pedologic Classification

CATEGORY	BOTANICAL CLASSIFICATION		PEDOLOGIC CLASSIFICATION	
VI	Phylum	*Spermatophyta*	Order	Zonal
V	Order	*Angiospermae*	Suborder	Forest
IV	Class	*Dicotyledoneae*	Great Soil Group	Gray-brown podsolic
III	Family	*Fagaceae*	Family	Miami
II	Genus	*Quercus*	Series	Coloma
I	Species	*alba*	Class	Loamy sand
0	Variety	....	Phase	Hilly
Common name	White oak		Coloma loamy sand	

the soil-type name, which must satisfy both scientific and practical requirements. This name is composed of two parts—the series designation and the class designation. A comparison between the botanical categories and the pedologic categories is shown more fully in Chart 1.

Another difference between the botanical system and the soil system is that the latter is much younger and is therefore still changing. It has developed only since the turn of the century. Not enough is yet known about soils to justify writing a book on their classification similar to Gray's *New Manual of Botany*. New soil types are being continually recognized. To get information about these types it is necessary to consult the soil survey reports that will be discussed in a later section. There is still considerable confusion concerning these soil types, however, because of the extreme youth of the science, which is evident in the fact that not all terms have yet been universally accepted nor are they all used in the same meaning.

Soil Orders. The largest category in the classification of soils is the *order,* of which there are three—*zonal, azonal,* and *intrazonal.* Zonal soils cover wide areas. Included are those great soil groups, possessing well-developed profiles that reflect the influence of the active factors of soil genesis. Topographically, zonal soils are situated on gently undulating, well-drained uplands in parent material not extreme in its texture or chemical composition.

The intrazonal soils have more or less well-developed characteristics that reflect the dominating influence of some local factor, such as the nature of the parent material, over the normal effect of climate and vegetation.

Azonal soils are without well-developed characteristics because some factor of parent material or relief has prevented their development. The relationships of the great soil groups are shown in Chart 2.

Division of Category V. It will be noticed that in category V soils are divided into groups on the basis of the factors which have been most instrumental in determining their characteristics or on the basis of the distinguishing characteristics of the soil group itself. For example, the cold and arid soils are direct products of climatic conditions. Grassland, transitional, and forest soils owe their properties to a combination of the influence of climate and vegetative cover. Excess moisture is responsible primarily for the nature of hydromorphic soils, while the accumulations of chlorides and of calcium salts are the distinguishing characteristics of halomorphic and calomorphic soils, respectively.

SOIL MAPPING AND THE SOIL SURVEY

Two subjects closely related to soil classification are soil survey and soil mapping. It was not until a systematic study of the soil resources of the United States was begun in 1898 that the need for a logical,

Chart 2. A Classification of the Great Soil Groups *

CATEGORY VI		CATEGORY V	CATEGORY IV	REMARKS
Zonal		Cold	Tundra	Subsoil continually frozen.
	Pedocals	Arid	Desert Red desert Gray desert Brown Reddish brown	Scarcity of rainfall limits vegetation. Organic content low. An horizon of calcium salts is found near the surface. Base content is high.
		Grassland	Chestnut Reddish chestnut Chernozem	Moderate rainfall. Moderate organic matter content under moderate grass vegetation. Calcium carbonate horizon at a greater depth than above.
		Transitional	Degraded chernozem Shantung or non-calcic brown Prairie Reddish prairie	Shows characteristics of both pedocals and pedalfers. Carbonate horizon may or may not be present at considerable depths. Organic content very high. Tall grass and forest vegetation.
	Pedalfers	Forest	Podsol Brown podsolic Gray-brown podsolic	Light-colored soils of the northern humid regions.
			Yellow podsolic Red podsolic Yellowish-brown lateritic Reddish-brown lateritic Laterite Terra rossa	Bright-colored soils of southern warm-temperate regions and tropics.

Order	Suborder	Great soil group	Description
INTRAZONAL	Halomorphic	Solonchak	Saline and alkali soils of imperfectly drained positions, chiefly in arid regions. Halogen elements conspicuous.
		Solonetz	
		Soloth	
	Hydromorphic	Hydrosol	Lakes and other water soils.
		Bog	Muck and peat.
		Half-bog	Waterlogged mineral soil with a deep organic surface horizon.
		Wiesenboden	Waterlogged mineral soil with a high proportion of organic matter in the surface.
		Alpine meadow	Poorly-drained mountain meadows of high altitudes.
		Ground-water podsol	Waterlogged mineral soils of northern humid regions. Ortstein layer conspicuous.
		Ground-water laterite	Waterlogged mineral soils of southern warm regions. Concretions conspicuous.
		Planosol	Soils characterized by thick massive hardpan.
	Calomorphic	Rendzina	Calcareous parent material remains dominant.
		Brown forest	
AZONAL		Lithosol	Fresh and slightly weathered rocks and rock fragments.
		Alluvium	Fresh sediments deposited on land by streams.

* Prepared from Table 2, "Soil Classification," M. Baldwin, C. E. Kellogg, and J. Thorp, 1938 Yearbook, U.S.D.A., pp. 993–1,001.

natural system of classification was encountered. And only through the processes of surveying and mapping have the many different soil categories been recognized and correlated. A few questions will assist in fixing in the mind of the student the primary points concerning soil surveying.

Questions:

1. What is a soil survey?
2. What are the steps in making a soil map?
3. What kinds of soil surveys are there?
4. What are the uses of soil survey reports and maps?

The Soil Survey. By the turn of the century there was an increasing awareness of the bonds between land and society. In an attempt to find the underlying causes of some agricultural problems and in an effort to build a solid foundation for future research, the United States Department of Agriculture in cooperation with the various state experiment stations began at that time a systematic investigation of our soil resources. This investigation assumed the form of a national inventory and survey.

At the present time soil surveying is the process of studying and mapping the earth's surface in terms of units called soil types. A soil survey report thus consists of two parts: (1) the soil map which is accompanied by (2) a description of the area shown on the map.

Soil Mapping. Surveys and maps are made for county areas. The actual process of mapping consists in walking over the land at regular intervals, generally about 80 rods, and taking notes by means of a "legend" and "field sheet" on soil differences and all related surface features, such as slope gradients, evidence of erosion, land use, vegetative cover, and cultural features. The field sheets are sometimes geological survey base maps, sometimes plane tabled by the soil surveyor himself, but in recent years contact prints of aerial photographs have become increasingly popular. Boundaries are drawn directly on the field sheets, representing in most places changes from one soil type to another.

Soil survey activities within a state are usually under the direction of the state experiment station. This organization shares the expenses of mapping with the Soil Survey Division of the United States Department of Agriculture, which inspects the work from time to time in order not only to maintain a certain working standard but also to correlate the mapping units in all sections of the country and to confirm the recognition of new soil types, phases, and series. After the field work has been completed the data are assembled, printed, and made available to the public by the United States Department of Agriculture.

There are over 2,850,000 square miles in the United States that are considered to possess reasonable possibilities for the production of

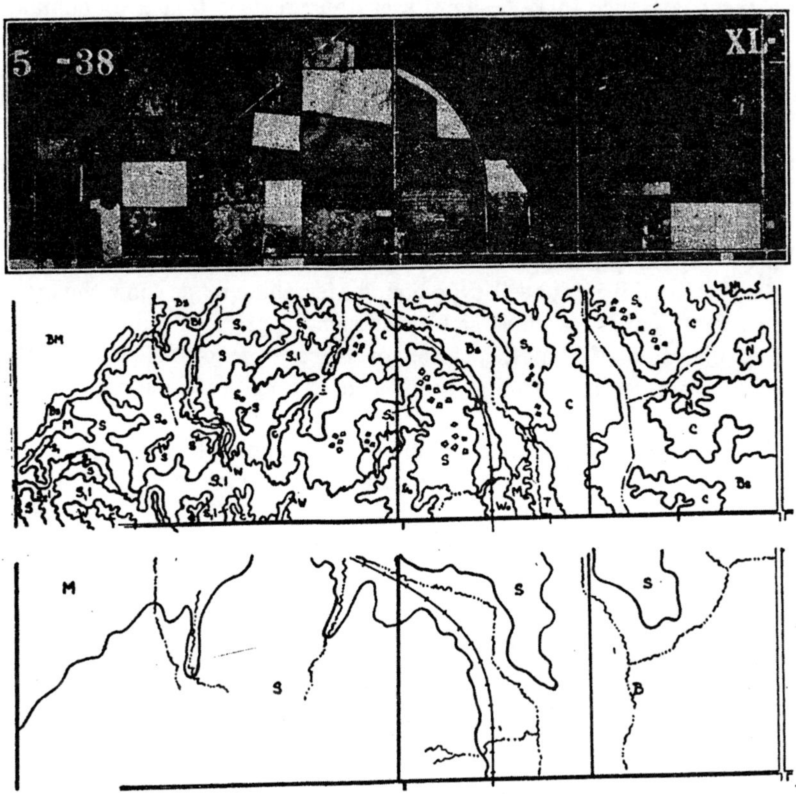

FIG. 16. Top: Aerial photograph of good farming land in glaciated area. Such photographs are used extensively for a base map in making soil surveys. Center: A detailed soil survey map of the same area. The symbols indicate soil types. Bottom: A land-type map of the area in which soil types of similar characteristics occurring in close association are mapped as a unit.

crops, grasses, and trees. Of this area, approximately 840,000 square miles have been studied and mapped in detail, the survey reports, including over 1,700 maps mostly on a county basis, having already been published. This figure includes surveys which are nearly complete of Puerto Rico and the Hawaiian Islands.

Types of Soil Surveys. Productive agricultural land is studied and mapped in *detail*. A *detailed map* is one in which the scale is one or more inches to the mile. This kind of work is expensive, costing from $25 to $40 for every square mile, that is, from 4 to 6 cents per acre. Large expanses of land are too poor, however, to justify this cost. These areas are mapped on small scales and the resulting study, called a reconnaissance survey, shows areas and regions that are dominated by a single soil type in contrast to the detailed map where all types and phases are separated. Detailed maps are still needed for about 1,250,000 square miles of land, whereas reconnaissance maps are needed for an additional 880,000 square miles.

Another kind of soil mapping is accomplished by the Soil Conservation Service. These maps are super-detailed, with scales often reaching 8 to 12 inches to the mile. Their purpose is to supply information necessary in planning land-use operations on individual farms and fields. In addition to soil-type boundaries, these maps show the extent of removal of the surface soil by erosion, location of gullies, gradient and direction of slopes, location of fences, size of fields, and other details needed in planning a soil-conserving, practical management plan.

In recent years still another kind of soil mapping has been employed, particularly in land of low agricultural value. Called *land-type mapping,* it consists in studying *soil-type associations* in contrast to merely separating areas in which a single type is dominant, as in reconnaissance surveying. *Land-type maps* are useful in studying extensive land-use problems, and they possess the desirable characteristics of being made quickly at a comparatively low cost.

Purpose and Value of Soil Surveys. The general purpose of a soil survey is to obtain an inventory of the soil resources of the country at large, of individual states, and more locally of counties and communities. More specifically, a soil survey gives information concerning soil quality, including topography, natural drainage, natural fertility, susceptibility to erosion, general crop adaptation, need for lime, and so forth, of any given community. In many cases this information can be applied to individual farms.

The soil survey is used by federal and state agencies, including colleges and universities, to obtain dependable information concerning soil conditions in given areas. Life insurance companies, banks, and other money-lending agencies use soil surveys in determining security for loans. Real estate companies and individuals interested in buying or selling land make extensive use of soil surveys. They are also used by highway and drainage engineers, by various kinds of manufacturers in selecting suitable locations for factories, and by merchandising and advertising companies in selecting areas for intensive campaigns. County agricultural agents and other extension workers find soil surveys helpful in their work. And last but by no means least, the farmers themselves are making increasing use of soil survey maps and reports in planning their management programs and in interpreting modern agricultural research into terms of their own farm conditions.

SOIL GROUPS IN RELATION TO CLIMATIC CONDITIONS

The information available from soil maps and soil survey reports makes possible broad geographic studies on the relationship between soils and their environmental conditions. Of these relationships none is more evident or of more academic interest than that which exists between the Great Soil Groups and climate.

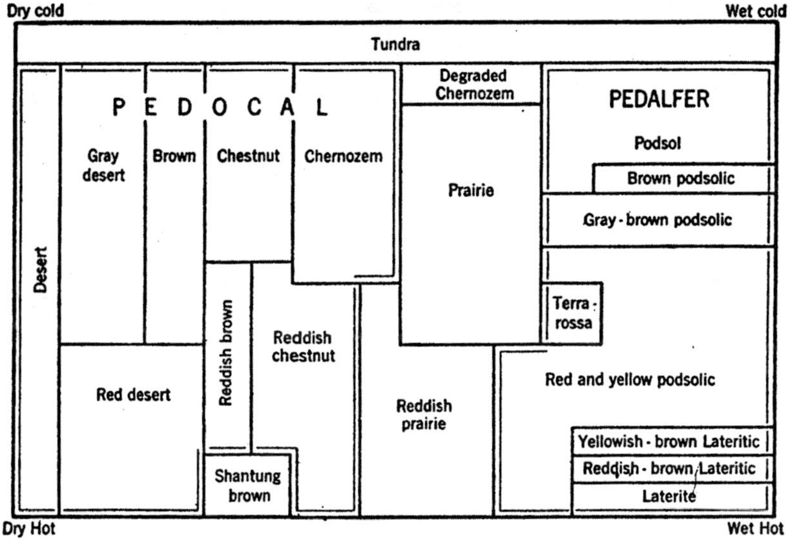

FIG. 17. A diagrammatic visualization of the relation between climatic conditions and the occurrence of the great soil groups. The relative size and position of each area indicates the extent and general geographic location of the soil group represented.

The rate of activity of the various forces of soil formation is greatly influenced by climatic conditions, as has been previously pointed out. Furthermore, the disposition of certain of the products of mineral decomposition is determined by the distribution, the quantity, and the nature of the precipitation. In view of these facts, it is inevitable that different climates must be related to the development of different profile characteristics. A study of Fig. 17 shows that there is a broad or general grouping of soils according to the climatic zones in which they were formed. The following questions emphasize important factors in these relationships. Limitation of space makes it necessary to confine the discussion to a few of the zonal soils.

Questions:

1. May soils be divided into groups on the basis of the leaching of salts from the surface layers?
2. In what soil groups was there a movement downward of iron and aluminum?
3. What soil groups contain an accumulation of salts?
4. What conditions gave rise to prairie soils?
5. In what soils do iron and aluminum accumulate in the surface layers?

Leached and Unleached Soils. Soils developed in a region where precipitation is sufficient to remove from the soil profile the very soluble salts of potassium and sodium, and the less soluble carbonates of calcium and magnesium differ in many respects from soils developed in a climatic zone where precipitation is insufficient to remove these compounds. A line drawn north and south across the United States near Lincoln, Nebraska, roughly marks the division between these two climatic zones.

Soils in Which Iron and Aluminum Have Leached. Not only are soluble salts leached readily from soils east of this dividing line but also a downward movement of iron and aluminum compounds takes place except where climatic conditions approach those of the subtropics. Suspended organic matter also moves downward. This leaching, which commonly results in a whitening of the soil layer immediately below the surface organic matter, is known as the *podsolization process* in soil formation. The more strongly leached part of the soil is known as the *horizon of eluviation.* The part in which iron, aluminum, and other compounds are concentrated is known as the *horizon of illuviation.*

The soils in this climatic zone are given a general or group name based on this tendency for iron and aluminum to accumulate in the profile. The first syllable of the Greek word *pedon,* meaning ground, is coupled with abbreviations of the Latin words *alumen* (aluminum) and *ferrum* (iron) to form the name *pedalfers.*

Gray-Brown Podsolic Soils. Eastward from the prairies to the Atlantic coast the soils were timbered. In the area bounded on the north by a line running east and west in the vicinity of Milwaukee, Wisconsin, and on the south by a line through Cairo, Illinois, are found the gray-brown podsolic soils. In these soils the horizon of humus accumulation is comparatively thin, ranging from 2 to 6 inches in thickness. The eluviated horizon is well developed, varying from grayish- to yellowish-brown in color. The horizon of illuviation is conspicuous because its color, varying from grayish-brown to reddish-brown, is generally darker than the color of either the parent material or the overlying leached horizon.

Podsols. North of the region of gray-brown podsolic soils, in a climate of lower temperatures and higher humidity (due to low evaporation), are developed the *podsols.* Here the humus horizon is of moderate thickness, the eluviated horizon is strongly developed, and the horizon of illuviation is conspicuous. The latter is usually brown to coffee-brown in color and is often characterized by a slightly cemented or indurated structure. Organic compounds together with iron oxides serve as cementing agents. The podsol profile represents the highest development of the podsolization process in soil formation.

Red and Yellow Soils. South of the region of gray-brown forest soils, high temperatures coupled with heavy rainfall accelerate the rate of mineral decay. The leaching of soluble products as well as the decay of organic matter are also increased. Soils developed in this climate are therefore characterized by a low accumulation of surface organic matter, a deep horizon of eluviation, and a deep, thick illuviated horizon in which the high rate of oxidation and hydration of iron produces bright red and yellow colors. The southern boundary of the red-and-yellow-soil area, where there is evidence in many places that the present soils were derived from old profiles developed under a more tropical climate than the existing one, merges into the lateritic zone.

Soils Contain Salt Accumulations. Soils in arid and semi-arid regions are characterized by an accumulation of calcium carbonate or by a mixture of $CaCO_3$ and $MgCO_3$ at some position in the profile. In recognition of this characteristic the Greek word *pedon* is combined with an abbreviation of the Latin word *calcis* or *calx* (lime) to produce the name *pedocal.*

In this group of soils the colloidal iron and aluminum oxides do not disperse and move downward in the profile because the soils retain an abundance of active bases. The neutral to alkaline reaction prevents dispersion.

Chernozems. In a wide north-south belt occupying the eastern part of the Great Plains of central United States is developed a group of soils containing little of the soluble salts of potassium and sodium but a striking accumulation of calcium salts occurring at a depth of approximately 24 to 36 inches. The natural vegetation of this region is composed of mixed prairie grasses, under which a relatively thick, granular, humus layer has accumulated. In places as much as 15 to 20 inches of dark surface material develop so that these soils are known as "black soils." They bear the name of *chernozem,* which

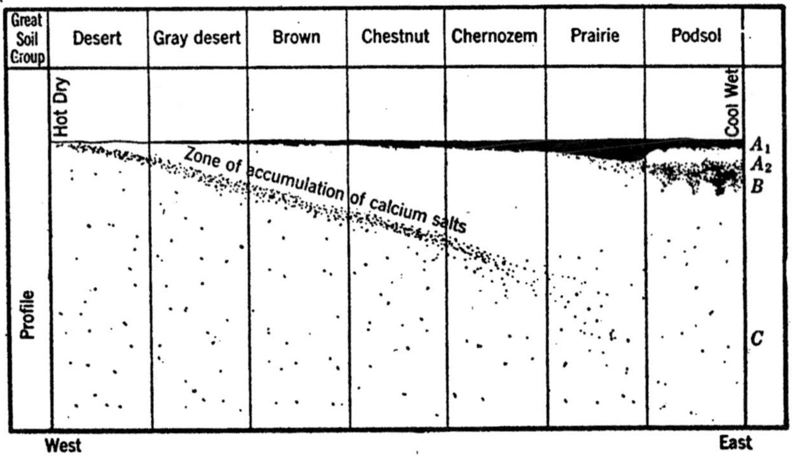

FIG. 19. Generalized profiles of the zonal soil groups in central and western United States. Note how the zone of accumulation of calcium salts increases in depth and diminishes with increase in precipitation. The thickness of the humus-rich layer also increases with increased precipitation until the grass vegetation of the prairies gives place to the forests of the podsolic region. The development of a clay-rich *B* horizon is prominent in the podsolic soils but disappears as one proceeds westward in the prairies.

was given to them when they were first classified on the steppes of Russia.

Chestnut Soils. As one travels westward from the chernozem belt into zones of less rainfall, the accumulation of calcium salts occurs nearer the surface (14 to 24 inches) and salts of sodium and potassium are present in increasing quantities. The vegetation becomes more sparse and shorter in stature. The very dark color of the soil gives place to a brown and the humus layer becomes thinner. This belt, lying just west of the chernozems, constitutes the *chestnut soils.*

Brown Soils. West of the chestnut soils is an area of light brown or grayish-brown soils. They are lower in humus content, support a still shorter and sparser grass vegetation intermixed with some shrubs, and are richer in soluble salt content. An horizon of accumulation of calcium carbonate occurs at a depth of 12 to 15 inches or less.

Desert Soils. Under desert conditions the humus content of soil becomes so low that the brown color of semi-arid regions is replaced by gray, which is intensified by a surface accumulation of salts. Usually a few inches of soil rest on the calcareous parent material. Shrubs replace grass and cacti appear in the warmer areas. Concentration of soluble salts is sufficient in some areas to produce "alkali" spots, which support only alkali-resistant plants.

Characteristics of Prairie Soils. Precipitation increases as one moves eastward from the chernozem belt, with a corresponding increase in height of the prairie grasses until, in what is now the heart of the corn belt, they stood at one time as high as a man's head. A deepening of the horizon of humus accumulation follows to some extent the increase in height of the grasses until a depth of 20 or more inches is reached. Accumulation of iron, aluminum, and clay in the illuviated horizon increases with increasing precipitation and the horizon of eluviation becomes more apparent. To a considerable extent the prairie soils developed in loessial parent materials and the topography varies from level to rolling with predominantly long and gentle rather than steep slopes.

Soils Showing an Accumulation of Iron and Aluminum. *Laterites.* As the humid tropics are approached the podsolic process gives way to one which is the reverse of it in one important characteristic. In the place of the hydrated oxides of iron and aluminum being removed from the horizon of eluviation, thus increasing the percentage of silica in the soil, the silica is carried downward, leaving a soil very high in content of hydrated aluminum and iron oxides. This process is called *laterization.* The resulting soil is not sticky and difficult to till, as might be expected from the high percentage of colloids present, but is easily permeable to water and lends itself readily to the preparation of a good seedbed. Although some soils in Florida are of a lateritic nature, no true laterites have yet been recognized in the United States.

AGE, RELIEF, AND PARENT MATERIAL IN RELATION TO SOIL GROUPS

In the preceding sections the ideal soil profiles in some of the great soil regions have been briefly discussed. It must be remembered, however, that by no means all soils in a zone have profiles that are identical with or closely approach the ideal. Unfavorable drainage conditions, topography, time that soil-developing processes have been active, and the great resistance of some parent material to weathering are

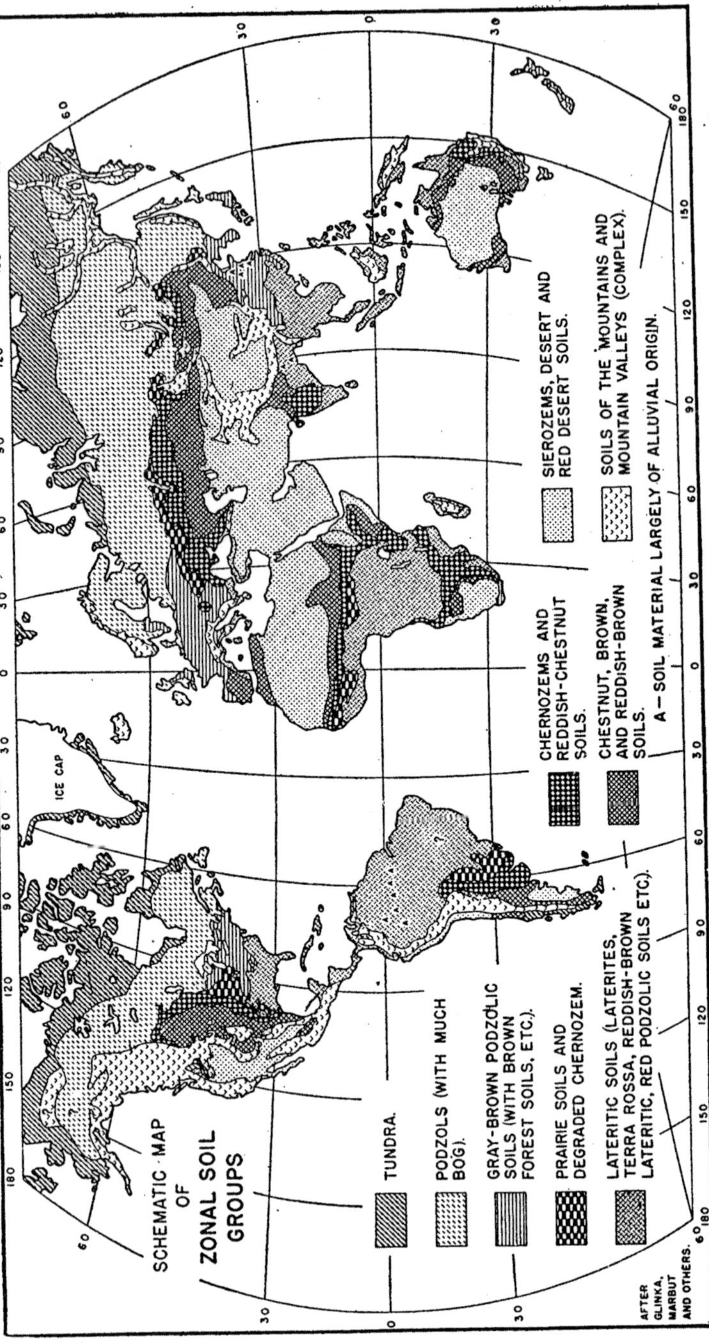

FIG. 20. A generalized map showing the approximate distribution of the zonal soil groups throughout the world. [From "The Soils That Support Us," by C. E. Kellogg. The Macmillan Company.]

among the factors which account for the lack of uniformity within a region. It is believed, however, that all soils within a group are gradually approaching the ideal in profile characteristics. And, lastly, in discussing soil areas it should be recognized that one group grades into another; just as there is no sharp line of division between soil types so there is no sharp line of division between great soil groups.

SOIL GROUPS IN RELATION TO VEGETATIVE COVER

Natural vegetative cover may be divided into the two general classes of trees and grass, and the soils supporting them are termed forest soils and prairie soils, respectively. There are several differences in soils developed in association with grass, which are of considerable agricultural significance. Furthermore, there are two types of forest, coniferous and deciduous, and two types of prairie grasses, the tall and short. The different effect of each kind of vegetation on the soil supporting it is brought out in the answers to the questions listed.

Questions:

1. Why is there a difference in humus content of prairie and forest soils?
2. Are equal quantities of nutrients returned to the soil annually by grass and trees?
3. Is action of microorganisms the same in prairie and timbered soils?
4. Does leaching proceed as rapidly under grass as under trees?
5. How are tundra soils developed?
6. By what process are organic soils formed?

Humus Accumulation. One of the first differences between prairie and timbered soils, when brought under cultivation, to attract the attention of a casual observer is the difference in color of the surface layer. Under similar drainage conditions prairie soils are much darker in color, owing usually to a higher humus content but in some instances to a difference in the quantity of calcium combined with the humus. The humus-rich horizon in these soils is also found to extend to a much greater depth. A partial explanation of these differences is found in the nature of the root systems of the two types of plants.

The fibrous-root system of grass completely fills the upper part of the soil with minute rootlets, which upon decaying leave the resulting organic matter thoroughly distributed within the surface layers. Tree roots, on the other hand, are large, with quantities of fine rootlets occurring very near the surface instead of being scattered in depth. In fact, many of them extend up into the more decayed sections of the litter. Only a thin layer of forest soil, therefore, is enriched in organic matter through rootlet decay. As the larger roots rot, the resulting organic matter is left in channels where it has no chance to be widely distributed throughout the soil.

The aerial portions of plants differ widely in their method of contributing organic matter to the soil. Grass falls more or less evenly

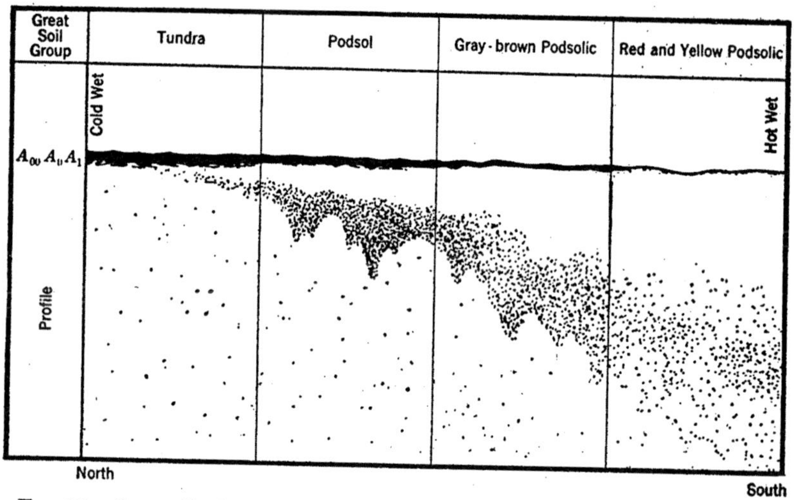

FIG. 21. Generalized profiles of zonal soil groups in eastern North America. Note how the accumulation of organic matter decreases with increase in temperature. The mineral material under the surface mat of organic matter in the tundra is very slightly altered. In the podsol zone eluviation is intense and a more or less indurated *B* horizon containing much humus and iron oxide is developed at shallow depths. In warmer climates eluviation extends to greater depths; the *B* horizon is thicker and is richer in clay with less iron and humus than in the podsol. The reds and yellows of the *B* horizon in the red and yellow podsolic soils are due to the state of oxidation and hydration rather than to an excessive quantity of iron.

on the ground, whereupon the dead stems and leaves may be dragged under the surface by worms and insects. The products of decay are also carried into the soil by water. Only a small part of the annual growth of trees, however, is returned to the soil surface in twigs and leaves. They contribute to the humus content of the soil in the same manner as grass tops if worm and insect action is equal, which frequently is not true. When large branches and tree trunks fall, the products of decay are deposited on limited areas instead of uniformly over the entire surface as with grass tops. When clearing forest land for cultivation, part of the leaf litter, dead logs, and branches are burned and hence never contribute their quota of humus to the soil.

Difference in Nutrient Cycle. Because of the completeness with which grass roots fill the soil and the annual habit of growth, large quantities of minerals are carried up to the leaves and stems and returned to the surface soil each year. This return of minerals to the

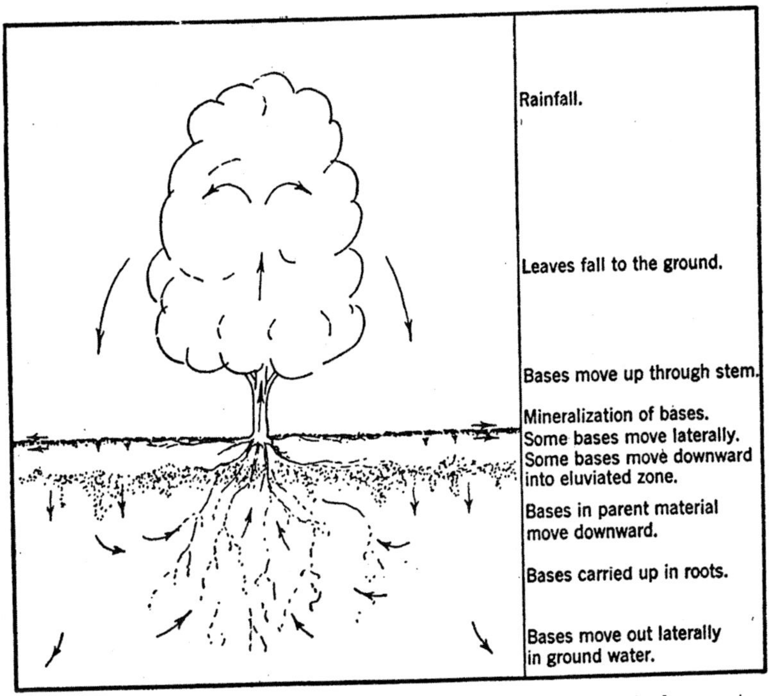

FIG. 22. The base movement in a forest soil. The base cycle has a strong influence on the properties of the soil developed under different types of vegetative cover. On the prairies all the leaves and stems of the grass with their content of bases are returned to the soil surface annually. In the forest a decidedly different cycle occurs since much of the nutrient supply taken up by the roots is stored in the tree trunks and branches.

surface layers results in a larger supply of available nutrients for the succeeding growth than would occur under timbered conditions, and it also delays the tendency for the soil to become acid. A grass cover also decreases the leaching away of soluble nitrogen compounds and hence leads to a further accumulation of humus because of the vital part played by nitrogen in humus formation (see Chapter VIII).

Much of the mineral and nitrogenous plant-nutrient content of leaves of trees is translocated to the woody parts of the plant before the leaves fall, and hence there is not the large annual return of these materials to the surface soil under a forest cover as there is under a grass cover.

Action of Microorganisms. In prairie soils both bacteria and fungi are active. On the other hand, under forest cover fungous action is thought to exceed bacterial action. This is especially true under coniferous forests. The products of the two types of decomposition are somewhat different, those resulting from the action of fungi being more acid in nature. This difference in the type of microorganic population may account in part for the difference in humus content of soils under the two types of cover.

Eluviation Is Slower in Prairie Soils. Because of the nature of the root system of grass which fills the soil completely, much soluble nutrient material is absorbed which would otherwise leach away. The high content of colloidal organic matter (humus) also provides for the retention or "fixation" of large quantities of nutrients. These factors, together with the large annual return of bases to the surface soil, are in part responsible for the slowness of development of an eluviated horizon in prairie soils. Such an horizon is found much more highly developed in timber land adjoining a piece of prairie than in the prairie soil.

Tundra Soils. Tundra soils are composed of a mat of living and dead grasses, mosses, sedges, and lichens. Little is known of the soil material underlying this mat, and apparently it has little effect on the accumulation of plant growth on the surface. The tundra occupies large land areas where the subsoil remains frozen throughout the year: those areas bordering the Bering Sea and the Arctic Ocean. The native vegetation varies, but on the average it consists of 30 per cent lichens, 25 per cent sedges, 25 per cent shrubs, and 20 per cent grasses, weeds and mosses. The tundra is capable of furnishing grazing for large numbers of reindeer, and some movement to utilize it for this purpose has been started.

Organic Soils. The characteristics of soils containing 20 per cent or more of organic matter are so dominated by the properties of this constituent that they are commonly called organic soils. These deposits are composed of partially decayed plant remains which have accumulated in bodies of water or on poorly drained areas where decay processes have been arrested through excess moisture. With these remains are mixed varying proportions of mineral soil which has blown or washed into the depression. In comparatively deep water various types of aquatic plants contributed the soil material. As the depth of the water lessened through the filling of the basin with plant remains or by improved drainage, reeds, rushes, sedges, grasses, and similar types of plants occupied the area and added their quota to the accumulating plant material. These plants were followed in turn by shrubs which ultimately gave place to trees as the water level lowered. During the various periods of accumulation mosses of various kinds, such as Sphagnum and Hypnum, grew on the surface of the water and sank to the bottom when their weight became too great, or grew over the swamp surface of the deposit.

When these organic deposits are yet in the crude, fibrous state, they are frequently designated as peat, but when decay has broken down the plant tissues until the material has something of a loamy consistency, it is called muck.

Deposits of organic soils are of common occurrence in the northern border states from Minnesota eastward, along the Atlantic seaboard, and in the Gulf coastal states. Considerable areas of these materials accumulated on bed rock near sea level in Florida or the Gulf coastal plain and, covered with tall sedges, grasses, and rushes, constitute the Everglades.

SOIL GROUPS IN RELATION TO POPULATION DENSITY AND PRODUCTION OF AGRICULTURAL PRODUCTS

Large cities generally develop in regions which produce large quantities of agricultural products. Some exceptions are due to the necessity of utilizing harbors for water transportation and of operating mines in non-agricultural areas. Not only do agricultural workers produce many of the articles required by factory workers but also they are themselves large consumers of the products of the factories. Moreover, a large number of persons are engaged in processing and marketing products of the soil. The mutual interests of the city and farm worker are many, and the common recognition of this fact will benefit everyone. The following questions will serve as guides in acquiring some knowledge of the agricultural products of each of the great soil groups and of the nearness of these areas to population centers where such products are consumed.

Questions:

1. What types of farming are followed on the great soil groups of the pedocal region?
2. What farming conditions prevail on the tall grass prairies?
3. How are the population density and agricultural development of the gray-brown podsolic area accounted for?

4. What type of agriculture do the podsols support?
5. Why are the red and yellow soils farmed so extensively?
6. How are organic soils farmed?

Chernozem Agriculture. Population is sparse in the chernozem belt because farming must be extensive rather than intensive, owing to limited rainfall. Chernozems are noted for their production of wheat not only in the United States but also in Russia and wherever climatic conditions have produced such soils. Other drought-avoiding and drought-resistant crops, such as varieties of the sorghums, are also grown. Corn and oats are produced to a limited extent, as well as considerable cotton in the South. Livestock raising is an important part of the farm program.

The soils are well supplied with mineral nutrients and nitrogen and are highly productive when supplied with moisture. The annual precipitation averages 20 to 26 inches. A mixture of tall and mixed grass is the prevailing native vegetation.

Agriculture of the Chestnut Soils. These soils constitute a considerable part of the spring wheat belt and also produce some sorghum crops for grain and forage and a little corn. Stock raising assumes a much greater relative importance than on the chernozems and extensive herds of cattle graze on the nutritious native grasses. The population is more widely scattered, for pastures must be large because of their relatively low carrying capacity. Ample water would make these soils highly productive, but the meager rainfall (15 to 22 inches) materially limits crop production.

Brown- and Desert-Soil Farming. On these soils low rainfall confines production of farm crops to irrigated tracts and to areas especially favored by location where dry-farming methods or above-average precipitation make crop production feasible. Wheat and corn are the chief crops. Grazing is extensive and large shipments of cattle move each year from this and the chestnut-soil zone to the feed yards of the corn belt. Dried buffalo and gramma grasses afford winter grazing for cattle. As precipitation decreases from about 18 inches to less than 5 inches, more shrubs, like greasewood and sage brush, and eventually cacti replace the grass.

Agriculture of the Tall-Grass Prairies. The tall-grass prairies lying east of the chernozems are the scene of extensive cereal and hay crop production and of intensive livestock feeding. Corn, oats, wheat, and soy beans are the principal grain crops. The clovers, alfalfa, and timothy, either alone or in some combination, are used for hay and pasture. Farms are comparatively large and are highly mechanized. Hog raising and cattle fattening are important enterprises. Dairying has developed extensively in the vicinity of larger towns and cities. Farming population is comparatively dense, and many moderate-sized cities have grown up to provide market places and processing facilities for farm products and to make implements and other manufactured goods for the farmer.

Annual precipitation ranges from 28 to 45 inches with a goodly proportion falling during the growing season. A midsummer drought is characteristic of the region, but nevertheless it contains a high percentage of the best farm land of the United States.

Farming in the Gray-Brown Podsolic Zone. Farming in this zone is more intensive with somewhat smaller farms than in the tall-grass prairie region. A larger variety of crops is grown, with less emphasis on corn and more attention to poultry raising, dairying, and truck crop production. Cattle and lamb feeding is followed and a large number of hogs are raised. The population density is greater, both urban and rural, than in the prairie regions, and a larger number of factories are found in both cities and small towns.

Precipitation ranges from 30 to 50 inches and although summer droughts are frequent, moisture relations are on the whole satisfactory. The growing season is long and summer temperatures are not excessively high except in the southern part of the zone. Soils are of moderate fertility, having suffered considerable leaching and varying losses from erosion. The humus content is medium. It is the combination of favorable climatic conditions and moderate to good soil fertility, permitting a highly diversified agriculture, that has made the gray-brown podsolic areas of the world the scenes of a highly developed agriculture and the location of population centers.

Farming the Podsols. Long winters and short, cool summers mark the podsol zone. Soil moisture is plentiful because of low evaporation. Soil fertility varies from low to high. Hay, potatoes, and roots are the main crops, with a limited production of grain. Because of the abundant forage, dairying is common. Butter and cheese are the main products because they are easily shipped considerable distances to market, which fact is important because the podsols support only a small local population. Cities in this soil zone owe their existence to lumbering and mining rather than to local agriculture.

Crop Production on the Red and Yellow Soils. The combination of ample precipitation, comparatively high temperatures, and long growing season, all of which permits the growing of many diversified crops, has resulted in a high degree of agricultural development in the red and yellow podsolic regions which support a dense population despite the mediocre productivity of the soils themselves. Farms are usually small to medium sized. An annual precipitation, varying from 40 to 60 inches, has caused severe leaching of the soil and the development of an acid reaction in addition to much soil deterioration through erosion losses. Rapid decay of organic matter results in only a small humus accumulation. These factors, coupled with the high value of many of the special crops grown, have led to the use of large quantities of commercial fertilizers.

Livestock farming is on the increase but it has not yet developed to the extent desirable, and more diversification in the cropping system is advisable. Aside from cotton, many special crops, such as tobacco, peanuts, vegetables, nuts, citrus fruits, and small fruits, are produced. There are many wood lots and forested areas.

Farming Organic Soils. When organic soils are drained in such a way as to remove excess water rapidly but yet maintain the water level at a relatively shallow depth, they may be used for very intensive types of crop production. In the northern states these soils are used for the production of onions, celery, mint, potatoes, cabbage, cranberries, carrots, and other root crops. Corn is produced to some extent and considerable areas are used as pasture. Late spring or summer and early-fall frosts are the greatest hazard to crop growth. A great variety of special crops is grown on the organic soils of the South and East. Special methods of tillage, coupled with careful and copious fertilization, are required to bring these soils to their highest state of productivity.

CHAPTER III

PHYSICAL AND CHEMICAL PROPERTIES OF SOILS

The physical properties of a soil have much to do with its suitability for the many uses to which man puts it. The rigidity and supporting power, both wet and dry, the freedom of drainage, moisture-storage capacity, plasticity, ease of penetration by roots, aeration, and retention of plant nutrients are all intimately connected with the physical condition of the soil. It is pertinent, therefore, that persons dealing with soil in any way should be acquainted with the physical nature of different soils, knowing to what extent and by what means man can alter those properties—whether it be as a medium for plant growth or as a structural material in the making of highways, dams, and foundations for buildings or other engineering purposes, in manufacturing brick and tile, or in building golf courses, athletic fields, and running tracks.

Aside from the chemical composition of soils as expressed in terms of the percentage of different elements present and the combination or compounds in which they occur, the most widely studied chemical phenomenon in soils is that of base exchange. In recent years the composition and structure of the exchange complex and the laws governing exchange reactions have received intensive study because of a growing realization of the determining role played by the colloidal fraction in soil productivity.

The study of the physical properties of soils and of base exchange will be presented under seven topical divisions.

Objectives:

A. Making a mechanical analysis.
B. Properties of the soil separates.
C. Soil structure.
D. Tillage operations and soil properties.
E. Porosity and weight of soil.
F. Soil color.
G. Soil temperature.

MAKING A MECHANICAL ANALYSIS

The physical properties of soils are dependent largely on the surface exposed by the particles and on the content of organic matter, although the chemical properties of the particles are also a contributing factor. The surface exposed by the particles is primarily a function of their size, and hence the study of the physical properties of a soil may well begin by determining the percentage of the different size groups or separates present; in other words, by making a mechanical analysis. The significant points relative to the making of mechanical analyses of soils are indicated by the following questions.

Questions:

1. How can the tendency for the particles to stick together be overcome?
2. How are the sands separated from the silt and clay?
3. By what means are the different size groups of sand separated?
4. What is the procedure for separating silt from clay?
5. What is the hydrometer method?
6. What is the procedure for the pipette method?

Dispersing the Soil. Before a soil can be divided into groups of particles on the basis of size, it is essential to overcome the tendency of the very small particles to cling to the larger ones and to each other. Every particle must exist as an individual. Organic matter is one of the chief agents which binds particles together, and hence the first problem is to destroy this cementing material. The commonly accepted procedure is to oxidize the organic matter by boiling the sample in hydrogen peroxide solution.

Other cementing agents are the oxides of iron and aluminum. They may be dissolved by suitable chemical reagents, but it is doubtful if such a procedure is advisable as the oxides are a part of the very fine or colloidal clay and should be determined as such.

The usual procedure is to shake the sample, after removal of the organic matter, in water in order to surround each small particle with a film of water which helps keep the particles apart. Usually a chemical such as sodium oxalate is added to the suspension in order to precipitate calcium, magnesium, and similar elements having a tendency to cause the fine particles to gather in clusters (floccules). The sodium also combines with the fine clay particles and decreases their tendency to cohere. The sample may be shaken moderately for several hours or it may be stirred violently for a few minutes in a stirring machine similar to those used in making malted milk drinks.

Separating Sand from Silt and Clay. The sand particles may be separated from the silt and clay by shaking the dispersed sample in a tall cylinder of water and decanting or siphoning off the turbid liquid when the sands have settled. Several decantations are necessary to effect complete separation.

Separation of Sands. The simplest procedure for separating the different-sized sand groups is by means of carefully calibrated screens.

Silt and Clay Separations. Usually clay is separated from the silt by the principle of sedimentation but settling is generally facilitated by means of centrifugal force. The suspension of silt and clay is placed in the tube of a special centrifuge which is run for a given time at a speed sufficient to throw the silt to the bottom of the tube, leaving the clay in suspension.

The Hydrometer Method. Bouyoucos has devised a method of determining the silt, clay, and sand content of a soil without separating

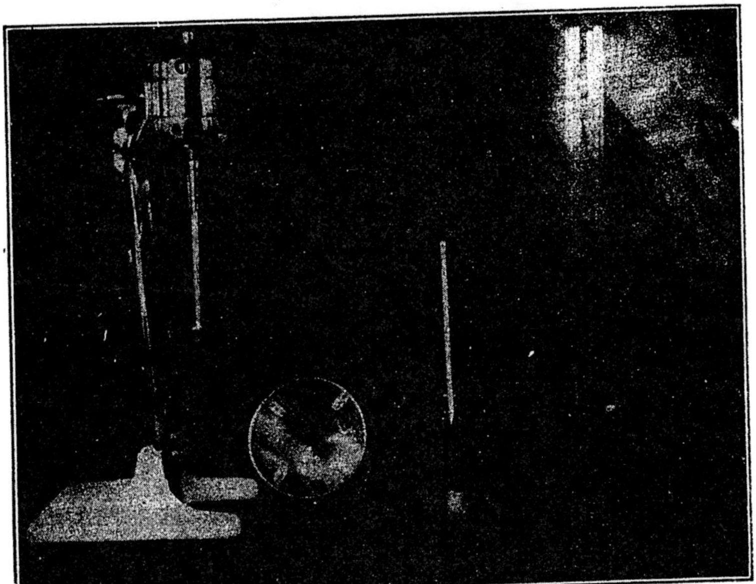

FIG. 23. Showing equipment used in making a mechanical analysis by the hydrometer or Bouyoucos method. The cup in which the soil is stirred is inclined to show the baffles which are essential if complete dispersion is to be obtained.

them. The sample is dispersed in a stirring machine by using a dispersing agent. The suspension is then poured into a tall cylinder and a specially designed hydrometer placed in it. By reading the hydrometer at the time interval (40 seconds) required for the sand (not less than 0.05 mm. in diameter) to settle, the quantity of silt plus clay in suspension can be determined, and hence the quantity of sand by difference. When sufficient time (1 hour) for the silt (not less than 0.005 mm. in diameter) to settle has elapsed, the percentage of clay may be determined by reading the hydrometer, and so the quantity of silt calculated. Likewise, the percentage of clay (0.002 mm.) can be found by taking a reading at the end of a 2-hour period. The very fine or colloidal clay is determined after a 24-hour period of sedimentation by the use of an especially sensitive hydrometer.

The Pipette Method. In this method the dispersed suspension is poured into a standard-sized cylinder and after a predetermined time a given volume of the suspension is drawn off from a given depth by means of a pipette. The withdrawn portion is evaporated and the sediment weighed. Thus the density of the suspension is determined at a given depth as a function of time. By repeating these measurements at given time intervals the percentage of particles present with different settling velocities can be determined. As the settling velocity varies with particle size, the percentage of particles of different sizes in the sample may be determined by calculation.

Many other methods for making a mechanical analysis of a soil have been suggested, most of them based on the principle that the size of particle which can be carried by flowing water depends on the velocity of flow; the methods described, however, are those in most general use.

PROPERTIES OF SOIL SEPARATES

Soils are divided into textural classes on the basis of the proportion of the different separates they contain. To understand the properties of the soil classes, it is essential to have a knowledge of the properties of the different soil separates. Except for the colloidal fraction the interiors of the soil particles are not exposed, and hence their chemical and physical activities must take place on the surface of the particles. The surface exposed by particles becomes, accordingly, of primary importance in studying their properties. The very fine particles, because of their crystal structure, have an active internal as well as external surface. The questions below are submitted as guides in the study of the soil separates.

Questions:

1. How do the separates vary in composition?
2. What are the chief functions of sand in soils?
3. Is silt an active soil constituent?
4. In what ways does colloidal clay differ from other clay?
5. How are the clay minerals constituted?
6. What are the chemical properties of the clay minerals?
7. What is the process of base exchange?
8. How is the colloidal fraction of the soil determined and what materials does it contain?

Composition of Soil Separates. As might be expected, the coarser soil separates contain a much higher proportion of minerals resistant to weathering, such as quartz, than do the finer separates. The content of several plant nutrients such as P_2O_5, K_2O, CaO, and MgO is appreciably higher in the finer than in the coarse separates, as is shown by the data in Table 2. There is, accordingly, a justification for the assumption that in general fine-textured soils are richer in total nutrients than are coarse-textured soils.

TABLE 2

COMPARATIVE NUTRIENT CONTENT OF SOIL SEPARATES *

Soil Origin and Number of Soils Analyzed	P_2O_5			CaO			MgO			K_2O		
	Sand	Silt	Clay	Sand	Silt	Clay	Sand	Silt	Clay	Sand	Silt	Clay
7 coastal plain soils	0.03	0.10	0.34	0.07	0.19	0.55	0.09	0.14	0.61	0.37	1.34	1.76
3 soils developed from crystalline and metamorphic rocks	0.07	0.22	0.67	0.50	0.82	0.94	0.48	0.86	1.24	1.60	2.37	2.86
10 glacial soils	0.15	0.23	0.86	1.24	1.30	2.69	0.54	0.88	1.80	1.72	2.35	3.08
5 soils developed from limestones and shales	0.19	0.17	0.49	7.55	6.82	6.67	0.44	0.52	1.84	1.46	1.95	2.67
2 arid soils	0.19	0.25	0.45	9.09	9.22	8.04	1.49	2.98	5.34	3.05	4.16	5.07
General average	0.12	0.18	0.61	2.61	2.57	3.07	0.47	0.77	1.70	1.41	2.15	2.78

* Diameter of separates—sand, 2.00–0.05 mm.; silt, 0.05–0.005 mm.; clay, 0.005–0 mm. Compiled from data in "The Mineral Composition of Soil Particles," G. H. Failyer, J. G. Smith, and H. R. Wade, Bul. 54, Bureau of Soils, U.S.D.A., 1908.

Functions of Sand. Sand particles are of comparatively large size and hence expose little surface compared to that exposed by an equal weight of silt or clay particles. Because of the small surface of the sand separates, they take an almost negligible part in the chemical and physical activities of a soil unless some of the particles are composed of $CaCO_3$, $MgCO_3$, or some other appreciably soluble compound. The smaller sand particles may contain a sufficient coating of the very small clay separate to attain some activity, but the action is due really to the clay and not to the sand. Because the sands are inactive, their chief function in soil is to serve as a framework around which the active part of the soil is associated.

Unless present in too small proportion the sands increase the size of spaces between particles, thus facilitating movements of air and drainage water. They also add considerably to the weight of the soil, as will be pointed out later in discussing soil structure and volume weight.

Silt in Relation to Soil Properties. The coarser silt particles are very similar to the finer sands in surface exposed and hence take very little part in the chemical activities of soils. The finer silt has sufficient surface to give it some slight chemical activity and if present in considerable quantity must be considered in determining the total activity of the soil. There is an appreciable tendency for the very fine clay to adhere to the surface of the silt particles which derive some activity from this source.

The minerals composing the silt particles have undergone little decomposition in certain soils, primarily those in areas of limited rainfall and in soil derived from recently deposited glacial or loessial material. In such soils the silt may yield appreciable quantities of nutrients to the soil solution or to the colloidal clay.

In themselves, silt particles have little tendency to stick together or to adhere to other particles and hence have little influence on the particle arrangement or structure of the soil.

Clay vs. Colloidal Clay. The term clay is rather loosely used. In early methods of mechanical analysis all particles with a diameter less than 0.005 mm. were considered clay. Further investigation showed that the largest of these particles, those with a diameter of 0.002 mm. and greater, had comparatively little activity and hence it was argued that they should be included in the silt separate. A given weight of particles with a diameter of 0.0001 mm. has 20 times the surface of particles with a 0.002 mm. diameter and, accordingly, is very much more reactive. On the basis of activity, it has seemed advisable to divide the clay fraction into two groups, calling the finer material colloidal clay (from the Greek words *kolla*, meaning glue, and *oid,* meaning like) and setting 0.0002 mm.[1] as the upper limit of diameter for this material. This size limit is the subject of considerable discussion and chemists are inclined to place it lower. The laboratory for chemical and physical investigation of soils, United States Department of Agriculture, uses 1 micron as the upper limit.

The coarser clay, that ranging from 2 microns to 200 millimicrons in diameter, has considerable reactivity and plays a definite part in the physical and chemical properties of soil. It is particularly effective when acting in association with the colloidal fraction.

Clay Minerals. The sand and silt of a soil are composed primarily of fragments of minerals and hence are dense or non-porous. Clay, on the other hand, and especially colloidal clay, is made up largely of minerals built up or synthesized in the soil. A chemical analysis of clay shows it consists largely of silicon, aluminum, and water, with varying quantities of iron, calcium, magnesium, potassium, and sodium. Small amounts of other elements may be present. The minerals are built up of layers or plates of SiO_2 and $Al_2O_3XH_2O$. In some cases portions of the Al are replaced by Fe or Mg. The kaolinite group [2] of minerals consists of 1 layer of silica and 1 of alumina, whereas the montmorillonite group is built of 2 plates of silica and 1 of alumina. The constitution of these minerals is discussed more fully on pp. 28–29. The X-ray shows these minerals to have definite crystal structure with a large amount of space between atoms composing the crystal. The crystal composed of three plates has much more internal space than that made up of two plates. The internal surface of these crystals added to the exterior surface gives them a much greater activity than they would otherwise possess.

The shape of the clay minerals is also a contributing factor to their activity. If they are platelike in structure, they have much more surface than if they are spherical or cubical. The shape of these minerals also determines the manner in which they pack together. Platelike forms adhere tightly when they are arranged flat side to flat side.

[1] 0.001 mm. constitutes 1 micron or μ; and 0.000001 mm., 1 millimicron (1 $m\mu$). 200 $m\mu$ would then be equal to 0.0002 mm. For convenience these designations of size will be used in referring to clay and colloidal clay.

[2] The minerals of the kaolinite group in addition to kaolinite are dickite, halloysite, metahalloysite, and nacrite. These have a 1 : 1 crystal lattice (1 plate of SiO_2 to 1 plate of Al_2O_3) and are fixed or non-expanding. The montmorillonite group includes also beidellite and nontronite. They have a 2 : 1 crystal lattice that expands on wetting. A third group of clay minerals with a 2 : 1 fixed crystal lattice is known as the illite group. Specific names have not yet been given to the members.

Chemical Properties of Clay Minerals. Chemically many clay minerals are similar to the negative radical of an acid, as PO_4^{---} or SO_4^{--}. In other words they are complex, negatively charged nuclei which will move toward the positive electrode when submitted to the action of an electric current. They may be pictured roughly as follows: The inner layer of negative charges is a part of the wall of the particle.

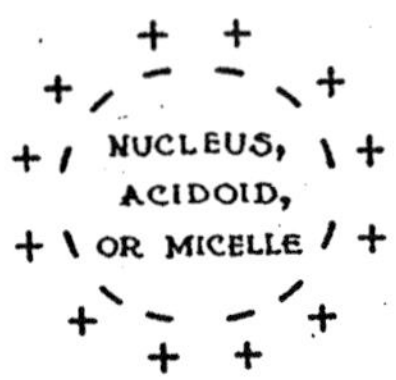

The outer layer is at a distance of molecular dimensions from the inner layer. In this diagram the inner layer is of negative charges and the outer of positive. In some cases the situation is reversed although the arrangement shown is the usual one. The outer layer may be replaced by other ions of a similar charge as K^+, Na^+, Ca^{++}, Mg^{++}, etc. As might be expected the behavior of a clay particle is greatly influenced by the nature of the ions in the outer layer.

Base Exchange. The fixation or adsorption of a positive ion by the colloidal nucleus, acidoid or micelle, and the accompanying release of one or more ions held by the nucleus is termed *base exchange*. For example, assume that the nucleus has one-half of its capacity satisfied with Ca ions, one-quarter with K ions, and one-quarter with H ions. Diagrammatically the situation would be as shown. Now suppose the

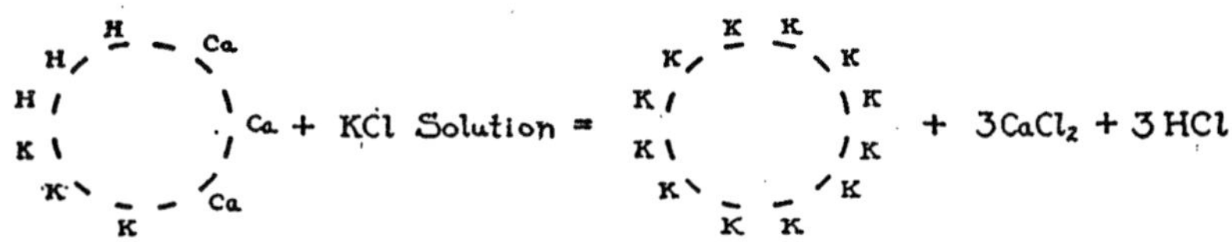

colloidal material is treated with a strong solution of KCl. First the K ions replace the Ca ions on the complex, and they in turn combine with the Cl ions to form $CaCl_2$. Upon continued treatment with the KCl solution the H is replaced. Thus by suitable procedure all the exchangeable bases on the nucleus may be replaced by any desired ion.

The efficiency with which ions will replace each other is determined by such factors as (*a*) relative concentration or numbers of the ions, (*b*) the number of charges on the ions, and (*c*) the speed of movement or activity of the different ions. The first factor is an application of the well-known chemical law of mass action. The greater the number of charges carried by the ion the greater its efficiency, other factors being the same. The speed or activity of an ion is primarily a function of its size, in which the degree of hydration must also be considered. If we consider the ions Li, Na, K, and Rb, we find that they are listed in the ascending order of size, and hence we would expect their efficiency of replacement to be Li Na K Rb. In reality, however, the Li ion associates itself with so many water molecules that its speed is much reduced and, furthermore, it cannot get near the micelle because of the shell of water. Likewise the Na ion is more highly hydrated than the K ion. The result is that the order of replacement is reversed, being Rb K Na Li. A listing of the more common ions in descending order of replacing power is as follows: H Sr Ba Ca Mg Rb K NH_4 Na Li.[3]

TABLE 3

IONIC RADII, HYDRATION, VELOCITY, AND EXCHANGE EFFICIENCY OF SEVERAL IONS

Ion	Radii of Ions in Angstroms (10^{-8} Cm.)		Velocity of Migration in Microns per Sec. per Volt per Cm.	No. of Charges Carried	Order of Base-Exchange Efficiency
	Dehydrated	*Hydrated*			
Li	0.78	10.03	3.45	1	8th
Na	0.98	7.90	3.31	1	7th
K	1.33	5.32	....	1	5th
NH_4	1.43	5.37	3.48	1	6th
Rb	1.49	5.09	3.25	1	4th
H	....	....	2.84	1	Ist
Ca	1.06	....	3.27	2	2d
Mg	0.78	....	3.18	2	3d

The total base-exchange capacity of a soil is expressed by the milliequivalents (m.e.) of ions which 100 grams of soil will adsorb. When the complex has its entire capacity satisfied by H^+, it is said to be "unsaturated." When other ions satisfy a part of the exchange capacity the degree of saturation is expressed as the percentage of the total capacity satisfied by such ions.

The expanding 2 : 1 crystal lattice of the montmorillonite minerals gives them a much greater base exchange capacity than have the kao-

[3] There are certain difficulties in accepting the idea that the hydration of ions is a determining factor in base exchange; yet there is much evidence in support of this theory and it is believed to be a useful concept. However, according to Grim, cation exchangeability cannot be explained satisfactorily on the basis of cation hydration. He points out that some cations, previously thought to be highly hydrated, probably do not hydrate at all and that other cations hydrate to a lesser degree than has been assumed. Modern Concepts of Clay Minerals, R. E. Grim, "Journal of Geology," Vol. 50, 1942, pp. 225–275.

linite minerals. The chemical composition of the minerals also affects their capacity to fix ions. In general the greater the ratio of silica (SiO_2) to sesquioxides ($Al_2O_3 + Fe_2O_3$), the greater the fixing capacity. This ratio is calculated as follows:

$$\frac{\%\ SiO_2}{\text{Molecular wt. of } SiO_2\ (60.1)} = A$$

$$\frac{\%\ Al_2O_3}{\text{Molecular wt. of } Al_2O_3\ (101.9)} = B$$

$$\frac{\%\ Fe_2O_3}{\text{Molecular wt. of } Fe_2O_3\ (159.7)} = C$$

$$\frac{A}{B + C} = \text{Silica : sesquioxide ratio}$$

Colloidal Fraction of Soil. In addition to the clay minerals the colloidal fraction of a soil usually contains a greater or less amount of organic matter and Fe_2O_3, Al_2O_3, and SiO_2 in various stages of hydration. The colloidal organic matter has the capacity to take up or adsorb ions in much the same way as do the clay minerals. The capacity of the colloidal fraction of the soil to adsorb or "fix" nutrients makes it a storehouse of available nutrients for plants. Soils which have little colloidal material have very low productivities.

The colloidal portion of soils also furnishes the binding materials which hold soil particles together in small groups or clusters or, on occasion, in small or large clods. Without the colloidal material soil particles would function largely as individuals with little coherence, as do sands.

The colloidal content of a soil may be determined by various methods. The hydrometer method, as previously described, may be used. Other methods are based on the capacity and avidity with which colloids adsorb water and on the assumption that the colloidal fraction is the only portion of a soil that adsorbs water. For example, the quantity of moisture adsorbed from a saturated atmosphere over 3.3 per cent H_2SO_4, by a given weight of colloid extracted from a soil and spread in a thin layer, is determined. A similar determination is made on a sample of the soil itself. The colloidal content is then calculated by the equation:

$$\frac{\%\ H_2O \text{ adsorbed by 1 gram of soil}}{\%\ H_2O \text{ adsorbed by 1 gram of colloid}} \times 100 = \%\ \text{colloid content}$$

Another method is to determine the heat evolved when a given weight of oven-dried colloid is moistened with water. The heat of wetting the soil is then determined. By substituting the heat of wetting in calories per gram of colloid and of soil in the above equation, the colloid content may be determined.

SOIL STRUCTURE

The term texture is used in reference to the size of soil particles, but when considering the arrangement of the particles the term structure is used. The structure of the different horizons of a soil profile is an essential characteristic of the soil just as are color, texture, or chemical composition. Structure influences moisture relations, availability of plant nutrients, action of microorganisms, and plant growth. Types of soil structure and the factors which influence the formation of them are suggested by the following list of questions.

Questions:

1. What is meant by single-grained structure?
2. What forms of compound soil structure are recognized?
3. By what process are secondary soil particles developed?
4. How do moisture and temperature conditions influence extent of soil granulation?
5. In what ways does plant growth influence granulation of soils?
6. How do wetting and drying and freezing and thawing affect the compound structure of soils?

Single-Grained Structure. If each particle in a soil functions as an individual, that is, if it is not attached to other particles, the structure is called single grained, and the individual particles are known as primary particles. Such a structural condition may exist in very sandy soils or in the eluviated horizon of podsolic soils from which the clay and organic matter have been leached. Some of the particles in many soils exist in the single-grained condition, but to find all the particles so functioning is unusual. If soils containing considerable clay are tilled or are tramped by grazing animals when too wet, a portion of the clay particles assume the single-grained condition.

Compound Structure. When particles adhere in clusters and the clusters then function as particles, a compound structure is developed. These clusters or compound grains may be called secondary particles. Some of the very small particles combine into clusters too small to be seen by the naked eye and so may be said to develop a microstructure. The small units together with primary particles unite into larger and visible units of various shapes and sizes to develop a macrostructure. In the last few years much study has been devoted to soil structure and particularly to the micro units. The terms granules and aggregates are used to designate compound or secondary structural units without regard to size or shape.

Various systems have been suggested for classifying the macrostructural units of soil, other than primary particles, but none has been universally accepted. The following system is as useful as any and sufficiently detailed for most purposes.

I. Particles or aggregates with all axes of approximately the same length, somewhat spherical or cubical in shape but exposing many rather than few faces, some of which are rounded.
   a. Nutlike structure—units of considerable size, greater than 6 mm. in diameter.
   b. Granular structure—units of smaller size, from 6 to 3 mm. in diameter.
   c. Crumb structure—units are small, porous aggregates which are quite easily crushed.

II. Particles or aggregates in which the vertical axis is appreciably longer than the others.
   a. Prismatic structure—units with faces moderately to well defined.
   b. Columnar structure—units are prismatic but have rounded tops.

III. Particles or aggregates in which the horizontal axes are appreciably longer than the vertical.
   a. Platy structure—units decidedly flattened but yet of considerable thickness.
   b. Laminar structure—units are quite thin.

IV. Particles or aggregates with no uniformity in length of axes, faces well defined but irregular in number and form.
   a. Blocky structure—units of considerable size, greater than 1 cm.
   b. Fragmental structure—units of smaller size, less than 1 cm.

Formation of Secondary Particles. An understanding of the causes for the development of compound structures in soils is of practical importance since structure has a great influence on plant growth and is also an important factor in soil classification.

Evidently there must be some agent which draws the particles together in order to produce aggregates and also some means by which they are bound rather firmly so that the structural forms may persist. Evidence points to the colloidal fraction as the active constituent in compound structural units, since without its presence single-grained structure prevails. It has been shown that there is a high degree of correlation between the amounts of clay (5 microns) in soils and the percentage of granules with a diameter greater than 0.05 mm. A similar correlation was found between percentages of aggregates of the same dimensions and the percentage of organic matter. Three groups of colloidal matter are thought to serve as cementing material in aggregate formation: (1) clay minerals, (2) colloidal oxides of Fe and Al, and (3) colloidal organic matter.

There are several theories concerning the process by which aggregation is brought about, but the following is probably as widely accepted as any. Because colloidal particles are charged bodies, water molecules attach themselves readily and firmly to them. As water molecules and colloidal nuclei carry both positive and negative charges, it is conceivable that oriented water molecules may connect colloidal particles. The linkage may include adsorbed cations because aggregation seems to be affected by the cations adsorbed.

Such water molecules are held very tightly, and as water evaporates from the soil the number of water molecules attached to two particles increases, and at the same time the length of each linkage becomes shorter and stronger and so pulls the colloidal particles close together. Larger soil particles to which the colloids are attached may be drawn into the cluster and some may be surrounded as the cluster or floccule develops. As more water is lost and the colloidal material becomes further dehydrated, it sticks or cements the particle group into a unit which is permanent until rehydration of the colloid reduces its binding power. Thus it is seen that water, acting in conjunction with the colloids, constitutes the force which produces granulation of soil and that colloidal material is the final binding agent.

Many soluble salts also exert a strong flocculating effect on colloids, and hence their presence may be a factor in inducing granulation in soils. Applications of lime to very fine textured soils in poor physical condition may noticeably increase granulation.

These processes of soil aggregation are greatly facilitated by other factors and processes which help to create lines of weakness in a soil mass and so permit groups of soil particles to be drawn around a new nucleus and finally cemented into a structural unit. In the following paragraphs some of the agencies which aid in soil granulation are discussed.

Influence of Climate on Aggregation. A relationship between the climatic conditions which have contributed to the formation of the great soil groups and the extent of aggregation or development of compound soil structure has been pointed out. Moisture conditions which have produced the chernozems and chernozem-like soils have led to a higher degree of aggregation than exists in the drier soils to the westward or in the forest soils to the east. The explanation lies in the fact that moisture supply was adequate for the development of considerable clay through mineral decomposition but was not sufficient to eluviate the clay as in the timbered soils. In the drier soils too small an amount of clay was developed to promote extensive aggregation. The large amount of organic matter in the prairie soils also promoted the development of a compound structure. The high organic content and lack of eluviation of the humid prairies led to a high degree of aggregation in these soils.

Low temperatures also promote the development of compound soil structures, as is evidenced by the much higher degree of aggregation in the northern than in the southern chernozems. The explanation of this situation probably lies in the greater content of organic matter accumulated under lower temperatures when moisture conditions are relatively constant. A reversal of the effect of temperature on degree of aggregation is found in the soils of the humid region in that the lateritic soils of the humid South show a high aggregation of the silt and clay (about 75 per cent). This is due to the binding power of the dehydrated oxides of iron and aluminum. The podsolic soils of the northern humid region show a much lower development of compound structure because of eluviation of clay and organic colloids, as previously mentioned.

Plant Roots Induce Granulation. It has long been observed that vegetative cover influences the development of compound structure in soils. Grass appears to be the most efficient type of vegetation in inducing aggregation probably because of the completeness with which its root system fills the soil. Several means by which plant roots cause aggregation have been suggested. They may be summarized as follows: (1) Roots and root hairs penetrating the soil produce lines of weakness along which the clod or soil mass may break into granules. (2) The pressure exerted by developing roots may induce aggregation. (3) Root secretions may flocculate colloids, stabilize, or cement aggregates. (4) Use of moisture by roots may cause dehydration of colloids, thus resulting in shrinkage and finally in cementation. (5) Organic colloids produced from decay of roots may be an aggregating agent.

Wetting and Drying Influence Structure. The granulating effect of the wetting and drying of soil is familiar to all who work with or till the soil. The disintegration of clods into granules under the influence of alternate wetting and drying is a common observation. The explanation of the process probably lies in the alternate swelling and shrinking of the colloids. Uneven expansion and contraction, which must occur because a portion of soil is never wet or dried uniformly and simultaneously, result in numerous cracks which separate the soil into smaller pieces. They may be drawn apart into units as shrinkage of the colloid occurs.

A contributing factor in disrupting a fragment or clod is the compression of the air trapped in the interparticle spaces as the exterior of the clod is wet. As the moisture moves inward the pressure becomes greater until finally the compressed air explodes the clod into fragments.

Aggregation Induced by Freezing and Thawing. Aside from the desire to get work out of the way so as to lighten the spring work schedule, farmers explain their interest in fall plowing on the basis of the improved tilth which it encourages in heavy soils. This increased granulation is due in part to repeated wetting and drying and in part to alternate freezing and thawing which take place much more readily in plowed land. Plowing also relieves the lateral pressures which oppose the granulating effect of the processes described.

As soil water freezes in a small crevice of the soil, it draws moisture from the surrounding soil as the ice crystals grow. Thus under slow freezing the ice crystals tend to break the clod and at the same time exert a pressure on the surrounding soil particles which presses them together and stimulates the development of new clusters or granules. Furthermore, as moisture is drawn from the adjoining soil, the colloidal material is partially dehydrated and shrinks and tends to cement the particles into an aggregate. The withdrawal of moisture in some cases also leads to the concentration of soluble salts which flocculate the colloids, and thus the foundation is laid for soil granulation. Repetitions of freezing and thawing from night to day accentuate these processes until they culminate in a thoroughly granulated soil.

TILLAGE OPERATIONS AND SOIL PROPERTIES

The texture of a soil can be altered by man only through the addition of soil separates, as is sometimes done by applying sand to clay soil or clay soil to sand. In irrigation sufficient quantities of silt are sometimes added in the water to alter appreciably the texture of the soil. Man spends much time, however, in improving the structure of the soil. The plow is the most effective farm implement for altering soil structure. It has been estimated that about one-third of all draft power used on farms is expended in plowing. Virtually all tillage practices, except those directed toward the killing of weeds, are carried out for the purpose of developing a desired soil structure.

If tillage operations are to improve soil structure, they must be carried on when the soil holds a suitable moisture content. When the soil is too moist the cementing material of the granules is so soft that the granules are destroyed. The particles readily slide over one another, smaller ones slipping in between larger ones to create a compact mass dominated by a single-grained structure. The soil is then said to be puddled. On the other hand, if the soil is too dry when tilled, it breaks up into large lumps. To restore a heavy soil, which has been worked when too wet or too dry, to a desirable structural condition frequently requires much subsequent tillage combined with wetting and drying or freezing and thawing. A year or several years may be necessary to undo the harmful effects. Some tillage implements and tillage operations are discussed under the headings suggested by a series of questions.

Questions:

1. How was the plow developed?
2. How does the plow induce soil granulation?
3. What types of plows are there?
4. What are the comparative merits of fall and spring plowing?
5. How deep should one plow?
6. For what purposes are harrows used?
7. What are the functions of rollers?
8. Why are row crops intertilled?
9. Is excessive tillage of the soil undesirable?

History of the Plow. Some type of plow for the purpose of loosening or turning over the soil was probably among the first tillage implements used. Paintings on the walls of ancient Egyptian tombs, dating back 5,000 years, depict oxen yoked together by the horns, drawing a plow made from a forked tree. The Greeks improved the Egyptian plow by adding a metal point. Similar points are still produced in a Connecticut factory for shipment to parts of South America, where plows made from forked trees are still used. There was little improvement in the plow until American Colonial days when a wooden-framed plow, having the moldboard covered with iron strips and carrying a curved iron point or share, was constructed.

Thomas Jefferson, who made the first scientific study of the plow, considered that it consisted of two wedges, one to cut the furrow slice free at the bottom and lift it and the other to cut the soil at the side and turn it. He calculated a correct mathematical combination of the two wedges. The cast-iron plow was made about 1800.

Granulating Action of the Plow. The design of the moldboard plow has been carefully worked out and the slopes and curvatures of its various parts calculated in order to produce the desired effect on the soil without requiring an excessive amount of power. As the furrow slice is cut free and passes over the sloping share, it is lifted and pushed sideways so that it is submitted to both a vertical and horizontal strain which tends to break it into layers perpendicular to the line of movement of the plow. At the same time a strain parallel to the passage of the plow is developed, owing to the friction between the soil and plow bottom. This stress results in a shearing of the soil into layers which are perpendicular to those previously mentioned. The result is that the furrow slice is fractured in three directions. As it passes over the curved moldboard, these fractures are amplified and the soil is thrown over in a completely shattered state so that the forces of granulation may draw the particles together and cement them into definite units.

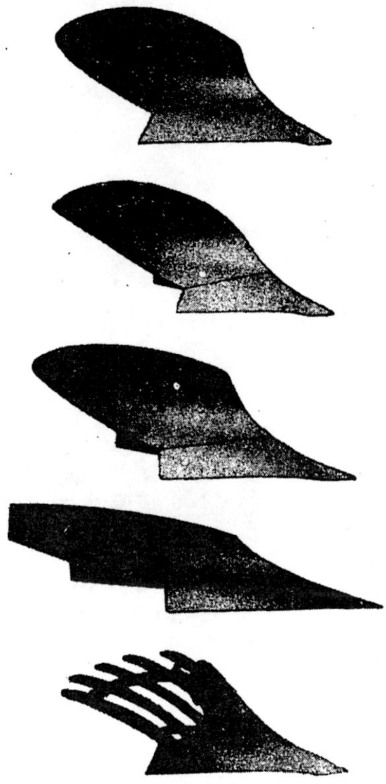

Fig. 24. Different types of plow bottoms showing variations in length and curvature of the moldboards. From top to bottom: Stubble, blackland, general purpose, prairie breaker, and slat moldboard. [Courtesy of John Deere, Moline, Illinois.]

Types of Plows. There are several types of moldboard plows, the most common of which are the (1) stubble, (2) general-purpose, and (3) sod, which are different in length and in the curvature of the moldboard. The stubble plow has a short, highly curved moldboard and hence accomplishes the maximum of soil pulverization. It is used primarily in light-textured soils and in those which are infrequently planted to grasses which bind the soil together by a dense growth of roots. The sod plow is used for breaking heavy sods in which the soil is permeated with roots and breaks up with great difficulty, the furrow slice tending to hang together as a continuous unit. This plow has a long, gently curved moldboard. The use of a sharply curved

FIG. 25. The lister has a double moldboard so that it turns a furrow slice in both directions, thus forming a broad trench. The seed of grain crops planted in rows is often planted in the bottom of the trench. [Courtesy of J. I. Case Company, Racine, Wisconsin.]

moldboard in such soil would require a great deal of power and would not accomplish desirable results. The general-purpose plow is intermediate between the two described, in length and curvature of moldboard. It is the most commonly used plow in general farming on the loam and heavier soil classes.

The lister is essentially a double plow with right- and left-hand bottoms mounted back to back. Hence it throws a furrow slice in each direction. The resulting ridges frequently are split, thus turning over the entire soil. The lister is used for ridging ground in the growing of special crops. On the great plains the lister is used for planting corn and various sorghums, the planting device being a part of the plow and dropping and covering the seed in the bottom of the trench.

Subsequent cultivations throw the soil from the ridges around the growing plants. Lister furrows, plowed at right angles to the prevailing winds, materially decrease soil blowing.

Various types of moldboard plows with either two bottoms or a reversible moldboard are made for use in hilly country or on experimental plats where it is desired to throw all the furrows in one direction.

The disk type of plow is convenient for use in hard or dry soils that are not easily penetrated by the moldboard plow and in loose, unconsolidated sandy soils. Sticky soils in which a moldboard will not scour and soils containing many larger roots and shrubs, are more easily plowed with a disk than with a conventional type of plow.

Fall vs. Spring Plowing. Fall plowing in many places is something of a habit and a tradition with farmers. In some agricultural sections of the country it was a custom to rate a farmer by the amount of fall plowing he was able to finish, and to some extent the same sentiment prevails today. When horses furnished the power on farms, plowing was a slow process and fall plowing was necessary if the spring work was to be done in time. The tractor has changed this situation and plowing which formerly required several weeks may now be done in a few days. With the passing of the horse has gone much of the necessity for fall plowing; yet it is practiced to a far greater extent than is good for the land. Fall plowing opens the way for loss by leaching of soluble plant nutrients, especially nitrates and lime. Fall plowing prevents the use of winter cover crops, which decrease erosion and the leaching of nutrients and add organic matter to the soil. And of more importance, fall plowing of all but very level land encourages loss of soil by erosion, which in many cases results in greater soil depletion than does crop production.

The beneficial effects on soil structure of plowing heavy land in the fall have been previously discussed. They can not be ignored on heavy, sticky soils, particularly those that are level. Furthermore, spring grain usually may be seeded earlier on heavy soils which have been fall plowed. But for crops planted in the late spring or early summer, spring plowing is just as satisfactory on the majority of soils, provided the seedbed is carefully fitted after plowing in order to granulate and compact it to the depth of the furrow slice. Many farmers experience unsatisfactory results from spring plowing because of insufficient preparation of the soil after plowing.

Depth of Plowing. Many comparisons have been made of different depths of plowing. In most cases little advantage has been found from plowing to a greater depth than 6 to 7 inches. Some benefit from deeper plowing of very compact soils may result occasionally, particularly in the growing of root crops. Shallow plowing has also been found desirable on some occasions when planting must follow so shortly after plowing that there is not time to prepare properly the deeper-plowed soil.

Harrows. With the exception of the disk, various types of harrows are designed primarily to stir the surface soil to a depth of 2 or 3 inches, thus killing weeds, breaking clods, leveling the surface, and drying out the surface of the plow layer by exposing it to the air. The harrow compacts the soil very little but may aid appreciably in inducing granulation. Numerous types of harrows are described in any good book dealing with tillage implements.

The disk harrow has diverse uses. With the gangs turned at a sharp angle the soil is moved considerably, loosening it and facilitating granulation. Manure or other organic matter can be incorporated into the soil by means of the disk. Use of the disk harrow before plowing cuts up crop remains on the surface and mixes them with the plow layer so that more complete mixing with the soil may be accomplished by plowing. In hard or compact soil the surface is also loosened, permitting the furrow slices to settle better and making quicker and more complete contact between plowed layers and the soil beneath. Indeed, on many soils unless there is considerable organic matter to work into the soil, a satisfactory seedbed for most small grains can be prepared by disking in place of plowing.

Rollers. Rollers are used to press down the furrow slices after plowing, to crush clods, to compact the soil around recently sown seed, to firm the soil around the roots of plants which have been loosened by heaving, and to settle and compact soil in seedbed preparation. The smooth roller has given place largely to those of the corrugated type because the corrugated roller penetrates the soil deeper, thus giving more subsurface packing. Corrugated rollers exert a side as well as a downward pressure and leave the soil in ridges rather than smooth, which is desirable. In some types the rear gang may be raised, thus increasing the weight on the front gang.

Intertillage of Crops. Numerous types of cultivators are available for use on different kinds of soils with different crops and under various climatic conditions. There are four primary purposes for which the intertillage of crops may be practiced: (1) to destroy weeds, (2) to conserve moisture, (3) to increase the absorption of moisture on sloping land, especially on soils which tend to bake or crust at the surface, and (4) to throw soil to or away from the plants as deemed advisable.

It is generally customary to make the first cultivation deeper than those that follow in order to loosen soil which later will be filled with roots and hence should not be disturbed. Later cultivations, and in some cases the first one, are made with a blade or with some type of broad flat shovel (such as the duckfoot) which stirs the soil to a shallow depth only, thus preventing the cutting of roots.

The most important objective in intertillage is the elimination of weeds that compete with crop plants for moisture and plant nutrients. After weeds are eliminated so that they do not use moisture, further cultivation generally conserves little more moisture on many soils. Considerable moisture may be prevented from evaporating, however, from soils which are very compact or heavy textured or which tend to crust and crack badly. This point is discussed more fully in the chapter dealing with soil moisture.

Many soils tend to crust on drying, allowing rain to run off in place of entering the soil. The granules in the surface of some soils are readily destroyed by rain, especially a beating rain, and the soil particles then run together, making a coating over the soil and permitting much of the water falling during the rest of the storm to run off. To overcome these tendencies and to roughen the surface and so impede the runoff of water, it is desirable to keep soils with an appreciable slope loosened and if possible somewhat ridged at right angles to the slope.

For crops planted or drilled in a continuous row, thus permitting cultivation in one direction only, it is desirable to throw a certain amount of soil toward the row so as to smother weeds between the plants. Soil is also thrown toward crops planted in lister furrows in order to conserve moisture and kill weeds. Some root and tuber crops need to be covered by soil as they develop in order to prevent sunburn and sometimes frost damage.

Intertillage or cultivation of the soil during the growth of a crop is considered advisable in some cases in order to hasten the decay of organic matter by microorganisms and thus produce nitrates and liberate other nutrients for use by the crop. The necessity for stirring the soil to increase the action of microorganisms varies with different soil and climatic conditions.

Adequate But Not Excessive Tillage. Sufficient tillage should be given a soil to prepare a good seedbed and to destroy as many weeds as possible before planting the crop. It is poor economy to slight the preparation of the seedbed. Subsequent tillage cannot compensate for the handicap imposed on crops by planting them in a soil insufficiently prepared. For most crops the soil should be thoroughly settled and firm to the depth of plowing. There should be only enough loose soil to insure planting at a proper depth with good coverage of the seed. In regions of limited rainfall seedbed preparation should cover a suitable period and include such tillage as is needed to provide for the storage of soil moisture.

After planting and before the plants have emerged or attained too great a height, many weeds may be destroyed with a harrow at a minimum of cost. Later tillage should be confined to what is necessary to attain the objectives set forth in the preceding section. More tillage not only adds needlessly to the cost of crop production but also is detrimental to the soil because it results in excessive decay of organic matter. Although many farmers may not fit good seedbeds and cultivate their crops adequately, there are also many who stir the soil to excess with tillage implements. Plowing when it is not necessary for seedbed preparation and excessive intertillage of crops are detrimental to soils.

POROSITY AND WEIGHT OF SOIL

That portion of a given volume of soil which is unfilled with solid matter is termed pore space. The proportion of a soil occupied by pore space depends on both the texture and structure of the soil and on the shape of the particles. A high content of organic matter also leads to much pore space. In soils containing little silt and clay the total amount of pore space is small even though the individual spaces are large. In soils containing a high proportion of silt and clay the relative volume occupied by pores is large, and if these soils are well granulated the total volume of pore space may be equal to or even greater than that occupied by the solid particles. This is because space exists not only between soil grains but also between granules.

The amount of pore space in a soil is expressed as a percentage of the total soil volume. The term porosity refers to the total pore space in a soil rather than to the size of the individual pores.

Pore spaces are important because they contain air and moisture, both of which play just as much a part in the crop-producing capacity of a soil as the mineral and organic matter. The amount of pore space in a given volume of soil greatly influences its weight. Several questions are listed which arise concerning the porosity and weight of soil and their interrelationships.

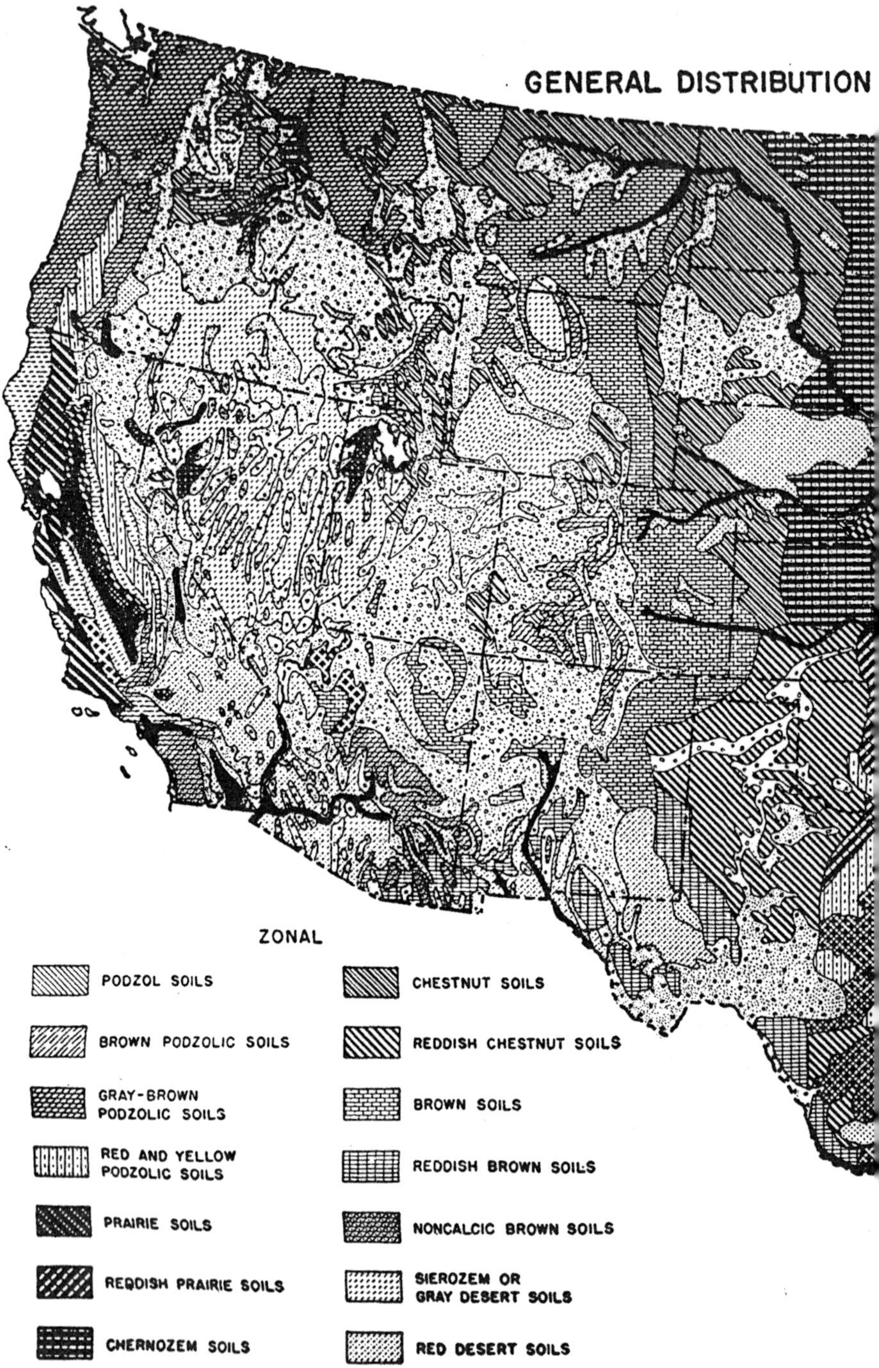

Fig. 18. General distribution of the great soil groups in th

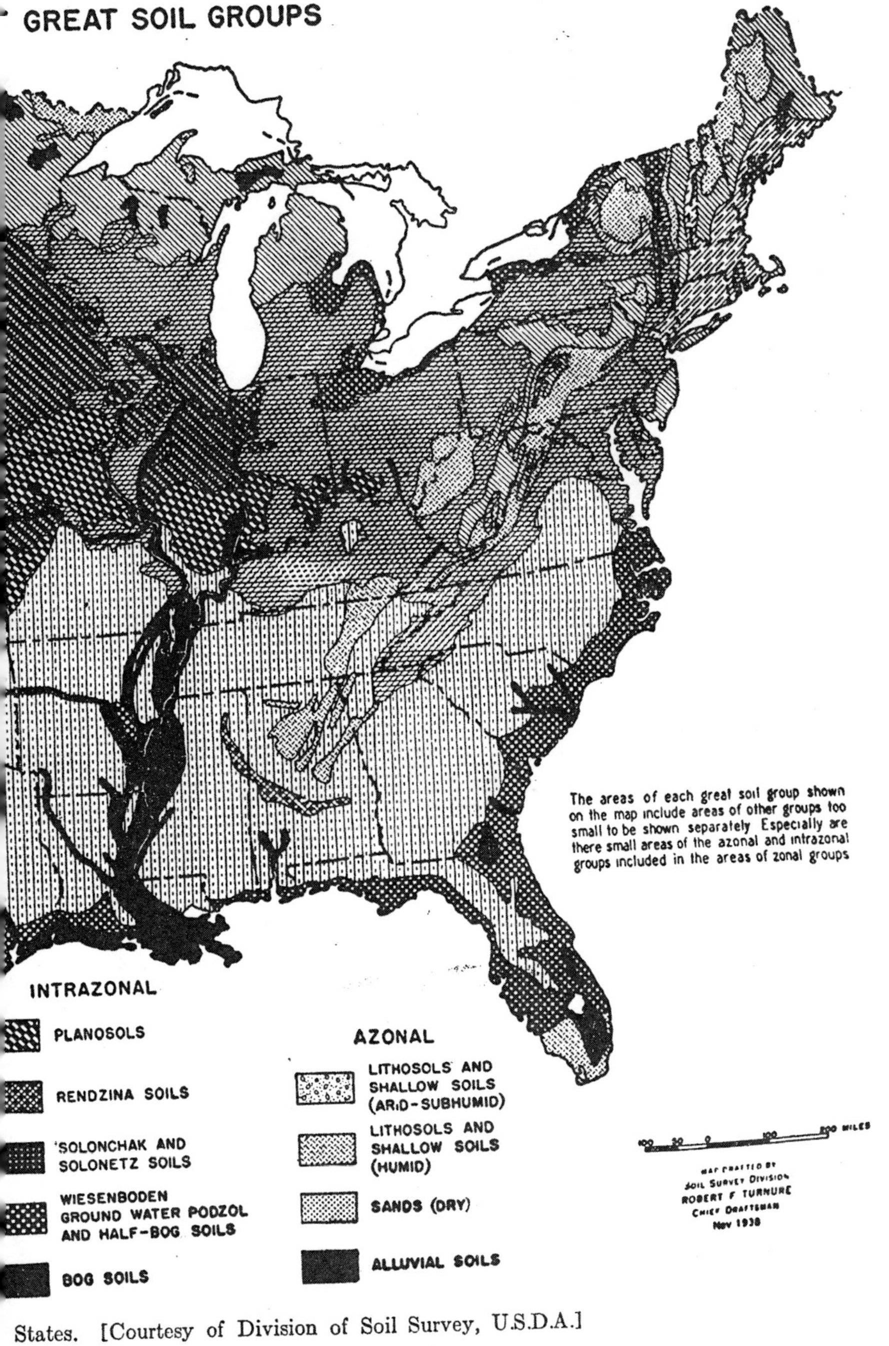

States. [Courtesy of Division of Soil Survey, U.S.D.A.]

Questions:

1. What is meant by the specific gravity of soil?
2. How does volume weight differ from specific gravity?
3. How much does an acre of soil, to a depth of 6⅔ inches, weigh?
4. How is porosity determined?

Specific Gravity of Soil. In determining the specific gravity of soil, consideration is given to the solid particles only. Thus the specific gravity of any soil is a constant and does not vary with the amount of space between the particles. Specific gravity is determined by dividing the weight of a given volume of dry soil particles by the weight of water that these particles will displace. Specific gravity is a weight ratio and does not vary a great deal for different soils unless there is considerable variation in content of organic matter. It averages about 2.65.

Volume Weight of Soil. In determining volume weight of soil, consideration is given to the pore space as well as to the solid particles. Volume weight of a given soil is, accordingly, a variable because the volume of pore space varies. It is calculated by dividing the dry weight of a given volume of soil in its natural structural condition by the weight of an equal volume of water. In determining the volume weight of field soils, great care is exercised not to destroy their structure. Samples are taken by driving into the soil a steel tube with a sharp cutting edge. The inside diameter of the tube is increased slightly immediately above the cutting edge so that the core of soil may be readily removed. The outside of the tube just above the cutting edge is also enlarged so that the soil is pressed outward as the tube is driven in, thus enlarging the hole and making withdrawal of the tube easy. As the dimensions of the tube are known the volume of the soil core removed is known, and so by drying and weighing it the necessary data are obtained for calculating the volume weight. Apparent specific gravity is a term used synonymously with volume weight. The volume weight of heavy soils may range from 1.1 to 1.6 and of sandy soils from 1.3 to 1.7.

Weight of Field Soils. As volume weight varies so greatly with soil structure so the weight of soils under field conditions must vary. A cubic foot of loam, sandy loam, or sandy soil in the dry condition weighs 80 to 110 pounds, whereas a cubic foot of silt loam, clay loam, or clay ranges from 70 to 100 pounds in weight. It is generally estimated that an acre of the finer-textured soils to the average depth of plowing (about 6⅔ inches) will weigh approximately 2,000,000 pounds when dry. This weight is used frequently in reporting the content of plant nutrients or of organic matter in terms of pounds per acre. Sandy and sandy loam soils have a somewhat greater weight. The weight of soils under field conditions, that is, with normal moisture content, is much greater than the above. As a result, in plowing an acre a very appreciable weight is turned over.

Determining Porosity of Soil. The percentage of pore space in a soil may be calculated from the volume weight and specific gravity. As specific gravity is assumed to be about the same for most soils, a volume-weight determination is all that is required for the calculation of pore space. If it is borne in mind that both specific gravity and volume weight are based on the weight of unit volume of water (1 ml. = 1 gram), it is readily seen that volume weight ÷ specific gravity × 100 gives the percentage of a volume of soil that is filled with solid particles. This percentage, taken from the total volume (100 per cent), will give the percentage of pore space; hence the formula

$$100\% - \left(\frac{\text{volume weight}}{\text{specific gravity}} \times 100\right) = \%\text{ pore space}$$

SOIL COLOR

The color of soil is a factor of service to both the farmer and soil scientist, providing one understands the causes of the various colors and is able to interpret them in terms of soil properties. Organic matter content, drainage condition, and freedom of aeration are soil properties related to color which are of interest to farmers. The investigator uses color as an aid in soil classification and draws from the color of the different horizons information as to conditions pertaining to and forces active during soil formation. The following division of the subject matter may be made.

Questions:

1. What are the colors of soil minerals?
2. What are the causes of gray and brown colors in soils?
3. Why are some soils red or yellow?

Colors of Soil Minerals. The minerals occurring in appreciable quantities in most soils are light colored. As a result, soils would be of a light-gray color if composed of crushed minerals which had undergone little chemical change. Accordingly, for an explanation of the dark-gray, brown, red, and yellow colors in soils we must look to chemical changes in the constituents (especially iron) of the minerals and to the addition of organic matter. There are a few instances, however, in which the proportion of colored minerals is sufficient to give the soil a decided color.

Gray and Brown Soils. The dark color of soils is generally due to the highly decayed organic matter they contain; in fact with some practice the percentage of organic matter in a soil may be judged with reasonable accuracy from its color. Although we speak of black soils, there are none which are truly black. Organic matter imparts a gray, dark-gray, or dark-brown color to soils unless some other constituent such as iron oxide or salt modifies the color.

If soils are poorly drained there is usually a greater accumulation of organic matter in the surface layers, thus giving a very dark color. The lower soil layers, which contain very little organic matter, on the other hand, are of a light gray color indicating the poorly drained condition. If drainage is intermediate the gray of the subsoil is likely to be broken by flecks of yellow.

Red and Yellow Soil Colors. When drainage permits aeration, and moisture and temperature conditions are favorable for chemical activity, the iron in soil minerals is oxidized and hydrated into red and yellow compounds. Highly hydrated iron oxides are yellow but as hydration diminishes, reds replace the yellows. Accordingly, we find shades of red in soils extending from the southern deserts across the semi-arid and subhumid belts to the humid states of the Southeast. The red and yellow colors of the subsoils in the southeastern states immediately catch the eye of the traveler. The low organic matter content of many soils in this area leaves undulled the brilliant colors of the iron oxides. With an appreciable humus content the red colors are converted into mahogany colors.

SOIL TEMPERATURE

As temperatures decrease the life processes of both plants and animals are slowed down until finally they cease altogether. Growth processes of most agriculturally important plants are very sluggish at temperatures about 40°F. and increase until temperatures ranging from 70°F. to 90°F. are reached. The chemical processes and activities of microorganisms which convert plant nutrients into available forms are also materially influenced by temperature. The farmer is much interested therefore in soil as well as in air temperatures, and a brief discussion of the topic follows under the heads indicated.

Questions:

1. What sources of heat warm the soil?
2. What factors determine the heat capacity of soil?
3. By what practices can man control soil temperature?
4. How does situation influence soil temperature?

Sources of Soil Heat. The movement of heat from the warm interior of the earth is counteracted by the loss of heat from the surface by radiation. Thus at a given depth, about 3 or 4 feet at the equator and 50 feet in the latitude of New York, the earth remains at an approximately uniform temperature, which is the same as the average annual air temperature, and the growth of plants is unaffected through the warming of the soil around their roots by the interior heat of the earth.

Likewise the soil is not warmed by the decay of organic matter within it, for the heat so generated is liberated so slowly that the effect on soil temperature is not measurable.

Man must depend, therefore, upon the sun to warm the soil sufficiently to enable crops to grow. Even the radiant energy of the sun would not perform this task satisfactorily if it were not for the tempering influence of the atmospheric envelope around the earth. Without the modifying influence of the atmosphere the sun's heat by day would be too intense and at night radiation would proceed so rapidly that life would be destroyed through freezing.

The Heat Capacity of Soil. Mineral particles require a comparatively small amount of heat to raise their temperature. The quantity of heat required to raise the temperature of a gram of soil particles 1°C. is only about one-fifth as much as is required to warm a gram of water the same amount. In other words, the specific heat of dry soil particles is 0.2 and it is evident from this fact that moisture content is the controlling factor in determining soil temperature. Soil high in water content will warm up slowly in the spring and will cool down slowly in the fall. Drainage therefore exerts a major influence on soil temperature.

Control of Soil Temperature. As has just been pointed out, the removal of excess water from a soil will facilitate changes in soil temperature. By providing drainage, man may exert some influence on the temperature relations of soils which are so situated that they hold excessive quantities of water. Aside from this there is little that man can do to affect the temperature of the soil he tills except to keep it in good tilth and increase the humus content. Soils high in organic content absorb more heat because of their dark color and hence tend to be warmer. But this effect is offset to some extent by the increased water-holding capacity. When in good structural condition soils also warm up more readily than when they are in poor tilth.

Location and Temperature. In the northern hemisphere soils which are located on southern and southeastern slopes warm up more rapidly in the morning than if they were located on the level or on the northern slopes. This is because they are more nearly perpendicular to the sun's rays, and hence a maximum amount of radiant energy strikes a given area. Soils with a southern or southeastern exposure often are selected for the growing of early vegetables and fruits.

CHAPTER IV

SOIL REACTION

Soils may be acid, neutral, or alkaline in reaction. The reaction of soils is significant in crop-production and soil-management practices because the various degrees of soil reaction are produced by the chemical conditions which exist in soils, each set of chemical conditions causing its corresponding degree of soil reaction. Each degree of soil reaction or set of chemical conditions in soils affects plant growth in a certain way owing to either a depressed solubility of some elements or to an increased solubility of others. The chemical conditions which accompany the different degrees of soil reaction, therefore, may be favorable to the growth of some crops, unfavorable to others, and in still other cases they may affect plant growth very little. Soil reaction, then, may be considered a *symptom of the particular chemical conditions which caused it,* and hence it may be used to indicate the possible effect of these conditions on plant growth. It is useful in diagnosing the fertility of soils. A knowledge of the conditions which cause different soil reactions, therefore, is of value to the student of soil science.

A Virginia farmer named Edmund Ruffin (1794–1865) was probably the first man in the United States to sense the prevalence of acidity and the need for lime in soils of eastern United States. In the latter part of the 19th and early years of the present century, Dr. Wheeler, of the Rhode Island Experiment Station, made soil tests for acidity over many of the eastern and central states and by his writings and lectures gave much publicity to the widespread need for liming soil.

The study of soil reaction and of the effects on plant growth of the chemical conditions accompanying various soil reactions may be taken up under three general heads or objectives.

Objectives:

A. Soil acidity and conditions which produce acid soils.
B. Properties of acid soils which affect plant growth.
C. Conditions of development and general effect of neutral and alkaline soils on the growth of plants.

SOIL ACIDITY AND CONDITIONS GIVING RISE TO ACID SOILS

There is a natural tendency for soils to become increasingly acid in humid climates and this trend is accelerated when soils are put under cultivation. Large areas of soil were acid when the land was cleared in preparation for farming, but the store of available plant-food elements in the virgin soils and the supply of bases liberated through rapid decay of the humus largely offset the detrimental effects of the acid condition. Consequently, farmers did not become conscious of the need of the soil for lime until clover began to fail. As most of the highly prized leguminous crops do not thrive on soils of strong acidity and as the yields of many other crops are decreased, the need of acid soils for lime becomes a very significant problem in the agriculture of humid regions. A series of questions commonly asked about soil acidity are listed.

Questions:

1. What is meant by an acid soil?
2. Into what two types may soil acidity be classified?
3. What kind of rocks give rise to acid soils?
4. How are acid soils developed from non-acid minerals?
5. What farm practices tend to increase soil acidity?
6. Where are acid soils found?
7. How is the acidity of soils expressed?
8. How acid and how alkaline may soils be?
9. Does the reaction of soil vary during the year?
10. How may soils be made more acid?
11. What is meant by buffering in soils?
12. In what ways may soil acidity be determined?
13. Does a soil acidity test tell how much lime is needed?

Acidity in Soils. Soil acidity is primarily a function of the colloidal fraction of soils, but soluble acids may contribute to the condition to a greater or less extent. It has been estimated that at least 95 per cent of the acidity of mineral soils is due to the reaction of the colloidal material. As previously pointed out, much of the colloidal material of soil reacts as a complex acid radical of very low solubility, around which are clustered positive ions at varying distances from the nucleus. These cations dissociate to some extent and their places on the acidoid are taken by H ions from the water as is shown in the illustration (page 96). The OH ions so formed produce an alkaline reaction in the soil solution. Although the H ions in the complex will also dissociate to some extent, there is reason to believe that they are held quite close to the acidoid, while other cations may be scattered at greater distances in the field of attraction.

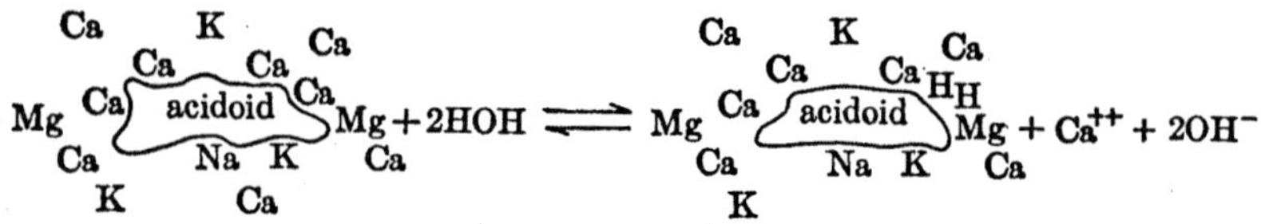

When the numbers of H ions and OH ions in the solution around the nucleus are equal, a neutral reaction exists. When the number of H ions exceeds the number of OH ions, the reaction of the soil is acid.

Traces of inorganic acids, such as HNO_3, HNO_2, H_2SO_4, and H_3PO_4, may be present, and in unusual soils which contain FeS a considerable concentration of H_2SO_4 may exist. Carbon dioxide is present in most soils and with H_2O produces a slightly acid reaction. Other organic acids, such as acetic, citric, and oxalic, are liberated during the decay of fresh plant tissue and may exist in the soil for a short period. Acids of this kind, however, are readily decomposed by soil microorganisms in well-aerated soils and so have only a temporary existence in the soil. Although inorganic and organic acids may contribute to some extent to the acidity of mineral soils, the reaction is due primarily to the relative proportion of H ions and other cations on the colloidal complex.

Studies have indicated that when the complex is saturated with bivalent ions, such as Ca^{++} or Ba^{++}, a slightly alkaline reaction results and, furthermore, that a soil with reaction approaching neutrality contains very little exchangeable hydrogen.

Types of Soil Acidity. For a soil to give an acid reaction there must be an excess of H ions over OH ions in the solution surrounding the colloidal complex. The concentration of these H ions can be readily measured and constitutes the *active* acidity of the soil. The concentration of these dissociated H ions is in equilibrium with those adsorbed on the complex. Should any process, such as the addition of lime or leaching, remove or neutralize a portion of the active H ions, more would immediately dissociate from the complex to restore the equilibrium. The H ions on the acidoid, therefore, constitute a reserve supply which is known as the *reserve* or *potential* soil acidity.

Soils from Acid Rocks. Rocks which contain an excess of quartz or of silica compared to their content of basic minerals or of basic elements are classed as acid rocks. Granite and rhyolite are examples. This classification is based on the fact that SiO_2, combined with various proportions of water, forms the different silicic acids of which orthosilicic acid (H_4SiO_4), metasilicic acid (H_2SiO_3), diorthosilicic acid ($H_6Si_2O_7$), and trisilicic acid ($H_4Si_3O_8$) are the most common. When rocks which are deficient in bases are disintegrated or decomposed in the process of the accumulation of soil material and soil development, the resulting material is acid even though no loss of base has taken place during the process. There are considerable areas of sandy soils which have been produced from acid rocks.

Acid Soils from Non-Acid Minerals. Aside from the oxides and carbonates most minerals are compounds of basic elements with some of the silicic acids or with aluminosilicic acid. As these compounds are broken down by the chemical processes of weathering, at least a part of the bases pass into the carbonate or bicarbonate forms which have an appreciable solubility and are leached away when rainfall is sufficient. On the other hand, the negative or acid radicals of the compounds pass over into hydrated aluminum silicate and acid salts of silicic acid usually with the elimination of some free SiO_2. The following reaction is typical:

$$\underbrace{(4NaAlSi_3O_8 + CaAl_2Si_2O_8)}_{\text{plagioclase}} + 6H_2O + 3CO_2 = 3H_4Al_2Si_2O_9 + CaCO_3 + 2Na_2CO_3 + 8SiO_2$$

The acid salts of silicic acids give acid reactions just as monocalcium phosphate [$CaH_4(PO_4)_2$] and potassium hydrogen sulfate ($KHSO_4$) react acid. The hydrated aluminum silicates are really aluminosilicic acids. These compounds are relatively insoluble and so accumulate in the soil. The result of the weathering process is the accumulation of material which grows constantly more deficient in bases.

During the decomposition of the soil minerals the clay minerals are developed which, it will be remembered, have a high base-exchange capacity. As bases are made soluble through mineral decomposition, they are adsorbed by the organic colloids and clay minerals, giving a complex with a high degree of saturation and an alkaline reaction. In soils used for crop production considerable CO_2 is liberated through the decay of organic matter and by the respiration of plant roots. Carbon dioxide combines with water to form the bicarbonate and hydrogen ions. $CO_2 + HOH = HCO_3^- + H^+$.

It will be remembered that the H ion is the most effective in base replacement, and hence the bases on the colloidal acidoid are being constantly dislodged by the hydrogen ions which are continuously being generated in the soil during the growing season. The bases so replaced may be used by plants or they may be leached out of the soil. It will be seen, then, that here is another force which continually operates to lower the base content of the soil and make it acid. Because of the small exchange capacity of sandy soils, they become acid before the heavier soils, other factors being equal.

Farm Practices That Increase Acidity. Under virgin conditions soils are covered with vegetation throughout the year, and during growth periods the use of water by the plants materially reduces the quantity which percolates through the soil and removes soluble bases. The intake of nutrients by plants during virtually all the warm part of the year also reduces losses by leaching. Furthermore, all the basic nutrients taken up by the vegetation are sooner or later returned to the soil. It is to be expected, then, that development of acidity through loss of bases is a relatively slow process in virgin soils.

When the land is used for farming, a rather different set of conditions prevail. Unless cover crops are used extensively, under most rotations the soil is devoid of vegetative cover during many months of growing weather. Even with a carefully planned use of cover crops, the soil is unprotected from leaching during the fitting of the seedbeds. Fall plowing leaves the soil in excellent condition for loss of bases by leaching during late fall and early spring. Summer fallowing also leads to much loss by leaching. Through the harvesting of crops most of the nutrients used by the plants are removed and a comparatively small proportion of them is returned in manure. It appears, therefore, that farming hastens the development of acidity and that practices which leave the soil free of growing vegetation and so subject to leaching are particularly objectionable.

Certain fertilizers tend to increase soil acidity through the development of mineral acids. This is particularly true of sulphate of ammonia. Other fertilizers, as calcium cyanamide, nitrate of soda, and calcium nitrate, tend to correct acidity. The effect of fertilizers on acidity will be discussed more fully in one of the chapters dealing with fertilizers.

The Distribution of Acid Soils. Acid soils will develop most rapidly in areas where rainfall is sufficiently abundant to leach away rapidly the bases liberated by ionic exchange and mineral decomposition. As the amount of water percolating through the soil diminishes, the rate at which acidity develops decreases. There are some soils in which acidity has become so pronounced and widespread as to require the application of considerable quantities of lime in order to permit the satisfactory growth of acid-sensitive legumes. These soils occur primarily east of a line drawn roughly north and south along the eastern boundary of Kansas. Many areas of soil west of this boundary are improved by liming, and in fact quite strongly acid soils are found near the mountains adjoining deserts. Large acreages of land east of the boundary line are adequately supplied with bases,

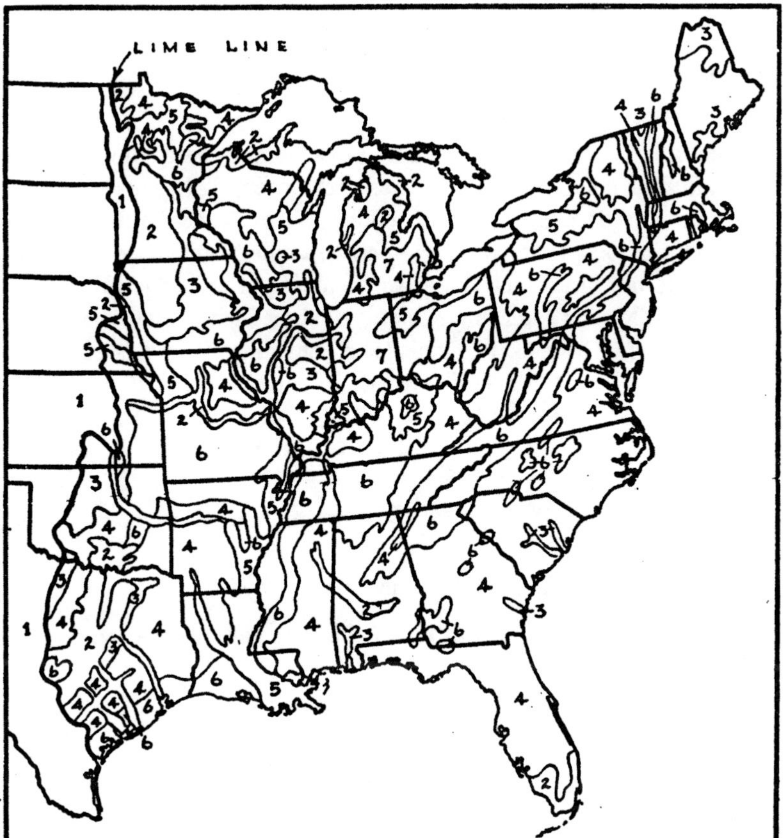

FIG. 26. Relative acidity and lime content of soil areas in eastern United States. [Prepared from a map in Farmers' Bul. 1845, U.S.D.A., by E. C. Shorey; revised by Oswald Schreiner.]

1. Areas of soils dominantly very rich in lime. 2. Areas in which soils and parent materials are dominantly fairly rich in lime, although the surface soil is somewhat acid in places. 3. Areas in which soils are acid but calcium in lime or other available form is within reach of deep-rooted plants. 4. Areas of acid soils low in lime and other available forms of calcium. 5. Areas in which some soils are high and some medium in content of lime or available calcium. 6. Areas in which some soils are medium and some low in lime content. 7. Areas comprised of soils with high, medium, and low lime content.

particularly in parts of the tall-grass prairie, in shallow or young soils derived from limestone, and in young soils developing on poorly drained locations and recently exposed lake beds. There are also areas in which leaching has not yet been sufficient to remove the limestone from the glacial till or other parent material.

Not all soils giving a moderately low *p*H value are in need of lime. In some soils of moderate to high acidity the exchange complex yet contains an abundant supply of bases which may supply the demand of crops for these nutrients.

Expression of Soil Acidity. There are several systems of expressing the acidity of soils. Within the last few years it has become customary to base the expression of acidity on the concentration of the dissociated H ions, that is, on the active acidity. This tendency has given rise to use of the *p*H values so frequently seen. The *p*H scale is derived as follows: Water is a neutral substance with equal concentrations of H and OH ions. The ionization constant of water is 10^{-14}. As concentrations of H^+ and OH^- are equal, each must be 10^{-7}.

$$\frac{\text{Conc. } H^+ \times \text{conc. } OH^-}{\text{Conc. } H_2O} = 10^{-14}$$

As concentration of H_2O (conc. H_2O) is infinite, it is ignored, so we may write

$$\text{Conc. } H^+ \times \text{conc. } OH^- = 10^{-14}$$

Considering only the concentration of the H ion as it is the cause of the acid reaction, we have

$$\text{Conc. } H^+ = 10^{-7} \doteq 0.0000001 \text{ gram H ions per liter}$$

The reciprocal of this is $\frac{1}{0.0000001} = 10{,}000{,}000$, the log of which is 7. Pure water is therefore said to have a *p*H value of 7, and as water is neutral any solution with a *p*H of 7 is said to be neutral. It is seen from the above that the *p*H value is the logarithm of the reciprocal of the H-ion concentration.

If the concentration of H ions in a solution were 0.000001 gram per liter (10^{-6}), the *p*H value would be 6. As the concentration of H ions in this solution is 10 times that in water, it is seen that a solution with *p*H 6 is 10 times as acid as a solution of *p*H 7. It should never be forgotten that *p*H values are logarithms.

As the product of the concentration of the H and OH ions is always equal to 10^{-14}, the concentration of the OH^- is known if we express the pH value. Thus pH 7 indicates a concentration of both H and OH ions of 10^{-7} ($10^{-7} \times 10^{-7} = 10^{-14}$). A pH of 6 would then indicate a concentration of OH ions of 10^{-8} ($10^{-6} \times 10^{-8} = 10^{-14}$). On the other hand a pH of 8 would require a concentration of OH ions of 10^{-6}, which is 10 times greater than 10^{-7}, the concentration of OH ions in a neutral solution. This solution therefore is alkaline. It is seen then that pH values above 7 indicate alkaline solutions.

Although pH values express the active acidity or intensity of acidity in a soil, they do not indicate the reserve acidity and hence are not measures of total acidity. For example, two soils may have the same pH value but have quite different total acidities or, in other words, may require decidedly different quantities of lime entirely to saturate the colloidal complex. Let us suppose that one soil, because of the quantity or nature of the colloidal material it contains, has a base-exchange capacity of 100 m.e., while another soil has a capacity of only 50 m.e. If the capacity of each soil is one-half supplied with calcium and one-half with hydrogen, then they would have about the same pH. Despite this fact, it would require twice as much calcium to saturate the complex of the first soil as it would to saturate the complex of the second soil. Because of this situation the relative acidities of different soils are sometimes expressed in terms of the quantity of limestone ($CaCO_3$) required to bring them to given pH values, as 6.5 or 7.0.

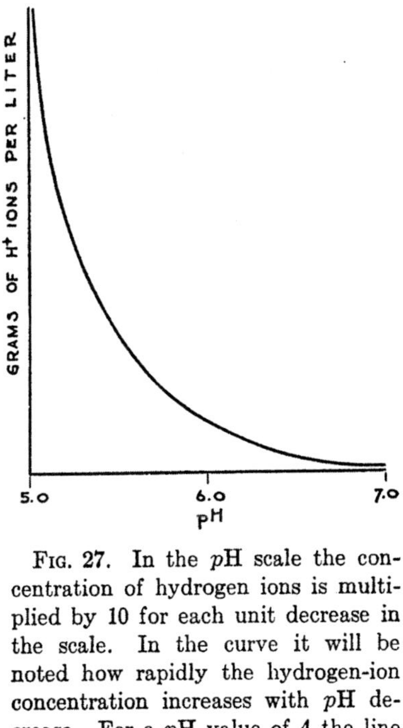

FIG. 27. In the pH scale the concentration of hydrogen ions is multiplied by 10 for each unit decrease in the scale. In the curve it will be noted how rapidly the hydrogen-ion concentration increases with pH decrease. For a pH value of 4 the line representing grams of hydrogen per liter would be over 4 feet long.

In other instances the degree of saturation is given in terms of percentage of the total base-exchange capacity in order to express acidity or lime requirement. Thus a soil may be said to have an exchange capacity of 80 m.e. with 75 per cent saturation.

Range of Reaction in Soils. Unless there is an abnormal development of strong acid such as H_2SO_4, it is unusual to find mineral soils with a *p*H reading of less than 3.5; in fact *p*H 4.0 might more correctly be given as the lower limit frequently encountered. In humid regions it is unusual to find a soil with a *p*H of more than 7.5 or 8, but in arid regions where soluble salts of sodium (Na_2CO_3) may accumulate, a *p*H of 9.5 to 11.0 is sometimes attained. The usual *p*H of agricultural soils in the humid region ranges approximately from 5.0 to 6.8.

In discussing soil acidity with those who are not familiar with the *p*H scale, it is customary to use descriptive terms to indicate the acidity and comparative lime requirement of soils. The scheme presented in Chart 3, with corresponding *p*H values, is representative.

Fluctuations in Acidity. It has frequently been observed that *p*H values of samples taken from a soil area at different times in the year show considerable variation. Several investigators have studied *p*H changes in soils by systematically sampling them during the growing season. Results show that acidity tends to increase from spring to midsummer in humid regions. The increase is accentuated by periods of drought and may amount to an entire *p*H unit. Rainy periods decrease acidity, and as winter approaches the soil gradually returns to the *p*H observed in the early spring. An increase in *p*H has also been observed to accompany the thawing of frozen soil. The seasonal decrease in soil *p*H has been found to correlate roughly with an increase in soluble salt content particularly of nitrates. A rapid growth of plants with accompanying absorption of nutrients has been found to restrict the accumulation of salts and so modify the increase in acidity. The dehydration of colloidal material has also been suggested as a cause contributing to increased acidity in warm, dry weather. Fertilizers appear to have little influence on the seasonal fluctuation of soil *p*H unless they contain $(NH_4)_2SO_4$ or other nitrogen compounds which produce acids in the soil.

The tendency of soils to increase in acidity in summer may be of small significance except in soil of low clay content, in which comparatively small applications of lime induce considerable changes in *p*H. When such soils are used for cropping systems in which small variations in acidity may affect quality, yield, or disease resistance of the crops, seasonal variations in acidity are more noteworthy.

Air drying of surface soil samples appears to change *p*H values very little in acid soils but to reduce the alkalinity of alkaline soils. The *p*H of subsoil samples decreases appreciably with air drying.

Because the mineralogical composition of a body of soil is not uniform, some variation in the *p*H of samples taken at various places is

Chart 3. A Summary of Soil Reaction, Lime Requirement, and Associated Plant-Nutrient Relationships

<table>
<tr><td>pH</td><td colspan="10">4.0 4.5 5.0 5.5 6.0 6.5 6.7 7.0 8.0 9.0 10.0 11.0</td></tr>
<tr><td>Acidity</td><td colspan="2">Very strongly acid</td><td>Strongly acid</td><td>Moderately acid</td><td>Slightly acid</td><td>Neutral</td><td>Weakly alkaline</td><td>Alkaline</td><td>Strongly alkaline</td><td>Excessively alkaline</td></tr>
<tr><td>Lime requirements</td><td colspan="2">Lime needed except for crops requiring acid soil</td><td colspan="2">Lime needed for all but acid-tolerant crops</td><td>Lime generally not required</td><td colspan="5">No lime needed</td></tr>
<tr><td>Occurrence</td><td>Rare</td><td>Frequent</td><td colspan="2">Very common in plow soil in humid climates</td><td>Plow soils when limed</td><td colspan="3">Common in subhumid and arid regions</td><td colspan="2">Black alkali occurs in limited areas in arid regions</td></tr>
<tr><td>Fertility conditions</td><td colspan="4">Phosphates fixed
Lime and potash leach
Iron, aluminum, and manganese are soluble
Fungi thrive
Bacteria languish</td><td colspan="3">Phosphates soluble
Lime present in optimum quantities
Desirable bacterial activity
Nitrogen freely fixed</td><td colspan="3">Phosphate fixed
Boron, iron, manganese, and potash may be deficient
Excess boron may occur in limited areas</td></tr>
</table>

to be expected. Likewise the uneven distribution of applied manure and of manure dropped by animals may cause local variation in acidity in a field for several years. The burning of weeds, grass, or brush, the digging of a fence post hole, or the removal of stumps or large rocks may also cause a soil sample to be unrepresentative of a field. As a result, several samples from a field should be taken for testing. In fields with rolling topography there frequently is considerable difference in reaction of soil in the depressions and on the higher ground. The soil occupying the lower sites is usually of higher *p*H than that on the knolls and slopes.

Ways of Increasing Acidity. There are a number of plants which grow well only in soils that are strongly acid. Some of those most commonly grown are cranberry, blueberry, rhododendron, and azalea. At times it becomes desirable to increase the acidity of the soil for the growing of some of these plants. Several procedures used to accomplish this purpose are (1) the addition of ground sulphur which oxidizes in the soil to sulphuric acid; (2) the application of ferrous sulphate or aluminum sulphate which hydrolyzes to form sulphuric acid; (3) the application of a solution of sulphuric acid or of phosphoric acid; (4) the mixing of very acid soil, usually peat, with the soil. The use of sulphur is probably the most common procedure.

Buffering in Soils. An acid or alkaline solution or system, like a soil or plant which requires more base or acid to neutralize it than its *p*H value indicates, is said to be buffered. Such a system resists a change in reaction, and hence the buffer capacity of a system is often designated as its capacity to resist change in reaction. Buffering comes about through the low ionization of weak acids, weak bases, and also salts, which give rise to these when strong acids or bases are added. For example, a solution of phosphoric acid (H_3PO_4), a rather weak acid, will require more base to neutralize it than would be calculated from its *p*H value because only a part of its hydrogen is ionized at any one time.

Soils may be buffered by the presence of carbonates, phosphates, and other salts, but generally the buffering capacity is due largely to the quantity of organic and inorganic colloidal material they contain. The colloidal soil complex functions as a slightly ionized acid or a slightly ionized salt of a weak acid. If the acid complex is represented by $H_{100}X$, then the calcium salt might be written $Ca_{50}X$. It would be assumed, therefore, that the addition of a small amount of acid, let us say $5H_2SO_4$, would give a marked decrease in *p*H. This result would not be attained, however, because we would now have present a partially neutralized weak acid ($Ca_{45}H_{10}X$) which still contains a large excess of calcium. On the other hand, the buffering effect of a strongly acid soil to the addition of alkaline material may be illustrated by the following reaction:

$$H_{100}X + Ca(OH)_2 \rightleftharpoons H_{98}CaX + H_2O$$

In this case as the dissociated hydrogen is neutralized, more will dissociate from the soil complex or weak soil acids, and hence little change in reaction will result.

As the buffering material of soils is usually present largely in colloidal form, it is to be expected that soils high in clay or organic matter content will be highly buffered and hence will require comparatively large quantities of lime to neutralize them if they are strongly acid. Coarse-textured soils, on the other hand, are usually only slightly buffered.

When a soil contains an appreciable quantity of salts of weak acids, such as the carbonates, bicarbonates, phosphates, or silicates of Ca, Mg, Na, and K, they contribute to the alkalinity by hydrolysis, as shown in the following reactions:

$$CaCO_3 + 2HOH \rightleftharpoons Ca^{++} + 2OH^{++} + HCO_3^{-} + H^{+}$$

$$Ca_3(PO_4)_2 + 2HOH \rightleftharpoons Ca^{++} + 2OH^{--} + Ca_2H_2(PO_4)_2$$

If a small addition of acid is made to such a soil, the OH ions neutralized will immediately be replaced through further hydrolysis of the salt, for only a small part of the salt will dissociate at any time. It can be seen then that the buffering capacity of the colloidal complex of a soil may be greatly supplemented by the action of salts. This situation prevails in many desert soils.

Methods of Determining Acidity. If it is desired to measure the active acidity of a soil, it is necessary to determine the H-ion concentration. This is the most commonly used measure of acidity both in soils and in many other materials. The most accurate procedure is by use of an electrode which sets up a potential with the H ions just as a copper electrode will set up a potential with the Cu ions in a solution of $CuSO_4$. The potential may be measured by balancing it against a half cell of known potential making use of a potentiometer. As the potential established is proportional to the concentration of H ions, this procedure offers the opportunity for measuring the active acidity. Various electrodes have been used, such as hydrogen gas adsorbed in finely divided platinum deposited on a platinum wire: a chemical, quinhydrone, which in water functions as a hydrogen electrode; a glass tube with the end sealed and blown out into a very thin walled bulb or into a spear-shaped point and filled with one-tenth normal HCl. The last, known as a glass electrode, is the most commonly used.

Various dyes which give distinct color shades at different *p*H values are used extensively, in the laboratory and especially in field kits, for determining *p*H values with relative accuracy. The soil may be placed either in a piece of paraffined paper, folded to make a trough, or in a depression in a porcelain plate, and the dye solution allowed to soak slowly through it. A drop of the clear percolate is drawn to the side for comparison with a standard color chart. Different dyes which change color within different *p*H ranges may be used separately or several of them may be included in one solution.

A test based on the occurrence of soluble iron in acid soils involves the use of a solution of potassium sulphocyanate (KSCN). This reagent gives a red color with ferric iron.

Some tests, formerly used much more than at present, involved the determination of active plus a portion of the reserve acidity. The soil was treated with a solution of a salt, such as KCl or Ca $(C_2H_3O_2)_2$, the basic ion from which replaced part of the H on the acidoid with the formation of HCl or $HC_2H_3O_2$. The acidity of the resulting solution was then determined by titration with a weak alkali solution or by other appropriate means.

Interpretation of Results of Acidity Tests. Because H-ion concentration determinations do not give a measure of total acidity, they must be used in association with other information in estimating the quantity of lime to apply in order to bring a given soil to a desired *p*H. In some states, the actual quantity of lime needed to raise each of the principal soil types one or more *p*H units has been determined, and these data are used in making lime recommendations from *p*H determinations. In other states the texture of the soil is assumed to correlate roughly with the base-exchange capacity. The quantity of lime recommended for a soil with a given *p*H is therefore increased progressively as the soils vary from sandy soils to loams, silt loams, clay loams, etc. Lime recommendations are usually made by experienced workers who have in mind the results of numerous field experiments involving the use of different quantities of lime on different soil types.

Several factors aside from *p*H values should be taken into consideration in deciding on the quantity of lime to apply to a given soil. For a discussion of them refer to Chapter V.

CONDITIONS IN ACID SOILS WHICH ARE BENEFICIAL OR DETRIMENTAL TO THE GROWTH OF PLANTS

As stated in the introduction to this chapter, soil acidity is an indication of a chemical condition existing in a soil. The components which contribute to this situation unquestionably vary in soils of different chemical composition, and hence the same plant may not be similarly affected in all soils showing the same *p*H. This point is well illustrated by the relative need for lime of organic and mineral soils occurring in the same locality. While a mineral soil with a *p*H of 5.5 may need two or more tons of limestone per acre for the growing of acid-sensitive crops, much less or possibly no lime would be needed on the organic soil. Furthermore, a scarcity of certain nutrients, as boron, in available form may occur in some soils with a high *p*H, whereas in other neutral or alkaline soils of the arid regions an excess of boron may be present.

In the following paragraphs are discussed the points raised by the list of questions concerning the conditions in acid soils which affect plant growth. This material is also presented in Chart 3.

Questions:

1. How does the concentration of H ions in acid soil affect plant growth?
2. Are soluble iron and aluminum salts detrimental to plants?
3. How does soil acidity affect the availability of phosphorus to plants?
4. Do plants get sufficient calcium in soils of low *p*H?
5. Does soil acidity influence the activity of microorganisms?
6. May plants be classified on the basis of their tolerance of acidity?

Effects of H Ions on Plants. Many studies have been made to determine the influence of H-ion concentration on the growth of various plants. The results show in general that, in a nutrient solution with a concentration of H ions corresponding to that of strongly acid soils, plants grow much better than in the acid soil itself. In other words, when separated from the accompanying conditions, which pertain in an acid soil, the H-ion concentration ceases to be especially harmful to many plants. Plants vary in this respect. It becomes necessary, then, to look for some cause in addition to H-ion concentration for the detrimental effect of acid soils on plant growth.

Soluble Iron and Aluminum. The quantity of soluble iron and aluminum in many soils increases as soil acidity increases. Although ferrous iron is rather toxic to many plants, ferric iron is not nearly so detrimental. Soluble aluminum, on the other hand, has a high toxicity for many commonly grown crops. In some soils the soluble aluminum present has been shown to be responsible for the larger part of the toxicity to certain crops. It is quite generally believed that soluble aluminum is one of the major contributing factors to poor plant growth in strongly acid soils.

Phosphorus Availability in Acid Soils. The chemical elements iron and aluminum both form compounds of low solubility with the PO_4 ion in acid soils. From about *p*H 5.0 up to the neutral point, soil phosphates have an appreciable solubility on account of the basic ions present that maintain the higher *p*H and at the same time form some calcium and magnesium phosphates in the soil. In soils from about *p*H 5.0 downward, however, complex phosphates of iron and aluminum may be formed which have a very low solubility and do not supply sufficient phosphorus for plants. The calcium and magnesium phosphates are more soluble in the presence of CO_2, which comes from decaying organic matter, but this solubility effect decreases up to about *p*H 8.0. Under strongly acid soil conditions, therefore, when soluble phosphate fertilizers are applied to the soil they revert strongly to the insoluble forms, and hence on these soils the efficiency of applied phosphates is greatly reduced. Applications of lime raise the *p*H of the acid soils and tend to increase the solubility of the soil phosphorus especially when CO_2 is present in the soil. Likewise, the plowing under of green manures and barnyard manures increases the CO_2 content of the soil when they decay and thus helps to make the soil phosphorus available to plants.

Phosphoric acid forms three common salts with calcium, namely, monocalcium phosphate [$CaH_4(PO_4)_2$], dicalcium phosphate ($CaHPO_4$), and tricalcium phosphate [$Ca_3(PO_4)_2$]. The monocalcium phosphate is water soluble, but the other two calcium phosphates are of rather low solubility. Magnesium and phosphoric acid form a similar series of phosphates, and the corresponding magnesium phosphates are more soluble than the calcium phosphates. Although the solubility of the dicalcium and dimagnesium phosphates in water is comparatively low, these salts furnish phosphorus to plants quite readily. The tricalcium and trimagnesium phosphates, however, have only a low degree of solubility. When the soluble monocalcium phosphate fertilizer is applied to soils having a *p*H representing a medium acid to neutral condition, a large portion of the phosphorus is reverted to the more insoluble forms, but some of the phosphorus is still held in a form available to plants. It should be remembered, also, however, that plant roots intimately contact the colloidal material of the soil and thus are able to obtain more phosphorus than that represented by the immediate water solubility.

Low solubility of phosphorus in alkaline soils results when the $Ca_3(PO_4)_2$ hydrolyzes in the presence of $CaCO_3$.

$$Ca_3(PO_4)_2 + 2CaCO_3 + 4H_2O = 2CaHPO_4 + Ca(HCO_3)_2 + 2Ca(OH)_2$$

The high concentration of Ca ions in the products of the reaction greatly depresses the solubility of the phosphorus. According to McGeorge and Breazeale,[1] this reaction takes place at a *p*H of 8.0 to 8.5. Through this *p*H range and, also somewhat below it therefore, the soil phosphorus may be highly insoluble and unavailable. At still higher *p*H ranges where the effects of the sodium and potassium ions may be appreciable, phosphorus is again held in the soil in a more soluble condition. The whole subject of phosphorus solubility in soils, however, is very complex and it is not always easy to predict how much phosphorus will be available to plants under any set of soil conditions.

Calcium Supply in Acid Soils. As soils become acid only when the supply of basic ions, principally Ca, on the colloidal complex has become depleted, there is a basis for the opinion that a calcium deficiency is one cause for the restricted growth of many plants on highly acid soils. Aside from the calcium used as a nutrient, the element plays several important roles in plant nutrition, discussed in Chapter V. It has been found that when calcium in a form unable to correct acidity, such as $CaCl_2$ or $CaSO_4$, is supplied, the detrimental effects of acid soils on certain plants are largely overcome. There is probably considerable variation in the response of different plants to a calcium deficiency and to the application of neutral calcium salts. Furthermore, the quantity of exchangeable calcium in different soils having the same *p*H is decidedly variable because of differences in base-exchange capacity. Nevertheless, a deficiency of calcium is undoubtedly a highly important factor in the growth of plants on many acid soils.

Activity of Microorganisms. The microorganisms which convert ammonia into nitrite and nitrite into nitrate are sensitive to acidity. The production of nitrates in strongly acid soils is accounted for, in part, at least, by the presence of decaying particles of minerals which

[1] W. T. McGeorge and J. F. Breazeale, "The Relation of Phosphate Availability, Soil Permeability, and Carbon Dioxide to the Fertility of Calcareous Soils." Ariz. Agr. Exp. Sta. Tech. Bul. 36, 1931, p. 375.

are liberating bases, thus establishing small areas of slightly acid to neutral reaction. As the *p*H value of a soil is the summation of the reaction of all parts, there may well be innumerable very small zones of neutrality in a soil of low *p*H. One group of bacteria (*Azotobacter*) which utilizes atmospheric nitrogen without association with a legume does not function appreciably below a *p*H of 6.0. Symbiotic bacteria which grow in conjunction with several of the most widely grown legumes, such as alfalfa, sweet clover, red clover, and to a lesser extent alsike clover and soy beans, are affected adversely by moderate to strong acidity. There is considerable range in the tolerance of symbiotic bacteria to acidity. Those forming nodules on the roots of lespedeza, crimson clover, lupines, and peanuts function in soils of relatively high acidity. This question is discussed more fully in Chapter VII.

Plants Vary in Tolerance of Acidity. The distribution of plants under natural conditions may well be considered as due to their ability to survive under those conditions rather than as an indication that the environment is the one in which they grow to best advantage. In other words, the existence of plants in a given environment is one of tolerance rather than choice. This fact is well illustrated by the rapid growth of some desert plants, such as mesquite, in a warm humid climate. Also some acid-tolerant plants like sorrel grow much more luxuriantly in a soil which has been limed. The intensity of acidity which plants will tolerate is also influenced by the supply of available nutrients and of moisture. When other conditions are favorable for the growth of a plant it will thrive in a much more acid soil than it would if the nutrient or moisture supply were limited. Experimental studies and observation have established the fact that most crops will grow more satisfactorily in a soil which is slightly acid than in one which is neutral or alkaline. This situation pertains to acid-sensitive crops such as alfalfa.

There are some plants which grow luxuriantly only in strongly acid soils. No satisfactory explanation has been offered for this phenomenon.

In Chart 4 a number of plants are listed on the basis of their tolerance of or requirement for an acid growth medium.

No satisfactory explanation for the variation in tolerance of acidity exhibited by different plants has been advanced. A useful exercise for the student of soils is to devise several rotations, using crops which are tolerant to medium or strong acidity. These rotations should meet all the requirements of a practical rotation so far as possible.

Chart 4. Tolerance of Plants for Acidity

VERY SENSITIVE TO ACIDITY	MODERATELY TOLERANT TO ACIDITY	HIGHLY TOLERANT OF ACIDITY	REQUIRE ACID SOIL
Alfalfa	Alsike clover	Rye	Cranberry
Sweet clover	Red clover	Buckwheat	Blueberry
Sugar beets	Soy beans	Crimson clover	Sphagnum moss
Table beets	Vetch	Sorrel	Labrador tea
Beans	Barley	Poverty grass	Leather leaf
Rutabagas	Wheat	Rhododendron	
Cauliflower	Oats		
	Corn		

CONDITIONS OF DEVELOPMENT AND EFFECT ON PLANTS OF NEUTRAL AND ALKALINE SOILS

A large proportion of the soils in the chernozem belt and westward has a neutral or an alkaline reaction. Exceptions are found in the soils occupying areas of high rainfall near the coast and in certain intermountain valleys. Some desert soils have a *p*H of slightly below 7. East of the chernozems many areas of soil are non-acid because they have one or more of the following characteristics: (1) location in which seepage water passing through lime-carrying soil material supplies lime to the soil; (2) young soils derived from limestone or lime-rich material and from which the Ca has not yet been leached; (3) soils developing on recently exposed lake beds or on poorly drained soil material containing $CaCO_3$; (4) lateritic soils in which the decay of organic matter is sufficiently rapid to maintain a supply of bases.

Neutral and alkaline soils occupy approximately half of the area of continental United States, and hence they deserve much consideration. The discussion of this important group of soils will be divided into sections, as indicated by the following questions.

Questions:

1. Does limited rainfall contribute to the neutrality and alkalinity of soils?
2. Why are "black alkali" soils so named?
3. How are "white alkali" soils developed?
4. What plants are tolerant of alkali in soils?
5. In what ways are high salt concentrations detrimental to plants?
6. What methods are used to improve alkali soils?

Low Rainfall and Soil Reaction. It has been shown previously that soluble salts, largely carbonates and bicarbonates of Ca, Mg, Na, and

K, are produced during the decomposition of rocks and minerals. When rainfall is sufficient to leach from the soil the Na and K salts but only a part of the Ca and Mg carbonates, the latter accumulate and keep the colloidal complex saturated. Hence a neutral or slightly alkaline reaction is maintained. With less rainfall a copious accumulation of Ca and Mg carbonates occurs and a decidedly alkaline reaction develops. When conditions permit some of the Na_2CO_3 to remain in the soil, a *p*H of 8.5 or higher may be attained and the term alkali is then usually applied.

Black Alkali. When Na_2CO_3 occurs in a soil it readily hydrolyzes, forming NaOH: $Na_2CO_3 + 2HOH = 2NaOH + H_2O + CO_2$. This strong base attacks organic material to form a dark-colored soil solution which may result in a black or brownish-black residue on the surface of the soil. As a result the term black alkali has been applied to such soils. Not only is the alkali corrosive to plant tissue but also the Na is adsorbed on the colloidal complex and causes a puddled soil structure which greatly impedes the development of roots and the absorption of irrigation water.

Saline Soils. In some desert soils an accumulation of the sulphates of Ca, Mg, and Na, and sometimes of NaCl and KCl, accompanies the carbonates of Ca, Mg, and Na. Such soils have a light-gray color and are given the name of white alkali. The source of the soluble salts in saline soils is sometimes difficult to explain. In some instances the problem is simplified through evidence that the area is the bed of an extinct saline lake or of an area of the ocean. There is no such evidence in many cases, however. Seepage water passing through rocks impregnated with salts is sometimes the explanation. The evaporation, over long periods, of drainage water containing small quantities of salts culminates in appreciable salt accumulations in some basins.

Plant Growth in Neutral and Alkaline Soils. Plants vary in their tolerance of alkalinity and salt content of soils just as they do in their tolerance of acidity. In Chart 5 are listed a number of crops and native plants which are especially tolerant of alkalinity and also several which are less tolerant.

Black alkali is harmful to the tissues of living plants, since it attacks them in the same way that it does the organic matter of the soil.

A concentration of soluble salts, such as occurs in white alkali soils, is not corrosive to plant tissue but may prevent the absorption of moisture by the plants because of high osmotic pressure. Likewise, the presence of some of the ions in considerable concentration may limit the intake of certain nutrients, as experiments have shown that the absorption of a given nutrient is influenced by the presence or absence of other ions in the nutrient solution. There are data which indicate that negatively charged nutrient ions are absorbed with difficulty by plants when the soil reaction exceeds *p*H 7.6. The use of a strongly acid fertilizer has been found to faciliate the intake of nutrients from strongly alkaline soils.

The solubility of the compounds of some elements which are required by plants is also suppressed by alkalinity. The insolubility of phosphates in a soil containing calcium carbonate is an example. Manganese is also usually found in a rather insoluble condition in alkaline soils, as is also iron. On the other hand, the salts of certain elements may be present in sufficient concentration to be toxic to planted crops.

Chart 5. Grouping of Plants Based on Their Tolerance of Alkali

VERY TOLERANT OF ALKALI	MODERATELY TOLERANT OF ALKALI	GROWTH INDICATES ABSENCE OF HARMFUL QUANTITIES OF ALKALI
Greasewood	Desert saltbush	Sage brush
Seepweed	Shadscale	Creosotebush
Pickleweed	Winterfat	Mesquite
Saltgrass	Alfalfa (established stands)	Little rabbitbrush
Salt sage	Sweet clover	Gallenta
White sage	Field peas	Giant wild-rye
Tussock-grass	Ryegrass	Bluegrass
Timothy	Sorghum	Corn
Orchard grass	Wheat	Potatoes
Brome grass	Barley	Oranges
Western wheat grass	Oats	Pears
Millet	Sugar beets	Figs
Rye	Grapes	Apples
Date palms		

Correction of Alkali. Under irrigation the alkalinity of soil may increase, owing to the upward movement of salts with capillary water. Several methods of treating soil to counteract the detrimental effects of black alkali have given more or less satisfaction. The application of ground sulphur which oxidizes in the soil to sulphuric acid and converts the Na_2CO_3 to Na_2SO_4 and H_2CO_3. The H_2CO_3 so liberated combines with $CaCO_3$ to make $Ca(HCO_3)_2$ which tends to develop a granular structure in the soil. Sulphuric acid and $CaSO_4$ produce the same results and both are used although the latter is used much more widely. Decaying organic matter has a very salutary effect since the CO_2 produced converts the calcium into the relatively soluble bicarbonate, in which form it is effective in inducing granulation. It is considered good practice to apply manure or to plow under a green-manuring crop in conjunction with the application of ground $CaSO_4$. Regardless of the soil corrective used in reclaiming alkali soils, the products of the reaction between the corrective and the alkali must be subsequently removed by leaching with irrigation water. An occasional heavy application of irrigation water is a good practice because it tends to prevent the accumulation of alkali salts. The subject of the management of alkali soils is discussed more fully in Chapter XV.

CHAPTER V

LIME AND ITS USE

Liming is an important practice in scientific soil management. Too frequently soil fertility is regarded chiefly as a matter of supply of nitrogen, phosphorus, and potassium, and insufficient emphasis is placed on calcium which performs many indispensable functions in plant growth and in modifying soil conditions. The liming of acid soils should be accepted by all as a fundamental and essential practice; it is frequently referred to as the "backbone" of permanent agriculture in humid regions.

In the discussion of soil acidity, it was pointed out that unfavorable soil conditions associated with acidity may be corrected by the application of lime. When the chemist speaks of lime, he usually refers to calcium oxide (CaO), but from the agricultural point of view a more general meaning is implied by the term. Agriculturally, lime may be said to be any compound of calcium or of calcium and magnesium *capable of counteracting the harmful effects of an acid soil.*

It is obvious why compounds of calcium and magnesium are used for liming purposes rather than compounds of such elements as sodium and potassium. In humid regions, the ions of calcium and magnesium are the usual predominating cations in soils, and it is natural that an attempt should be made to restore the supply when depleted. Moreover, materials carrying calcium and magnesium are relatively abundant, widespread, and cheap. These materials are effective in neutralizing all types of soil acids and have no detrimental effects on the physical properties of soils and are not toxic to plants in amounts normally applied. Potassium salts are too expensive to use for correcting acidity, and the use of alkaline salts of either potassium or sodium encourages the development of a poor physical condition of the soil. Furthermore, these salts are caustic to handle and are toxic to plants if applied in large amounts. Liming materials are used to bring about in the soil chemical, physical, and biological changes that are beneficial to plant growth. Used intelligently they bring satisfactory results, but unwisely used they may prove detrimental to soils and crops. This subject of lime and its uses will be discussed under the following topics.

Objectives:

A. The need of soils for lime.
B. Functions of lime in the soil.
C. Forms of lime used for application to the soil.
D. Guarantees of chemical composition and fineness of lime.
E. Sources of lime for agricultural use.
F. Use of lime on acid soil.

THE NEED OF SOILS FOR LIME

It has been estimated that perhaps 70 per cent of the tillable land in the eastern half of the United States is acid and requires liming for maximum crop production. A large part of this area at one time contained sufficient calcium for optimum growth of legumes and other lime-loving crops. But this calcium supply has greatly diminished and in most of eastern United States the supply is no longer sufficient. Certain areas on the west coast, where there is considerable rainfall, are also acid. Liming has been practiced in some of the eastern states since the beginning of the development of agriculture, and the use of lime has gradually spread westward, even into regions of limited rainfall. There are several important reasons why the popularity of lime as a soil fertilizer and soil amendment is so widespread.

Questions:

1. Were the parent materials of all soils originally well supplied with lime?
2. Does lime readily leach from soils?
3. Do crops remove large quantities of lime from soils?
4. Do the acid substances normally added to soils tend to increase the need of soils for lime?

Some Soils Develop from Acid Rocks. Certain soils have developed from rocks and minerals which in themselves were naturally acid. Those rocks which contain more quartz or silica than basic constituents are known as acid rocks. Soils developed from such materials are acid even though no loss of bases may have taken place during the formation. (See p. 96.) There are considerable areas of siliceous sandy soils which have always been in need of lime. Most acid soils, however, have developed as a result of leaching losses and crop removal of bases.

Leaching Increases Need for Lime. As has been indicated, the amount of rainfall and the lime needs of soils are closely related. No one factor tends to remove the basic materials from the soil in humid regions to as great an extent as leaching, and calcium is the element which usually suffers the greatest loss. As a rule, the basic elements in soils are more readily soluble than the acid constituents and hence are more readily carried away in drainage water as it percolates through the soil. With a continuation of this process, it is obvious that a deficiency of bases and the development of acid soils will occur sooner or later. Since calcium and, to a less extent, magnesium are readily lost from soils and since calcium and magnesium are the active constituents of lime, soils which do not supply them in sufficient quantities for plant growth are designated as lime-deficient.

The leaching losses of calcium and magnesium from soils depend to a large extent on the quantities of these elements present and the forms in which they are held, the amount of water percolating through the soil, the vegetative cover, and the production of carbonic acid. Calcium and magnesium for the most part are held in the soil in either the silicate or the carbonate form or as exchangeable ions on the soil colloids, and both forms are subject to leaching losses. Owing to the continuous production of carbon dioxide through organic matter decomposition and plant excretion and to the consequent development of carbonic acid, there is a constant tendency toward a replacement of calcium and magnesium ions on the soil colloids by the hydrogen ions of the carbonic acid. These basic ions are thus released and are susceptible to loss by leaching. Furthermore, the carbonate forms of calcium and magnesium in the soil may be gradually converted to the more soluble bicarbonate forms. Leaching losses are much greater from fallow fields than from comparable fields growing crops. This difference is due to the smaller quantities of percolating water in cropped land and to the intake of bases by the plants.

Crop Removal of Bases. The removal of bases through cropping is greater than the removal of acids, thus leaving the soil more acid. Such crops as the legumes, particularly alfalfa and clover, require considerably more calcium and magnesium than any other crops and consequently make a heavy demand for bases. An average crop of alfalfa will remove about 100 pounds of calcium per acre; clover, about 60 pounds; and soy beans, 30 pounds. Cropping, therefore, tends to deplete the lime content of the soil. (See p. 98.)

Acid Substances That Increase the Need for Lime. Near industrial areas large quantities of sulphur are carried into the soil with the precipitation. This sulphur is readily changed to the sulphate form and as such encourages the loss of soil bases. Decaying organic matter also produces various acids and the process of nitrification results in the formation of nitrates which greatly facilitate the loss of bases. These acids produce salts which on the whole are quite soluble. The use of certain fertilizers (see p. 331) is known to increase the need for lime.

The need of soils for lime, then, is caused by the loss of the basic constituents, calcium and magnesium, through leaching and cropping and is aggravated by other processes or practices which add acid constituents to the soil, although as pointed out a need for lime may exist even though no loss of bases occurs during soil formation.

Lime was applied as a soil amendment in ancient times, long before the Christian Era, and the same factors which created lime-deficient soils at that time are still operating today. Soils in humid areas will always be affected by these lime-depleting processes.

FUNCTIONS OF LIME IN THE SOIL

Lime produces several specific effects, resulting in the improvement of soils and increased crop production. Since soils may vary greatly in their physical, chemical, and biological properties, it is not to be expected that a lime deficiency will produce exactly the same unsatisfactory conditions in each case, and, as a result, some soils may give a much greater response to lime than do other soils. Injurious effects may result from the use of lime under certain conditions, especially when applied in excess. Lime has several important functions in the soil, as indicated by the questions which follow.

Questions:

1. Are all forms of calcium effective in neutralizing soil acids?
2. Does the calcium ion have a direct effect on plant growth?
3. What is the effect of liming on the solubility of soil mineral elements?
4. Are the desirable microbiological soil processes favorably influenced by the presence of lime?
5. Does the presence of lime affect the chemical nature of soil colloids?
6. What are the effects of lime on soil structure?
7. Does the application of lime increase the effectiveness of fertilizers and manure?

Neutralization of Soil Acids. Since the accumulation of hydrogen ions in soils is accompanied by the loss of basic elements, chiefly calcium, an addition of lime will correct the calcium deficiency and at the same time neutralize soil acidity. Acid soils are beneficially changed in character by the application of calcium compounds in the oxide, hydroxide, or carbonate form but not by the addition of calcium in the sulphate (gypsum) or chloride form. When the last two compounds are added to acid soils, strong, stable mineral acids (H_2SO_4 and HCl) are formed, and unless they are leached from the soil no decrease in acidity can result although the quantities of soluble calcium in the soil may be increased by the application of these substances.

Lime in the forms of oxide, hydroxide, and carbonate reacts rather rapidly with moist, acid soils. The reactions of these forms vary somewhat, but the final result is the same. When lime is added, the soil water becomes charged with calcium ions which have the ability to replace (by ionic exchange) the hydrogen ions on the colloidal complex. The hydrogen ions are released and they unite with hydroxyl ions to form water. When the colloidal clay becomes saturated with calcium, it is no longer acid; the acidity of the soil has been neutralized or corrected. In other words, the process of neutralizing acid soils is the reverse of that by which they become acid, as explained in the chapter on soil reaction (p. 95).

The Influence of Calcium on Plant Growth. Calcium is an indispensable element for the growth of all crops. Many soils are too low in lime content to supply the needs of crops, especially certain species of leguminous plants. All the functions calcium performs within the plant are not definitely known, but it is generally believed that calcium has an influence on the translocation of carbohydrates and certain mineral elements within the plant and on the development of roots. It appears to be essential for cell-wall construction and perhaps aids in neutralizing organic acids within the plant or in regulating the acid-base balance in plants. Calcium may also influence, either favorably or unfavorably, the absorption of other elements. Within certain limits, for example, an inverse relationship has been found between the intake of calcium and potassium by plants. It may counteract to some extent the toxic effects of high concentrations of magnesium and sodium.

The nutritive value of animal feeds is to a large extent influenced by the calcium content of plants. It is important in the development of the bones of animals and of the shell of eggs. That lime should be present in sufficient quantities in all properly balanced livestock rations is now an accepted fact. The importance of the calcium content of food for growing children and for treating certain types of human disease is well recognized by physicians.

The Effect of Lime on the Solubility of Soil Mineral Elements. As a general rule, a relatively high percentage of the phosphorus in soils well supplied with lime is available for plant use, and the phosphorus of calcium-deficient acid soils is in relatively unavailable forms even though the total phosphate content is comparatively high. Soil phosphorus is generally most readily available to plants in neutral or slightly acid soils, and with increasing acidity its availability decreases. In strongly acid soils, in the presence of aluminum and iron compounds, soluble phosphates combine with these elements, forming relatively insoluble aluminum and iron phosphate compounds. In alkaline soils in the presence of excess calcium carbonate (about 2 per cent), phosphorus combines with calcium, forming tricalcium phosphate. This form of phosphorus is of low solubility but is more soluble than the phosphates of aluminum and iron.

Liming acid soils has a tendency to make the phosphate compounds in the soil more available. Increasing the calcium content of soils will convert a part of the phosphorus that is present as aluminum and iron phosphate to the more available calcium phosphates. The more desirable forms from the standpoint of plant use are the monocalcium and dicalcium phosphates. It is also likely that liming results in the liberation of the organic phosphorus in the soil through stimulation of decomposition processes. It is obvious, however, that lime alone will not solve the problem of phosphorus availability because many soils are so depleted of phosphorus that lime has little effect in increasing crop yields unless accompanied by applications of phosphate fertilizer.

The effect of lime on the solubility of soil potassium cannot be stated with any degree of certainty. Theoretically, calcium by the process of ionic exchange should displace or liberate some of the exchangeable potassium of the soil. Under field conditions this is difficult if not impossible to demonstrate because of so many complicating chemical and biological factors. There is experimental evidence indicating that the solubility of potash is decreased in certain soils that have been overlimed. Marked crop response to potash fertilizers is often secured on such soils. It is perhaps safe to say that lime does not greatly influence the availability of potassium.

When soils become deficient in bases, the solubility of aluminum, iron, and manganese increases; in strongly acid soils, the high concentrations of these elements may be toxic to crop plants. Excess quantities of these elements may become available at *p*H values below 5.5. Soils having reactions between *p*H 5.5 and *p*H 7.0 usually supply plants with sufficient quantities of both iron and manganese, but at *p*H values above 6.5 or 7.0, especially in sandy soils, they may become insoluble to such an extent that plants are unable to satisfy their needs. Caution is therefore necessary in the use of lime. A normal application may prevent toxic concentrations of these substances; yet too much lime may create an iron and manganese deficiency.

Effect of Lime on Microbiological Processes of the Soil. Calcium is closely associated with certain important microbiological processes. The more important effects of calcium on the soil population may be grouped as follows: (1) promoting the decomposition of organic matter; (2) making conditions favorable for nitrification and sulphofication; and (3) providing favorable conditions for the growth and functioning of both symbiotic and non-symbiotic nitrogen-fixing bacteria. Each group of organisms which contributes to the above processes functions most efficiently in soils well supplied with lime. In many of these processes, it is not necessarily a matter of changing the *p*H but one of supplying soluble calcium.

Lime may be used as a preventive of certain types of plant diseases which occur only in acid soils. However, an alkaline soil, caused by the use of too much lime, is likely to depress certain desirable microbiological processes, such as nitrification and the decomposition of organic matter. Furthermore, it is generally believed, although it does not always occur, that the application of lime in amounts sufficient to make a soil neutral or alkaline favors potato-scab disease.

The Effect of Lime on the Chemical Nature of Soil Colloids. As has been explained, acid substances of soils are, for the most part, not soluble in the ordinary sense, but rather they are colloidal. Furthermore, calcium is usually the dominant exchangeable cation in neutral soils in humid regions. Under leaching conditions, the calcium and other bases are gradually replaced by hydrogen ions from the surface of colloidal particles. As these particles become more acid, they tend to become more chemically unstable. Silica splits off from the inorganic colloidal materials and their base-exchange capacity is greatly reduced. The maintenance of a high base-exchange capacity in soils is thus largely dependent upon the maintenance of a high calcium content. Aluminum may also come into play as an exchangeable cation. The remedy for this is lime. Not merely does it act as a neutralizing agent but also it restores calcium on the soil particles, producing significant effects on various chemical as well as physical properties of the soil.

The Effect of Lime on Soil Structure. The replacement of calcium ions on the soil colloids with hydrogen ions results in the production of a dispersed condition of the colloids. If these dispersed colloids are carried downward in the profile, dense subsoil horizons may gradually develop. But as long as the colloids remain essentially saturated with calcium, dispersion will not occur.

A soil with a high content of lime is more likely to remain in good tilth than one low in lime. This is especially observed in heavy soils; they are usually more porous and granular and less likely to puddle when wet or cake when dry if large amounts of calcium are present. The presence of sufficient lime aids in making soils loose and friable, permitting better aeration, assists in drainage, and is important in plowing and seedbed preparation.

Lime may affect the physical properties of the soil indirectly by regulating to some extent the kind and amount of organic matter returned to the soil, the rate of its decomposition, and the nature of the humus that is formed.

The Influence of Lime on the Effectiveness of Fertilizers and Manure. The full benefits of fertilizers and manures are realized only after ample supplies of lime have been provided. Fertilizers and manures are more effective when all soil conditions are favorable. Seldom do they produce maximum results in sour soils. Lime is not a substitute for fertilizers and manure, but when lime is needed its application will greatly increase their effectiveness.

In soils deficient in lime, various substances, as iron and aluminum compounds, are in a condition to combine with soluble phosphorus, applied in fertilizers and manure, and change it into less soluble forms. Since the solubility of the natural phosphorus compounds is also low in lime-deficient soils, it is evident that crops are likely to suffer from the need of phosphorus in acid soils.

Nitrogen and potassium of fertilizers and manure are not made insoluble as a result of a lime deficiency, but since crops do not grow so well as in lime-rich soils, the plants cannot utilize the nitrogen and potassium as well as they could if growing conditions were made more favorable by the presence of lime. If nitrogen is applied in forms other than nitrate, the presence of lime will hasten the production of nitrates.

FORMS OF LIME

Lime is generally placed on the market in three chemical forms, namely, the oxide, hydroxide, and carbonate. They all have the advantage of leaving no harmful residue in the soil. The oxide and hydroxide change to the carbonate and bicarbonate forms in moist soil at a fairly rapid rate. As the calcium and magnesium of the bicarbonate form is adsorbed by the colloidal clay particles, hydrogen ions are displaced, producing carbonic acid. Carbonic acid is an unstable acid and readily breaks up into water and carbon dioxide, leaving no undesirable residue. Calcium and magnesium salts of strong acids, such as the chlorides and sulphates, are not considered liming materials, as is explained on p. 119.

Calcium silicate, which is a constituent of blast-furnace slags, can be classed as a liming material. It supplies calcium, neutralizes acidity, and leaves no harmful residue. It has not been used extensively in the past, but in the future it may be classed as an important form of lime. In discussing the different forms of lime the following questions will be answered.

Questions:

1. What are the different forms of lime, how are they prepared, and what are some of their outstanding properties?
2. What is the relative neutralizing power of the different forms of lime?
3. Are the different forms of lime equally effective in the soil?

Carbonate Forms of Lime. Calcium carbonate is commonly referred to as carbonate of lime, under which name several lime compounds are sold. Of the various chemical forms of lime, this is the most widely used for liming purposes; and most of the naturally occurring liming materials exist in the carbonate form.

By-product lime from various industries contains what is frequently referred to as precipitated carbonate. Such carbonates are usually produced from either calcium oxide (CaO) or calcium hydroxide ($Ca(OH)_2$). Calcium carbonate ($CaCO_3$) may be produced by combining calcium oxide and carbon dioxide (CO_2). In this reaction, 56 pounds of calcium oxide combine with 44 pounds of carbon dioxide to produce 100 pounds of calcium carbonate. The steps in this reaction are as follows:

$$CaO + H_2O \rightarrow Ca(OH)_2$$

$$Ca(OH)_2 + CO_2 \rightarrow CaCO_3 + H_2O$$

The calcium in limestones is largely in the carbonate form. Many such stones are nearly pure calcium carbonate. Certain other limestones contain various percentages, up to about 45 per cent, of magnesium carbonate with the calcium carbonate. These latter limestones are often referred to as dolomitic or high-magnesium stones. The term high calcium is applied to limestones containing little magnesium. Any stone containing less than about 80 per cent of carbonates is usually considered low grade, while those stones having more than

95 per cent of total calcium and magnesium carbonates are preferred for liming purposes.

Marl, chalk, marble, coral, and shells also contain lime in the carbonate form in varying proportions.

Calcium carbonate is only slightly soluble in pure water, but if the water is saturated with carbon dioxide at ordinary temperature, it is soluble to the extent of about 1 part by weight in 1,000 parts of water, forming calcium bicarbonate [$Ca(HCO_3)_2$].

Oxide Forms of Lime. Calcium oxide is known under several commercial names, such as lime, burned lime, quicklime, caustic lime, stone lime, lump lime, and unslaked lime. Commercial oxide of lime is usually marketed in paper bags and in a pulverized form, although some is still sold in the granular or lump condition. The oxide form of lime is highly caustic and disagreeable to handle.

Commercial oxide of lime is prepared by heating any form of carbonate of lime ($CaCO_3$ and $MgCO_3$), the carbon dioxide being driven off as a gas and the calcium oxide and magnesium oxide remaining as a solid residue.

$$CaCO_3 + \text{heat} \rightarrow CaO + CO_2$$

$$MgCO_3 + \text{heat} \rightarrow MgO + CO_2$$

On burning 100 pounds of pure $CaCO_3$, 44 pounds of CO_2 are lost and 56 pounds of CaO remain. There is no loss of calcium in burning; the calcium oxide contains all the calcium of the original material.

The impurities commonly found in commercial oxide forms of lime consist of the original impurities of the limestone or other carbonate which is burned. These impurities usually consist of clay, sand, and compounds of iron and of aluminum. The oxides readily take up water and in turn carbon dioxide to form hydroxides and carbonates, respectively. Furthermore, if the burning has been incomplete (semi-calcined lime), a considerable amount of the original carbonate may remain. While commercial oxide of lime varies greatly in composition, the ratio of calcium oxide to magnesium oxide is usually 4 or 5 : 1.

Hydroxide Forms of Lime. When calcium or magnesium oxides come in contact with water, they undergo what is known as slaking; water combines with the oxides, forming compounds known chemically as calcium hydroxide and magnesium hydroxide or commercially as slaked lime, caustic lime, hydrated lime, hydrate, agricultural hydrate, or water-slaked lime.

$$CaO + H_2O \rightarrow Ca(OH)_2$$

$$MgO + H_2O \rightarrow Mg(OH)_2$$

According to the first reaction, 56 pounds of calcium oxide combine with 18 pounds of water to form 74 pounds of calcium hydroxide.

Commercial slaked lime is thus impure calcium and magnesium hydroxides. It is a mixture of hydroxides and carbonates together with such impurities as were present in the original lime before slaking. Furthermore, if the slaking process was not carried to completion, it will contain a corresponding amount of oxide. Thus six important chemical forms of lime may be present, in hydrated lime, the oxides, hydroxides, and carbonates of both calcium and magnesium.

When burned lime combines with water, marked physical changes occur. The lime increases in bulk, breaks down to a very fine powder, and generates much heat. One volume of burned lime of good quality, on slaking, increases two or three times in volume.

Commercial hydrate appears on the market as a white powder and is usually handled in paper bags, although if produced locally, bagging is not always necessary. Slaking sometimes is done in small piles in the field and the product spread directly from the piles. The hydrate is similar to the oxide forms in that it is caustic and disagreeable to handle.

Air-slaked lime is calcium oxide (together with magnesium oxide) that has absorbed moisture from the air and changed into hydroxide. As the slaked lime is formed, it slowly absorbs carbon dioxide from the air, forming carbonate. (See p. 140.) Air-slaked lime usually contains larger quantities of carbonate and less of hydroxide than water-slaked lime.

Silicate Forms. Blast-furnace (basic) slag is a by-product of pig-iron manufacture from iron ore and limestone. In the smelting process, silica of the ore reacts with calcium of the limestone to form calcium silicate. Carbon dioxide is driven off during the burning process, and there are also some magnesium and calcium oxides in the slag. It is essential that this material be finely ground in order to be effective. It is used as a liming material to a considerable extent in some of the southern states.

The Relative Neutralizing Power of the Different Forms of Lime. The neutralizing power of liming materials is always calculated on the basis of pure calcium carbonate, taken as 100 per cent. It has been pointed out that when pure calcium carbonate is burned in a kiln, 100 pounds of the dry material gives off 44 pounds of carbon dioxide gas, leaving 56 pounds of calcium oxide. When this 56 pounds of calcium oxide is moistened, it reacts with 18 pounds of water to form 74 pounds of hydrated lime. It is obvious, then, that 100 pounds of pure calcium carbonate, 74 pounds of pure calcium hydroxide, and 56 pounds of pure calcium oxide all contain the same amount of calcium and all have the same power to neutralize soil acidity because it is the calcium ions that displace the hydrogen ions in correcting acidity.

The relative ability of different liming materials to correct acidity frequently is expressed on a percentage basis. Thus, the neutralizing power of the different forms of lime in the pure state is determined by their molecular weights. The molecular weight of calcium carbonate is 100; of calcium hydroxide, 74; and of calcium oxide, 56. By dividing 100 by 74 the figure 1.35 is obtained, which means that 1 pound of calcium hydroxide supplies the same amount of calcium as 1.35 pounds of calcium carbonate. In other words, if expressed on a percentage basis (1.35×100), pure calcium hydroxide has a neutralizing value of 135 per cent relative to calcium carbonate. Likewise, calcium oxide $\left(\frac{100}{56} \times 100 = 178\right)$ has a neutralizing power of 178 per cent. Pure magnesium carbonate with a molecular weight of 84 has a neutralizing value of 119 per cent $\left(\frac{100}{84} \times 100 = 119\right)$. A limestone containing 80 per cent calcium carbonate and 20 per cent magnesium carbonate would have a neutralizing value of 103.8 per cent [$80 + (20 \times 1.19) = 103.8$]. Frequently magnesium limestones have a neutralizing power of 107 or 108 per cent. The molecular weights, neutralizing values, and calcium carbonate equivalents for the common chemical forms of liming materials are given in Table 4.

TABLE 4

RELATIVE NEUTRALIZING POWER OF DIFFERENT FORMS OF LIME

Form of Lime	Molecular Weight	Neutralizing Value (Percentage)	Pounds Equivalent to 1 Ton of Pure $CaCO_3$
Calcium carbonate	100	100	2,000
Magnesium carbonate	84	119	1,680
Calcium hydroxide	74	135	1,480
Magnesium hydroxide	58	172	1,160
Calcium oxide	56	178	1,120
Magnesium oxide	40	250	800

As indicated above, pure hydrated lime made from a pure calcium limestone would have a neutralizing power of about 135 per cent, but if made from a stone high in magnesium, the neutralizing power may run as high as 150 per cent. If all the oxides of calcium or magnesium were not converted into the hydroxide form during the process of hydration, the so-called hydrated limes may have a neutralizing power of about 170 per cent. Theoretically, pure calcium oxide has a neutralizing value of 178 per cent, but if commercial oxide of lime is made from a magnesium stone, the value may be higher if the stone is completely burned; and if incompletely burned, leaving considerable quantities of carbonate, the neutralizing value may be considerably less than the theoretical value for calcium oxide.

Effectiveness of Different Forms of Lime. In considering the effectiveness of a liming material, it is to be remembered that the total neutralizing power is not the only factor of importance. The density, hardness, and fineness of grinding of limestone, for example, may greatly influence the speed of its action in the soil. Dolomitic stones are harder and dissolve more slowly than calcium stones, and for this reason it is not customary to consider the higher neutralizing value of magnesium stones in determining the amount of lime to apply. In other words, a ton of a high-calcium stone seems as effective as a ton of a high-magnesium stone in correcting an acid soil.

Burned lime (CaO) is changed to hydrated lime under ordinary moist soil conditions before it has time to act chemically with the soil; consequently it would not be expected to be more effective than hydrated lime applied directly. Theoretically, the hydrate (and oxide) form of lime can be expected to react in the soil even somewhat more rapidly than finely ground limestone because the former is a very fine powder and can therefore be more thoroughly distributed in the soil. Moreover, freshly slaked lime consists mostly of calcium hydroxide, which is more soluble in water than is the calcium carbonate of ground limestone. This difference in solubility may result in its being more thoroughly and uniformly distributed through the soil if rain follows soon after the application of lime. Consequently, because of greater solubility and more thorough distribution in the soil, calcium hydroxide should be more active than the carbonate in producing chemical changes in the soil. It is to be emphasized, however, that calcium hydroxide will sooner or later change into the carbonate form in the soil. Under certain conditions this change may require several weeks, and during that time the hydroxide form can be considered more chemically active than the carbonate. But over a period of years, or even a rotation, there seems to be little difference in the effectiveness of the various forms of lime when applied in equivalent quantities.

Therefore, the conclusion is made that the effectiveness of a liming material in correcting an acid soil depends on its content of calcium and magnesium and is little influenced by the particular chemical form (hydroxide, oxide, or carbonate) in which these bases occur.

LIME GUARANTEES

Several states have laws regulating the sale of commercial lime materials. The regulations deal with both the guaranteed chemical composition and the fineness. Although there is little uniformity in the provisions of the lime-control laws of the different states, they all have as their aim the establishment of certain standards or methods of expressing guarantees that may be used for comparative or evaluation purposes. The topic is discussed under two headings, as indicated by the questions.

Questions:

1. What is included in the chemical guarantee and what are the various ways in which it may be expressed?
2. Of what value is the guarantee of fineness?

The Chemical Guarantee. Since the potential capacity of a liming material to neutralize soil acidity is determined by its content of calcium (and/or magnesium) and since these materials are sold on the basis of their neutralizing power, chemical guarantees are of great importance. The chemical guarantees for commercial oxide and commercial hydrated lime may be stated in several different ways. The guarantee may be phrased in terms of oxide content, calcium oxide

TABLE 5

DIFFERENT METHODS OF EXPRESSING THE CHEMICAL GUARANTEE OF COMMERCIAL OXIDE

Material	Oxide Content (Percentage)	Elemental Ca or Mg (Percentage)	Calcium Oxide Equivalent	Neutralizing Power
CaO MgO	75 20	53.3 12.0	103	183.3

equivalent, neutralizing power, or the percentages of elemental calcium and magnesium. In order to clarify the meaning of these different methods of expression, the chemical guarantee of a sample of commercial oxide is expressed in the four ways in Table 5.

The figures representing the oxide content are the forms in which chemists usually express the quantities of calcium or magnesium in any material. In other words, in this particular sample the calcium in all its forms, expressed as CaO, was equivalent to 75 per cent of the material. The magnesium compounds, whether they were in the oxide, hydroxide, or carbonate forms, expressed as MgO, made up 20 per cent of the material by weight. In arriving at these values, the chemist makes a determination for total calcium and magnesium (53.3 per cent and 12.0 per cent, respectively, for this sample) and expresses the results in terms of CaO and MgO. The calcium oxide equivalent is nothing more than a term expressing both the CaO and MgO in terms of CaO. Since 1 pound of MgO is equivalent to approximately 1.4 pounds of CaO, the CaO equivalent of the MgO in the sample is 28 or $(20 \times 1.4 = 28)$. The total CaO equivalent becomes 103 or $(75 + 28 = 103)$. As already explained (p. 125), the neutralizing power is obtained by expressing the strength of the lime in terms of its $CaCO_3$ equivalent; in this case it is 183.3 $(103 \times 1.78 = 183.3)$.

In most states having chemical lime control laws, it is customary to state the guarantee of commercial hydrates and oxides in terms of oxide content. This shows the amount of magnesium as well as calcium present, which is an important consideration particularly where magnesium may be needed in the soil. In addition, the guarantee may carry one or more of the other forms of statement. The statement of the chemical analysis in terms of CaO equivalent or in terms of neutralizing power is convenient in that it is made in one figure and thus renders comparisons between different materials quite simple.

The method of stating guarantees for ground limestone is somewhat different from that for commercial oxides and hydrates. The quantities of calcium and magnesium carbonates are usually expressed separately and in terms of total carbonates on a percentage basis. These may also be expressed in terms of CaO equivalent, in terms of neutralizing power, or even in terms of elemental calcium and magnesium. A representative sample of ground limestone may be expressed in one or more ways, as illustrated in Table 6.

Of the various methods of expressing guarantees, the neutralizing power (from which the CaO equivalent easily can be calculated) is the easiest and most rapidly determined. From the analytical point of view, this is an important consideration when many samples are involved. Yet the meaning of the carbonate method of guarantee is perhaps easier for the farmer to understand, and it gives the quantity of magnesium present, which is an advantage of this method of presentation. However, the differences between these two values (carbonates

TABLE 6

VARIOUS METHODS USED IN EXPRESSING THE CHEMICAL GUARANTEE OF COMMERCIAL GROUND LIMESTONE

Lime	Separate Carbonates	Total Carbonates	Neutralizing Power	Calcium Oxide Equivalent	Percentage Ca or Mg
Ground Limestone	80% $CaCO_3$ 15% $MgCO_3$	95%	97.9	174.3	32.00 7.14

and neutralizing power) are usually so small that practically they are of little or no significance.

Fineness Guarantee. A fineness guarantee is based on a mechanical analysis of a particular sample of ground limestone made by the use

FIG. 28. Alsike clover is more tolerant of soil acidity than are alfalfa, sweet clover, and red clover. Nevertheless, on this field at the Tennessee experiment station the crop was greatly increased by an application of lime. [From Farmers' Bul. 1845, U.S.D.A.]

of standard screens of definite mesh or size openings. A 10-mesh screen, for example, has 10 openings per linear inch or 100 openings per square inch. The quantities of stone that will pass through each of the various screens are stated in the guarantee. The following is an example: 100 per cent through an 8-mesh screen, 50 per cent through a 60-mesh screen, and 40 per cent through a 100-mesh screen. Various grades of ground limestone appear on the market, based on the percentage of the stone that will pass through screens having a specified number of meshes per linear inch.

With all other factors equal, the finer a limestone is ground, the more rapidly it will dissolve and the more thoroughly can it be mixed with the soil. The effectiveness of a ground limestone of a given neutralizing power is determined not only by its rate of solubility but also by its contact with the colloidal particles. However, the finer the stone is ground, the greater will be its cost and the less will be its lasting qualities. Furthermore, a very finely ground limestone is not only difficult to handle but also unpleasant to distribute. Therefore, it is generally recommended that a ground limestone of medium fineness be purchased. Such a grade can be ground rather cheaply, and it contains a sufficient quantity of fine material to give immediate effects and a sufficient amount of coarse material to give it lasting qualities. Such a ground limestone would be one in which all would pass a 10-mesh screen and 30 to 50 per cent would pass a 100-mesh screen.

SOURCES OF LIME

In the introduction of this chapter, it was pointed out that one reason why calcium compounds rather than other compounds are used for liming purposes is that there is a plentiful supply of calcium compounds and that they are widely distributed. In this connection, mention should be made of the different sources of lime and of the nature and supply of such materials. Questions are proposed to call attention to pertinent points relative to supplies of liming materials.

Questions:

1. What is limestone? Is there a sufficient quantity to meet the needs of the soil?
2. What is marl? How is it formed and where are the deposits found?
3. Is sugar-factory lime of any value for liming purposes?
4. Do wood ashes contain significant amounts of calcium?
5. Can lime from gasworks, acetone plants, acetylene plants, and paper mills be used for liming purposes?
6. Is blast-furnace slag of any value for correcting soil acidity?
7. Are oyster shells considered a source of lime?

Limestone. Limestone is a sedimentary (deposited in water) rock formed by the precipitation of calcium and magnesium in water in the form of carbonates with subsequent consolidation of the deposits. Limestone deposits are widely distributed, and they constitute an important source of lime. Many limestone quarries are operating in the humid regions of the United States and are in a position to meet the demand of agriculture, as well as of all other industries, for lime. It might be added that only a small portion of the output of these quarries is used in agriculture at the present time.

Limestone is the most important single source of commercial liming materials. The approximate consumption of liming materials on United States farms during 1940 is presented in Table 7. A very large

TABLE 7

APPROXIMATE CONSUMPTION OF LIMING MATERIALS ON UNITED STATES FARMS DURING 1940 *

FORM OF LIME	TONS USED
Ground limestone	2,819,060
Limestone screenings and meal	9,384,886
Burned lime	185,535
Hydrated lime	221,499
Marl	482,869
Miscellaneous materials †	339,766
Total liming materials	13,433,615
Effective lime oxides ‡	6,004,181

* From data compiled by National Lime Association.

† Consists of ground shells, by-product lime, etc.

‡ Computed on the following basis: 50 per cent for ground limestone, miscellaneous materials, and commercial marl; 35 per cent for farm-dug marl; 42 per cent for limestone screenings and meal; 70 per cent for hydrated lime; and 85 per cent for burned lime.

percentage of the commercial lime that is used for liming purposes is used in the form of ground limestone or in some form made directly or indirectly from limestone. The calcium of commercial hydrate, commercial oxide, and calcium silicate of the blast furnaces all comes from limestone.

Marl. Marl[1] or bog lime is formed by lime which was dissolved from the soil in drainage water and carried to swamps, lakes, and other bodies of water where it was thrown out of solution. This precipitation of calcium may be a purely chemical reaction, the loss of carbon dioxide from calcium bicarbonate, or it may occur as the result

[1] From an agricultural point of view bog lime is usually spoken of as marl. Geologists frequently refer to marl as a calcareous clay of variable composition. In certain areas unconsolidated calcareous rock is called marl; the calcium carbonate is finely disseminated and is also in the form of shells and fragments of shells. Glauconite (green sand) is also sometimes referred to as marl.

of the action of certain plants or of shell-forming animals. Frequently it accumulates on the bottom of these bodies of water as a soft, mush-like material, or it may occur in nodular form. Often these two forms are intermixed. It may be found underlying areas of muck and peat and along the shores and in the beds of lakes. The marl may be covered by a few inches or several feet of muck. Marl beds are also found on dry land where an old lake has been drained by a change in the drainage level. The thickness of these marl deposits may range

FIG. 29. A marl pit in Michigan.

from less than one inch to several feet, and the areas are quite variable in size.

The lime in marl exists principally in the calcium carbonate form with varying quantities of magnesium carbonate. The value of a marl to correct acidity is usually expressed as though all the carbonates were present as calcium carbonate; thus a marl testing 95 per cent has the same neutralizing power as a ground limestone with a neutralizing value of 95. The purity of marl may vary considerably, even within the same deposit. The common impurities found in this material are sand, silt, clay, and organic matter. Of these, clay is perhaps the most objectionable. A marl containing a rather high clay content is very sticky when wet and becomes very hard and lumpy when dry; furthermore, it is difficult to handle and cannot be applied uniformly.

Owing to the variable moisture content of marl, it is usually spoken of and dealt with in terms of cubic yards. On the average, a cubic yard of good marl (testing about 95 per cent on the dry basis) weighs about 2,300 pounds in the wet condition and contains approximately 1,600 pounds of calcium carbonate. Since it is usually difficult to spread marl uniformly, two cubic yards of a good marl frequently are recommended for each ton of limestone needed.

In some states, marl is the most important source of liming materials. In Michigan, for example, approximately 80 per cent of all liming materials used is in the form of marl.

Sugar-Factory Lime. In the manufacture of sugar from sugar beets, lime water is added to clarify and purify the juice from the beets. Carbon dioxide gas is then bubbled through the juice to precipitate the lime as calcium carbonate, which is subsequently filtered out and discarded. This material contains insignificant quantities of organic matter and plant nutrients. This lime frequently is washed out to the waste pile with the water used in washing the beets and may thus contain considerable dirt. On a dry basis, sugar-factory lime will usually test about 80 per cent calcium carbonate. Owing to its variable moisture content, it usually is handled on the cubic-yard basis.

Some of the sugar factories prefer to burn their own lime, and when the kilns are cleaned, impure burned lime or stone lime is obtained. This material is mostly calcium oxide when fresh, but after a short time it takes on water and forms hydrate, and in turn some of it may react with carbon dioxide, forming calcium carbonate.

Wood Ashes. In certain localities wood ashes may be available in sufficient quantities for use on the soil. The composition will depend not only on the composition of the original wood but also on the completeness of burning, leaching that has occurred, and impurities with which it may have become mixed. The lime in freshly produced ashes is in the oxide form, but on exposure hydroxide and carbonate will be formed. Wood ashes, as ordinarily used, may be expected to contain from 20 to 50 per cent of lime, expressed as calcium carbonate. They will also contain about 5 per cent or less of potash and very small amounts of phosphorus but are usually regarded as more important as a liming material than as a source of potash.

Lime from Acetone Plants, Gasworks, Acetylene Plants, and Paper Mills. There are various manufacturing plants which produce or discard lime as a by-product. Such material is usually of rather high grade, and when it is available locally and at a low cost, its use may be advisable. The principal disadvantage of this lime is that it usually contains a high content of water which makes spreading difficult. The cost of transportation and the difficulty in distributing must be taken into consideration when comparing costs with other sources of lime. These materials generally can be obtained at little or no cost since they are regarded as waste products.

Calcium oxide is used at gasworks to remove impurities from the gas, after which it is either discarded or sold to farmers. This lime is largely in the hydroxide and carbonate form and it contains many impurities, some of which, such as sulphites and sulphides, may injure seeds. On exposure to the air, these sulphur compounds soon change over to the sulphate form which is harmless.

Acetylene lime is a solid residue formed in the manufacturing of acetylene gas from calcium carbide and consists largely of the hydroxide and carbonate forms of lime. This material is wet and contains some acetylene gas which is injurious to seeds, but this gas will escape within a few weeks on exposure to the air.

Blast-Furnace Slag. As mentioned previously (p. 125), blast-furnace slag contains calcium in the form of calcium silicate. The slag is a by-product in the smelting of iron ore. This may be considered a liming material inasmuch as it supplies active calcium, leaves no objectionable residue (the silicate ion produces a very weak acid), and has the ability to reduce soil acidity. Basic slags (in which materials are added to remove acid impurities) usually contain magnesium oxide and also calcium oxide. A typical grade of basic slag produced in the United States contains 6 to 8 per cent of calcium silicate, 5 to 7 per cent of magnesium oxide, and 45 to 48 per cent of calcium oxide. This is an important source of lime in some of the southeastern states. Basic slag is also an important carrier of phosphorus (see p. 316).

A relatively new type of slag is now becoming available in substantial quantities from rock phosphate reduction furnaces. This product consists largely of calcium silicate and contains most of the original constituents of phosphate rock, except the phosphorus, together with the silica which was added. This product constitutes a satisfactory liming material.

Oyster Shells. Clean oyster shells contain about 90 to 95 per cent of calcium carbonate. When finely ground they are a very satisfactory liming material which is used to some extent on farms located near the source of supply. Oyster-shell lime sometimes is burned and used in the calcium oxide form.

THE USE OF LIME

After a discussion of the general importance of lime in agriculture, its functions in the soil and the forms, composition, and sources of lime, it is important that certain points dealing more specifically with the practical use of lime be clearly understood. May it be emphasized again that lime, if needed and wisely used, is indispensable to a permanent agriculture. Several questions are suggested as guides in this discussion.

Questions:

1. How can the need of soils for lime be determined?
2. Are all crops benefited by the application of lime?
3. What kind of lime should be used?
4. What changes does lime undergo when applied to soil?
5. How much, and how frequently, should lime be applied?
6. When should lime be applied?
7. What are the different methods of applying lime?

Determining the Need of Soils for Lime. In arriving at a conclusion as to the desirability of applying lime and the amount needed, it is necessary that soil tests be made. The general need of soils for lime may be made by observation within certain limits, but the amounts to apply are not so indicated. Some common indications of an acid soil are: (1) a prevalence of red sorrel, (2) the repeated failure of alfalfa and clover, (3) thin or patchy stands of alfalfa, or clover on soils which produce good yields of other crops, or (4) alfalfa and clover may turn yellow in early spring. These symptoms may be caused by some factor other than a lack of lime; yet these conditions frequently are associated with a deficiency of calcium. From such observations it is obvious that the amount of lime needed cannot be ascertained, and it may be practicable to use it long before any of the deficiency symptoms are apparent. Furthermore, an excess of lime is to be avoided; consequently soil tests must be used. The idea should be discarded that lime is a cure-all and that, if a little is good, more is better.

Soil tests for determining lime requirement are discussed in Chapter IV, p. 105, and will not be repeated here. It may be mentioned, however, that the simple colorimetric tests are sufficiently accurate for practical purposes. It is much more desirable from the practical standpoint to test a large number of samples from a particular field with the simple tests than to test only one or two samples by one of the more accurate but time-consuming laboratory methods.

The different fields on any particular farm frequently vary a great deal in acidity; likewise each field may vary in acidity from place to place. Consequently, when the whole farm or an entire field is given a uniform application of lime, some parts may get too little and others too much. It is important, therefore, that a field be properly sampled and tested in arriving at the lime needs. Several tests should be made of the surface soil (plow depth) and at least one or two tests of the subsoil (at a depth of 2 to 3 feet) of each soil type in each field. It is sometimes recommended that the surface soil of each acre of a field be tested. The number of tests which should be made in a field of a given size will depend largely on the topography. On very level fields fewer tests need be made. It is necessary that several tests be made of each soil type in order to avoid such irregularities as may be due to droppings of manure, ashes from burned weeds, brush, or crop residue, and straw-stack residues.

Soil-acidity tests, together with an understanding of the chemical factors associated with the test results and an understanding of a plant's requirements, furnish a basis for making lime recommendations.

The Response of Crops to Lime. Crop response to lime depends upon a number of factors, among which the *p*H of the soil is an important one. It is, however, generally accepted that *p*H as such is rarely the limiting factor in plant growth within the range normally existing in agricultural soils. Plant growth frequently is retarded in acid soils because of unfavorable chemical conditions associated with a low *p*H; for example, the degree of base saturation is definitely related to *p*H and, since calcium, magnesium, and potassium are all involved, plant growth must be affected by the supply of these elements. The solubility of aluminum, iron, manganese, and perhaps other minor elements is also related to *p*H, as is also the activity of microorganisms. The experimental evidence relative to crop response to lime is, as would be expected, rather confusing because of the wide range of soil and climatic conditions under which many of the lime experiments have been conducted.

In general, the legumes respond more to lime than do the non-legumes. The fixation of nitrogen by legumes is much more effective where high levels of available calcium are present. Of the legumes commonly grown, alfalfa, sweet clover, red clover, and Dutch clover respond to the lime most markedly. Garden and field peas, soy beans, and alsike clover give noticeable response to lime, but less than the legumes previously mentioned. Likewise, cowpeas, vetch, field beans, and lespedeza are benefited to some degree by lime, but less than the preceding group. Contrary to popular belief, all legumes do not respond favorably to lime. For example, serradella and the lupines may actually be injured by liming.

Of the non-leguminous crops, sugar beets, table beets, cabbage, and cauliflower are particularly responsive to lime even though the soil may not be strongly acid. Corn, oats, wheat, barley, rye, sorghum, Kentucky bluegrass, and timothy appear to be benefited by liming, although the benefit may be brought about in part indirectly. The indirect effect is produced through the stimulation of legume crops in the rotations which in turn benefit the non-legume crops. Lime permits greater growth of the legumes and thus leaves more organic matter and larger quantities of active nitrogen in the soil. Crops like cotton, tobacco, Irish potatoes, redtop, strawberries, buckwheat, bent grass, fescue, and millet will make satisfactory growth on soils low in active calcium and appear to be little influenced directly by the addition of lime. However, indirectly (through its effect on legumes in the rotation) lime will usually have a beneficial effect. Certain non-leguminous plants may be injured by lime; included in the list are such plants as watermelons, blueberries, cranberries, laurel, azaleas, and rhododendrons. Furthermore, lime may injure potatoes by encouraging a disease known as scab.

FIG. 30. Although soy beans are somewhat tolerant of soil acidity, the crop responds remarkably to an application of lime on strongly acid soil. Left: Plants from limed soil. Right: The same number of plants from unlimed soil. [From Farmers' Bul. 1845, U.S.D.A.]

It is thus apparent that it becomes a matter of great importance not only to know the condition of the soil with respect to its content of active calcium and magnesium but also to understand the response of crops to these elements. As has been indicated, crop response to liming is a complicated phenomenon, and only rather broad, general conclusions can be drawn.

Kind of Lime To Use. In selecting the liming material to use, several factors must be taken into consideration which will vary with the conditions attending each individual farmer. The following factors are among those which should be considered in choosing the form or grade of lime that should be used: (1) relative cost of calcium in the different forms, (2) fineness, (3) rapidity of action, (4) convenience in handling and storage, (5) character of soil, and (6) crops to be grown.

Since equivalent quantities of the different chemical forms of lime are essentially equally effective in neutralizing soil acidity and in increasing crop growth, obviously *cost* is one of the most important factors to consider in choosing a liming material. The cost of equivalent quantities of limes applied to the soil may be determined by simple arithmetic calculations. From these calculations it is easily learned which gives the greatest quantity of lime per dollar spent. Total cost is determined by the cost of the material at the plant or at the source of supply and the distribution costs, including delivery to the farm and application to the soil. If the material has to be transported a considerable distance, it may be less expensive to purchase the more concentrated materials (burned or hydrated lime). Whereas if the material is close at hand, the forms containing less calcium (ground limestone, marl, refuse lime, etc.) may be the least expensive.

Where immediate or first-year effects are especially desired, particular attention should be given to the degree of fineness, and under such circumstances allowance must be made for this factor in comparing costs of ground limestone, shells, etc. Properly slaked lime is probably in the finest state of subdivision of any of the liming materials. Hydrated lime is somewhat more soluble than the carbonate form, and until such time as it may change to the carbonate form, the hydrate form must be considered more active.

The carbonate forms of lime are generally more convenient to handle and less disagreeable than the caustic forms, even when the latter are bagged. Furthermore, limestone in storage, if kept dry, will not change form as will the caustic forms. The difficulties in scattering materials, such as marl and certain industrial by-product limes, have been mentioned (p. 134) and should be considered in selecting the form of lime to use.

Only in rather special cases does the character of the soil affect the form of lime that should be used. Obviously, on certain soils containing large quantities of magnesium, it is desirable to avoid the use of those forms of lime rich in magnesium. Furthermore, where the caustic effects of oxide and hydrate forms are likely to occur, such as on light, sandy soils, it is safer to use the carbonate form.

Changes Lime Undergoes When Applied to the Soil. The immediate changes that lime undergoes when added to the soil depend among other things on the form of lime, the chemical nature of the soil, and the moisture supply. Lime in the form of either the oxide or hydroxide undergoes rather rapid change when applied to moist soils. The attraction between burned lime (oxide form) and water is very intense. Water combines with the oxides of calcium and magnesium to form the corresponding hydroxides or water-slaked lime. In moist soil this reaction occurs so rapidly that perhaps the oxides as such do not have time to react chemically with the soil.

Slaked lime has a strong affinity for carbon dioxide and the two combine to form carbonate. In moist soil the reaction probably occurs as follows:

$$Ca(OH)_2 + Mg(OH)_2 + 2H_2CO_3 \rightarrow CaCO_3 + MgCO_3 + 4H_2O$$

This reaction would proceed rapidly in moist soils well supplied with active organic matter because of the abundant production of carbon dioxide.

It is thus evident that lime applied to soils in the form of oxide or hydroxide is sooner or later changed to the carbonate form. It might thus be assumed that little difference would be found in the activity of the various forms since they all change to the carbonate form. This is essentially true; yet as has been mentioned before, the slaked limes are more soluble than the carbonate forms, and until they are converted to the carbonate forms they will be more active.

However, under field conditions over a period of years differences in the effectiveness of the various forms of lime generally have not been observed and practically are of little significance. Lime (in its various forms) may be considered as acting principally in the carbonate form in the soil.

The carbonate form is only slightly soluble in water; yet in the finely divided state it appears to displace hydrogen ions of the colloidal soil particles, thus reducing the potential acidity. The reactions of the carbonate in neutralizing acidity may be illustrated as follows:

$$2H(\text{soil}) + Ca(HCO_3)_2 \ (\text{or } CaCO_3) \rightarrow Ca(\text{soil})_2 + 2H_2CO_3$$

First the carbonate may react with carbonic acid to form calcium bicarbonate [$CaCO_3 + H_2CO_3 \rightarrow Ca(HCO_3)_2$], which in turn may react with the soil, or else the calcium carbonate may react directly with the soil colloids. The colloids are thus charged with bases accompanied by the liberation of carbon dioxide.

It is obvious, therefore, that lime added to soils exists principally in two forms, as carbonates and as exchangeable ions in association with the organic and inorganic colloids. In both forms the bases are susceptible to the action of carbonic acid and form bicarbonates which may readily leach out of humid region soils. The reactions involved are essentially the reverse of those given above.

$$CaCO_3 + H_2O + CO_2 \rightleftharpoons Ca(HCO_3)_2$$

$$Ca(soil)_2 + 2H_2O + 2CO_2 \rightleftharpoons 2H(soil) + Ca(HCO_3)_2$$

Quantity and Frequency of Lime Application. The amount of lime to be used depends primarily on the (1) degree of acidity of the soil, (2) buffer capacity of the soil, (3) acidity of subsoil, (4) crops to be grown, (5) form or grade of lime to be used, and (6) frequency of application. There are various tests (p. 105) which are used to aid in determining the amount of lime to apply to soils, but because the quantity needed is dependent on so many factors, all lime recommendations are at best only approximations.

Under most general farming conditions, it is usually advisable to regulate the application of lime in order to maintain the soil reaction at a pH of about 6.5 or 7.0. Under other farming conditions and in special cases, it is sometimes desirable to regulate the quantity of lime so as to maintain either a higher or lower pH.

Sandy soils usually require less lime to bring them to any desired pH than do heavy soils having the same degree of acidity. This difference is due largely to the much greater buffer capacity of the latter soils. Furthermore, the amount of lime required to change the pH of a strongly acid muck or peat soil is much greater than for mineral soils. On mineral soils it is usually not necessary and frequently impractical to apply more than the equivalent of 2 or 3 tons of finely ground limestone per acre. Five to 10 tons of lime per acre are sometimes recommended for certain acid organic soils. Perhaps the most generally recommended application of lime considered necessary for mineral soils is 1,000 to 2,000 pounds of finely ground limestone (or its equivalent) per acre, once in a rotation. There is some difference of opinion as to whether a large application is better than smaller and more frequent applications. When finances available for liming are limited, the latter is generally advisable.

In some instances, the use of a relatively small amount (200 to 500 pounds per acre) of very fine lime, drilled in the row at seeding time, has proved satisfactory. When this method is used, it is done as a regular part of the legume seeding whenever the crop comes around in the rotation. This method has an advantage in decreasing the initial cash outlay, gives a better distribution of labor, and permits the lime to be transported greater distances. It is not to be inferred that this method should displace the ordinary plan of liming, but the use of relatively small amounts of fine limestone may permit liming under many conditions where heavy liming is not commonly practiced.

The question of overliming has recently received considerable attention. It is generally believed that if the *p*H of the soil is raised above 7.0 or 7.5, the availability of phosphorus, potassium, iron, manganese, and perhaps copper, boron, and zinc is decreased. There is much evidence to indicate that especially on sandy soils, where the quantities of manganese, zinc, and iron are low, excess quantities of lime will convert these into less available forms. There is also evidence to indicate that overliming disturbs temporarily the utilization of phosphate in plants. No doubt, the question of overliming has in some instances been overemphasized, yet it is well to keep in mind some of the possible dangers resulting from such a practice.

Time of Applying Lime. Generally speaking, lime may be applied any time during the year when it is most convenient. In many cases, however, the type of the rotation, system of farming, and form of lime used will be the deciding factors.

If the caustic forms are to be used for spring-seeded crops, it is usually best to apply the lime in the fall. These forms will change over to the carbonate form by seeding time and will lose their caustic properties and have no injurious effect on seed germination. Lime in the carbonate form may be applied any time without danger of injury. Regardless of the form of lime used, it is best applied when the soil is not too wet; otherwise it is difficult to get an even distribution. Under wet-soil conditions, the caustic forms and very finely ground carbonate tend to ball badly when spreading.

It is advisable, of course, to apply lime where it can be used to the best advantage in the rotation, for example, preceding the legume crop or in connection with a green-manure crop. It is usually best to apply lime a considerable time in advance of seeding legumes, in order that the lime will have time to correct the acid condition of the soil.

To insure best results, lime should be in the soil at least six months prior to seeding legumes, although successful legume seedings sometimes are obtained by applying lime immediately preceding or with the legume seeding.

If the legume is to be seeded with a nurse crop, such as wheat, frequently the lime is applied just before the wheat is seeded in the fall. In a rotation of corn, oats, wheat, and clover, it is usually most conveniently and satisfactorily applied before the wheat. If a sod crop is to be plowed under for such crops as corn or beets, to be followed by one or two grain crops in either of which a legume is to be seeded, lime probably can be most conveniently applied to the sod before plowing.

When potatoes are included in the rotation, lime should follow this crop if scab is prevalent, for it is generally believed that lime makes conditions more favorable for the development of the organisms which produce scab. Many potato growers have found that this crop gives greater yields when it follows a leguminous sod. Generally the increase in yields are sufficient to justify using lime on sour soils to make the growth of leguminous crops possible. This, therefore, usually necessitates the use of a minimum amount of lime directly after the potato crop is harvested.

In general practice it is usually recommended that lime be applied when most convenient since the most important consideration is that of applying it regularly in the rotation as needed. Late summer or fall application is usually most convenient. At this time of the year, there is a lull in farm work, and the roads and fields are in a good condition for hauling and spreading.

Methods for Applying Lime. The principal requirement of any method of applying lime is that it should be distributed evenly and, except when applied to pastures, it should be thoroughly mixed with the soil. Lime even as it dissolves moves to no appreciable extent horizontally and only to a limited extent vertically. Movement is not sufficient to distribute the lime evenly over the field or to mix it thoroughly with the soil. Since soil acidity is due largely to the colloidal clay acids, it is essential that lime comes in contact with all the soil particles so far as possible. This requires a thorough and even mixing of the lime with the soil. An even distribution is best accomplished by the use of any of the standard lime spreaders on the market. The only way of mixing the lime thoroughly with the soil is by tillage operations.

There are three types of spreaders in common use: (1) trucks with specially designed hoppers and distributors, (2) the two-wheeled box type of spreader which can be operated by one man and a team, and (3) the endgate spreader which is attached to the rear end of the wagon box. The distributing device of this attachment is operated by a chain that is running on a gear bolted to one of the rear wagon wheels. It requires two men to operate the endgate spreader, one to

FIG. 31. The two-wheeled box type of lime spreader is an efficient implement for applying ground limestone and hydrated lime. [From Farmers' Bul. 1845, U.S.D.A.]

drive while the other shovels the lime into the spreader. Similar types of attachments for manure spreaders are also on the market.

Grain drills sometimes are used for applying lime. The lime is put into the fertilizer compartment and may be drilled either at the time of seeding or as a separate operation. This is an excellent method of applying small amounts of finely ground limestone.

The old method of spreading lime from a wagon with a shovel has the advantage of not requiring any additional investment in equipment. Sometimes the lime is placed in piles over the field and later scattered with a shovel. These methods require more labor, are usually more expensive, and are more disagreeable; also, it is more difficult to spread the lime evenly than to apply with a machine, but such methods can be used if a mechanical spreader cannot be obtained.

Marl, sugar-factory lime, and similar types of liming materials which are usually wet and difficult to handle are generally applied by means of a manure spreader or truck equipped with a spreading device.

A thorough mixing with the soil can perhaps be best accomplished by applying the lime evenly on plowed land. The disking, harrowing, and other tillage operations permit a thorough mixing of lime and soil particles.

CHAPTER VI

SOIL MOISTURE

Moisture supply receives more general attention than any other factor concerned with crop production. The bringing of water to otherwise fertile land has entailed the construction of immense dams and elaborate canal and pumping systems and has resulted in the conversion of huge expanses of desert into highly productive farms. The use of irrigation to supplement natural moisture supply in humid regions also has been developed to a considerable extent, largely for the production of special crops. On the other hand, in humid regions the moisture problem is more often one of removal than of supply. In many places, extensive drainage systems have been necessary to remove the excess water from countless acres of farm land in regions of moderate to high precipitation.

An inadequate moisture supply frequently limits crop production to a greater or less extent in virtually all regions where natural precipitation is depended upon to supply the moisture needs of plants. The problem of getting sufficient water to enter the soil confronts many farmers. On irrigated land the farmer is interested primarily in increasing the permeability of his soil to irrigation water, and the farmer in humid areas endeavors to increase the percentage of the precipitation that enters the soil instead of running off over the surface. The farmers of both regions are concerned with the conservation of soil water for crop use.

As the moisture problem under irrigation will be discussed in another chapter, let us now consider moisture relations in humid and semihumid regions. Water which reaches the earth in the form of precipitation may be classified under headings of (1) that which does not enter the soil and (2) that which does enter the soil. It is suggested that the study of soil moisture be taken up under four headings or objectives.

Objectives:

A. Soil water which yields to the pull of gravity.
B. Soil water which is retained against the pull of gravity.
C. Water in relation to plant growth.
D. Loss of moisture from the soil.

SOIL WATER WHICH YIELDS TO THE PULL OF GRAVITY

Water which the soil is unable to retain against the force of gravity is known as gravitational, free, excess, or drainage water. As this moisture occupies the larger spaces in the soil which should be filled with air, if plants are to grow well, the rapidity with which gravitational water passes below the main root zone of crops is of interest. Soil through which free water passes slowly is referred to as tight or cold. Planting frequently is delayed on such soils, with the resulting hazard of frost damage in the fall in some climates. In other soils the movement of free water is very rapid, thus producing warm or early soils. The escape of gravitational water usually is influenced by the underlying soil horizons to a greater extent than by the surface soil, and hence when investigating soil drainage an examination should be made to a depth of three or more feet. In moving downward through the subsoil, the drainage water as a whole follows well-developed drainage channels and does not seep slowly around particle after particle. There is, of course, some downward and oblique seepage as the moisture moves toward the channels of freer flow. Likewise, as the channels fill there is seepage out from them so that the entire body of the soil is wet. These drainage ways are not large and are comparatively close together. They consist of channels left by decaying roots and of those made by worms and insects. Small crevices between the structural units of the soil also serve as drainage channels, as do numerous cracks formed by drying of the soil. An idea of the methods used in removing excess water from the soil and of the benefits to crops of its removal may be gained by seeking answers to the questions proposed.

Questions:

1. Why do many crops grow poorly in inadequately drained soils?
2. How should tile lines be placed in the field?
3. How deep and how close together should tile lines be placed?
4. What size of tile is advisable?
5. Should careful attention be given to outlets?
6. Is tile drainage always preferable to open ditches?
7. Are appreciable quantities of nutrients lost in drainage water?

Detrimental Effects of Poor Drainage. Poorly drained soils warm up slowly in the spring and cool slowly in the fall because water has an heat-absorbing capacity approximately five times greater than soil particles and so much heat energy must be added or lost to bring about a change in temperature. The early-season coldness of poorly drained soils is one of the limiting factors in their use for crop production. The high moisture content of these soils also delays the fitting of the soil for planting and thus necessitates their use for late-planted crops or else for the seeding of early crops in poorly prepared soil.

The roots of the most commonly grown crops require a supply of oxygen for the process of respiration, and as poorly drained soils contain an abnormally low proportion of air the roots tend to suffocate in them.

With an inadequate oxygen supply, decomposition of organic matter must be largely of an anaerobic nature with the production of compounds, some of which are toxic to plants, in a reduced chemical state. The disagreeable odors from stagnant pools during warm weather bear witness to the production of unoxidized compounds in waterlogged soil. Under such conditions oxidized mineral compounds may be also reduced to poisonous combinations such as sulphides from sulphates, nitrites from nitrates, ferrous iron from ferric iron, etc. Plants may also suffer from an insufficient supply of nutrients in poorly drained soils because the normal decay of organic matter with the accompanying liberation of available nutrients is hampered. Furthermore, it is doubtful if the plant roots can effectively carry on the processes by which nutrients are liberated and absorbed from the surfaces of the soil particles.

Systems of Tile Drains. Different soil situations require different types of drainage systems. A few of the conditions frequently encountered will be discussed. First, there are the level lands with a high water table which require a complete, systematically placed tile system. The main tile lines should be placed at such intervals that the secondary lines, commonly called laterals, flowing into them will not be more than 80 rods long. Furthermore, the laterals should enter the main lines at a small angle to the direction of flow rather than at right angles to it. Two commonly used systems for such land are shown in Fig. 32.

Rolling land of clay loam or silt loam soil may need drainage only in the valleys. The first inclination is to place the tile lines near the center of the valley as that is where the need for drainage is most apparent. It is usually found, however, that the water seeps down from the adjoining hills, and the proper place to install the tile is near the foot of the slope where they will intercept the water and prevent it from reaching the main valley floor. The aid of an experienced drainage engineer is needed for the proper location of effective drains in such land. The depth at which such drains are laid is highly important.

A covering of sandy soil of varying depth over clay frequently offers an interesting drainage problem. The portion of the field where the

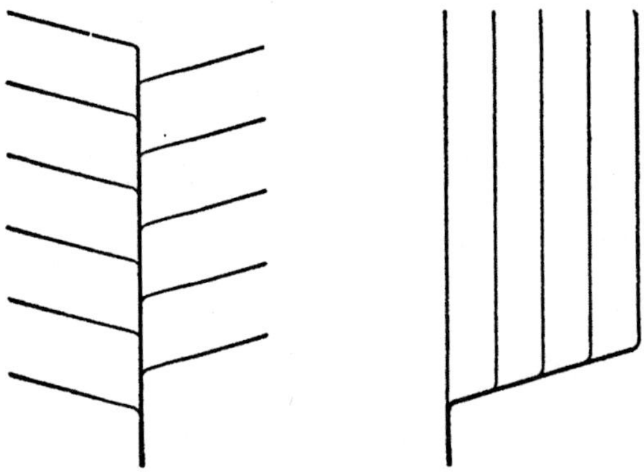

FIG. 32. Two systems used in laying tile. On the left is shown the herring-bone method and on the right the gridiron system. The larger tile line into which the smaller ones or "laterals" empty is called a "main."

sand is of considerable thickness may need no drainage, but as the sand covering gets thinner excess water is likely to cause trouble. Again in this case the tile should be placed not in the main body of the wet sand but near the edge of the thicker sand covering, for it will be found that the water is moving over the clay floor from the deep sand. The tile should also be set so that they lie largely in the clay.

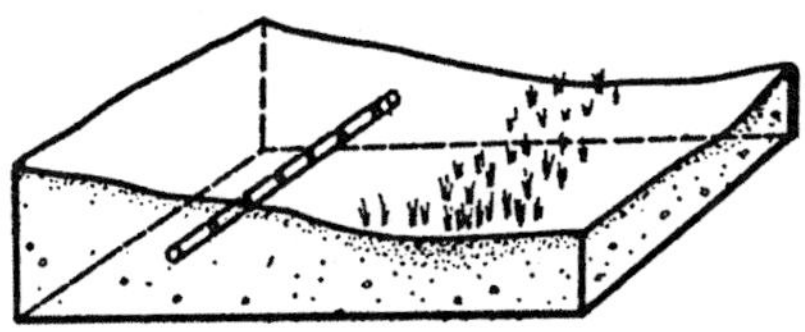

FIG. 33. In draining a wet or "springy" area near the bottom of a slope, the tile is not put through the poorly drained spot itself but is put some distance up the slope so that it will intercept the water as it seeps downward.

In some fields comparatively small depressions are wet although the soil as a whole is sufficiently well drained. It would seem that lines of rather small tile would be adequate to remove the excess moisture from such small areas. This conclusion is incorrect since the water from a considerable portion of the field may slowly accumulate in these depressions. Accordingly, it is necessary to use a sufficiently large size of tile in each case to carry the water from all the land that drains into the depression.

Depth and Spacing of Tile Lines. Unless some unusual condition prevails, tile should be laid at least 2½ feet deep. The depth of the outlet has some bearing on the depth at which the drains may be placed and still provide a satisfactory fall. A fall of at least $\frac{1}{10}$ foot per 100 feet is considered necessary, and more is desirable. The water level is lowered by the lines as shown in Fig. 34. In order, therefore, to have a sufficiently low water table between tile lines, to permit crops to develop an adequate root system, it is essential that the tile be 2½ or more feet deep.

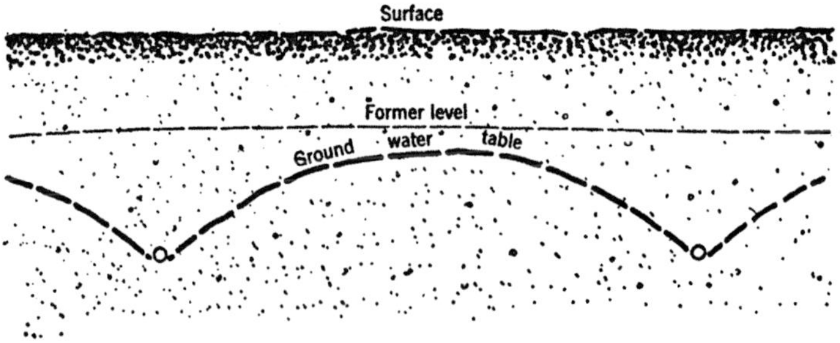

FIG. 34. The effect of tile lines in lowering the water table or ground-water level. In very heavy soil it takes several years for the water table to reach the proper adjustment. The benefit of the drainage is first evident immediately over the tile and gradually spreads to the soil area between them.

The freedom with which water moves through the soil determines the distance apart drains may be placed and yet afford adequate drainage. If the soil through which the water must percolate to reach the tile contains a high percentage of clay and the structural condition is such as to make water movement slow, the tile lines should not be more than 4 rods apart. In more pervious soil a spacing of 5 rods is permissible, and in soil through which water drains rapidly the drains may be placed at 6-rod intervals. If the entire field needs drainage, tile lines should not be farther than 6 rods apart even in very pervious soils.

Sizes and Kinds of Tile. Experience and study have led to the conclusion that the small 2- and 3-inch tile formerly used extensively for laterals are not advisable. Nothing smaller than a 4-inch tile is now recommended even for short lines, and if the drain is to be of considerable length, a 6-inch size is preferable. Main lines into which the laterals empty should be of 6-inch, 8-inch, or larger tile, depending on the volume of water to be carried. In the long run it is better to use a size larger tile than is thought necessary rather than a smaller size. A drainage engineer will have a table for calculating the size of tile to use according to volume of water to be carried, fall, rate of movement of water into the tile, and similar factors.

Three kinds of tile are usually found on the market. The soft clay tile are of varying quality. If well burned so that they give a clear, ringing sound when struck, they are quite satisfactory. On the other hand, if they are poorly burned they are usually not very durable. Tile which are glazed, generally by the use of salt, are of very good quality and although more expensive are more economical than a poor grade of soft clay tile when length of service is considered. For a number of years concrete tile have received considerable attention. They vary greatly, owing to the difference in curing procedure. Unless the manufacturer has facilities for keeping the tile moist for a time sufficiently long to permit thorough setting of the cement, they are not durable. This uncertainty in quality has brought concrete tile more or less into disrepute. Also, in highly acid soils they are likely to disintegrate prematurely. They are seldom recommended for use in muck or peat soils.

In laying drains in organic soils or in other soils which are not sufficiently rigid to form a firm base and so keep the tile in line, it has been customary to place them on boards to prevent settling. A better procedure is to use a somewhat larger tile and procure them in double length (2 feet). Each of these tile will present much more bearing surface than the short ones of smaller diameter, and hence they will stay in line much better.

Good Outlets Are Important. Before installing a drainage system a careful investigation should be made of outlets which will provide for disposition of the water removed from the land. Frequently, when there are extensive areas of land which need drainage, the natural streams which can serve as outlets are so far apart that tile lines must be too long. They would be required to carry such a large volume of water that very large and expensive tile would be needed, and also it would be impossible to get sufficient fall and yet place the tile at the upper end of the system deep enough to provide efficient drainage. Also, many such natural streams are too shallow to afford good outlets for tile lines and too small to carry the volume of water which would be quickly thrown into them from a tile system. To overcome these limitations dredging of the natural streams is frequently resorted to and, in addition, a quite complete system of dredge ditches is sometimes necessary.

In some land areas lying close to lakes, bays, or large streams into which the drainage flows, there is not sufficient difference between the land and water levels to allow for satisfactory tile drainage. Such situations require the most careful investigation by competent drainage engineers before steps are taken to develop the land. Some attempts have been made to dyke areas of such land when the quality of the soil appeared to warrant so expensive an enterprise. A drainage system would be installed which carried the water to concentration points along the dyke, from which it would be pumped over the dyke into outlet streams or ditches. The expense involved in the construction and operation of such developments makes their feasibility questionable except under very unusual conditions.

The openings of tile lines into streams and ditches should be constructed in such a way as to prevent the undermining of the outlet by water and the slumping of the tile out of line. Several lengths of sewer tile at the end of the drain are often suggested. The opening of the tile should also be protected against the entrance of rabbits and other small animals.

Tile Lines vs. Open Drains. A tile drainage system has several advantages over drainage by open ditches, some of which are (1) when properly installed, little cleaning or other maintenance work is required; (2) the land over the tile is available for crop production and the inconvenience of turning at the ditch when working in the field is avoided; (3) there is no ditch bank to serve as an area for production of weed seed; and (4) tile lines may be placed at closer intervals in the field and so provide more complete drainage. Two disadvantages of tile systems are (1) the initial cost and (2) their inability to carry an excess volume of water during periods of unusually heavy rainfall.

In fields which are inadequately drained with tile or open ditches, surface drains made by running dead furrows across the field at frequent intervals help to remove the excess water during heavy spring rains. These drains are usually closed by cultivation later in the season.

Loss of Nutrients in Drainage. As previously mentioned, drainage water tends to displace the soil solution from the surface soil and small cracks and channels as it passes downward through the soil. This phenomenon may be demonstrated by filling a percolation cylinder with moist fertile soil and allowing it to stand for several hours to permit the soil solution to come to equilibrium with the soil. An inch or so of distilled water is now added to the surface and the percolate caught at the bottom of the tube in successive portions of 25 ml. Conductivity determinations on these portions will show them to contain appreciable and quite uniform quantities of soluble salts until the distilled water added to the surface approaches the bottom of the percolator. It becomes evident then that drainage is a source of loss of plant-food materials and would be entirely objectionable were it not for the fact that the space occupied by excess water in the soil is needed for air.

FIG. 35. Exterior view of lysimeters in use at the Virginia agricultural experiment station.

The quantities of nutrients lost in drainage water are extremely variable. Efforts to determine the quantities of different nutrients lost from a particular soil under given climatic conditions and when a specific cropping sequence and soil treatment system are followed have been made by use of lysimeters. They are concrete or metal tanks of various sizes set so that the top extends a few inches above the ground and having a bottom constructed so that the drainage water will flow through a tin tube into a receptacle. Two rows of tanks usually are set a suitable distance apart so that a passageway may be constructed between them to hold the drainage receptacles and allow space for the attendant to measure and sample the percolate. The inside of the lysimeters is coated with asphalt or other water-proofing paint which is not toxic to plants. The tanks are filled with soil taken from the field in 6-inch or 1-foot layers and placed in them in the same order that the layers occurred in the field. Some tamping is necessary to get the full depth of soil into the lysimeter.

The lysimeter method has been used to study the quantity and composition of drainage water by a number of experiment stations. Some of the larger installations are at the experiment stations in Illinois, Tennessee, New Jersey, and New York. Data in Table 8 from the New York station, at Ithaca, illustrate the type of results obtained.

FIG. 36. Interior view of lysimeters at the Virginia station. Note receptacles for catching percolate which comes from lysimeter tanks through tin tubes.

The inside dimensions of these tanks are 4 feet 2 inches square by 4 feet deep. They are filled with 3½ tons of silty clay loam soil which was removed from the field and placed in them in 1-foot layers. Some of the lysimeters are cropped to different rotation systems and others are uncropped.

The effect of cropping on the loss by leaching is worthy of consideration, as is also the large quantities of certain ions lost, especially Ca^{++}, and NO_3^-. Some of the factors which affect loss of nutrients in drainage water have been found to be (1) whether or not the soil is cropped; (2) the cropping system; (3) annual rainfall; (4) application of fertilizer, lime, or manure; (5) soil texture; and (6) soil composition.

The effect of cropping land, compared to keeping it cultivated but not planted to crops, on the total quantities of the various nutrients removed from the soil is illustrated by the data in Table 9, which are

TABLE 8

NUTRIENTS IN POUNDS PER ACRE PER YEAR CONTAINED IN DRAINAGE WATER. AVERAGE FOR 15 YEARS. AVERAGE RAINFALL 32.52 INCHES

	Percentage of Rainfall	Ca	NO_3	Mg	K	S	Bicarbonates
Cropped to Various Rotations	48.15	203.7	22.5	37.0	55.4	37.3	882.8
Not Cropped	66.43	369.0	301.4	58.1	68.3	46.0	1,091.2
Difference	18.28	165.3	278.9	21.1	12.9	8.7	208.4

from the New York experiment station. It is seen from these results that the production and harvesting of a crop may actually result in the removal of appreciably less Ca, Mg, and NO_3 than if the land were clean-cultivated (fallowed) throughout the season. These results suggest that summer fallow should only be resorted to when necessary for the storage of moisture for a following crop or for the eradication of noxious weeds.

SOIL WATER WHICH IS RETAINED AGAINST THE PULL OF GRAVITY

If the pore space in a soil is filled with water and then drainage is permitted until no more water yields to the pull of gravity, the quantity remaining in the soil is known as the *water retained* or *field capacity*. Plants draw very largely on the retained water for their moisture supply, and hence the capacity of a soil to hold water against the pull of gravity becomes of practical significance. Not all the retained water is available to plants because the mutual attraction between soil and water is so great that a certain portion of moisture is held with a force greater than the absorptive power of plants. Several questions are suggested as guides in the study of this section.

Questions:

1. What forces hold water in soil?
2. How is the moisture content of soil expressed?
3. By what methods may soil moisture content be determined?
4. Into what classes may retained water be divided?
5. How does moisture move in the soil?

TABLE 9

EFFECT OF CROPPING ON THE REMOVAL OF NUTRIENTS FROM THE SOIL *

	Potassium		Calcium		Magnesium		Sulphur		Nitrate (NO_3)	
	Cropped	Uncropped	Cropped	Uncropped	Cropped	Uncropped	Cropped	Uncropped	Cropped	Uncropped
Removed in the harvested crop	78.8	....	13.2	.....	7.1	....	10.1	....	77.0	....
Lost in drainage	45.9	61.0	173.3	367.4	37.0	65.2	38.2	48.5	5.4	92.8
Total	124.7	61.0	186.5	367.4	44.1	65.2	48.3	48.5	82.4	92.8
Reduction of loss in drainage as a result of cropping	15.1	....	194.1	.....	28.2	....	10.3	....	87.4	....
Decrease in total loss to soil as a result of cropping	−63.7†	....	181.9	.....	21.1	....	0.2	....	10.4	....

* Values expressed in pounds per acre per year.

† This means that 63.7 pounds more of potassium are removed from the cropped soil than from the uncropped soil.

Forces Holding Water in Soil. Two forces are primarily responsible for the retention of water in the soil against the pull of gravity: (1) the mutual attraction between soil and water, which may be designated as adhesion; and (2) the attraction of water molecules for each other, which is a manifestation of cohesion.

If there were no attraction between soil and water, there would be no force to hold water in the soil and it would all yield to the pull of gravity and drain away. In other words, the retention of water in the soil is primarily due to the force of adhesion. With this retaining force in operation, the quantity of water held and the forms or shapes assumed by the water films are due largely to cohesion, as manifested by surface tension, which is a phenomenon resulting from unbalanced cohesive forces. Surface tension is illustrated in Fig. 37. In this illustration, any molecule in the main body of the liquid, such as molecule *A*, is being attracted equally by molecules all around it, and hence its position is the result of balanced forces of cohesion. On the other hand, molecules at the surface are attracted only by molecules to each side and below them. Consequently, the resultant force is downward, which results in a compression and so a higher concentration of water molecules at the surface. These molecules act as though they formed a tightly stretched elastic membrane over the surface of the water. This is the phenomenon known as *surface tension.*

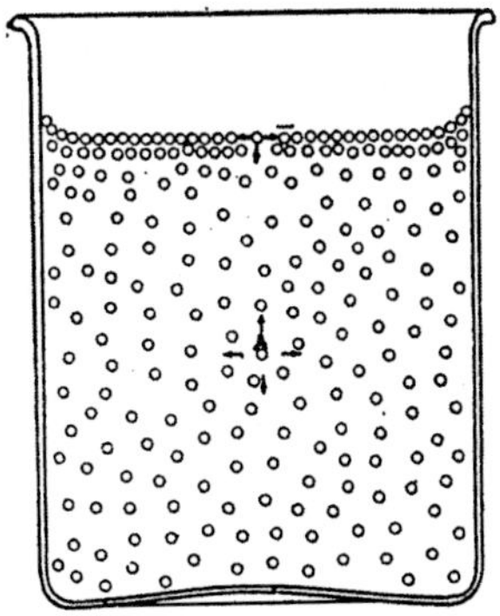

FIG. 37. A concentration of molecules at the surface of a liquid, as a result of unbalanced cohesive force, acts as a tightly stretched elastic membrane.

The apparent inclination of molecules to climb up the side of the beaker is evidence of adhesion and shows that the attractive force between glass and water is somewhat greater than the attraction between the water molecules themselves. A similar relationship exists between water and soil particles.

Temperature has a marked influence on cohesion in water and hence on the quantity of water retained. This fact may be demonstrated by permitting a column of saturated soil to drain to the field capacity in a low temperature and then removing the column to a warm room and noting that drainage will again take place. The presence of soluble organic substances in the soil solution or of soluble salts will alter the surface tension of the solution and so affect the quantity of water retained.

There has been some effort to express the force by which water is held in soils at different moisture contents in terms of the centimeters in height of a column of water required to produce a tension of equal magnitude. It has been proposed that the logarithm of the number of centimeters be used as an expression of the energy involved and that the value be called the *pF* of soil moisture. The *p* is to indicate that the value is a logarithm, and *F* is a symbol of free energy. The expression has some similarity to the designation of soil reaction in terms of *p*H.

The energy with which water is held in soils at different moisture contents may be measured by several methods. A very high-velocity centrifuge is made which will submit the soil moisture to pulling forces as high as 300,000 times gravity. Perhaps the most widely used apparatus is the tensiometer. This consists of a porous cup made of clay and attached to a mercury manometer. The cup and connecting tube are filled with water, and the cup is placed in the moist soil. When equilibrium is reached between the moisture in the cup and the soil, the attractive force of the soil for water at that moisture content may be determined by reading the height of the mercury in the manometer. The values so determined are sometimes called the *capillary potentials* of the soil. Determination of the freezing point of a soil at different moisture contents is another way of determining capillary potentials. The capillary potentials of a sand, loam, and clay at various moisture contents are given in Table 10. The great

TABLE 10

THE CAPILLARY POTENTIAL OF SOILS AT DIFFERENT TEXTURES AND AT DIFFERENT MOISTURE CONTENTS *

Soil	Moisture Content in Percentage of Dry Soil—Capillary Potential in Centimeters per Gram										
Sand	Moisture content	20.9	11.7	7.6	6.2	5.1	4.8	4.7	4.1	3.9	3.7
	Capillary potential	4	26	82	150	301	452	492	684	692	801
Loam	Moisture content	39.6	34.2	30.2	22.8	19.4	17.5	16.4	15.5	15.0	14.6
	Capillary potential	5	29	55	144	269	371	489	556	715	796
Clay	Moisture content	66.5	60.4	51.5	46.3	40.2	31.6	29.0	28.5	28.3	27.7
	Capillary potential	6	22	58	85	196	438	650	664	784	776

* "The Usefulness of Capillary Potential to Soil Moisture and Plant Investigators," L. A. Richards Jr. Agr. Res., Vol. 37, No. 12, 1928, p. 732.

variation in moisture content of the three soils with exertion of similar capillary potentials may be noted in the first vertical column. Likewise it will be noted that when capillary potentials of similar magnitude, namely, 492, 489, and 438, are exhibited by the sand, loam, and clay, respectively, the moisture percentages are 4.7, 16.4, and 31.6.

Moisture Content of Soil. It is a common practice to express the quantity of water in a soil in terms of percentage based on the weight of the soil. Frequently the weight of the moist soil is used as a base for calculation, as is the common practice in expressing the percentage of moisture in articles in trade, as hay, corn, vegetables, ores, etc. The disadvantage of this method is that the base of calculation is different for each moisture content. To overcome this objection the weight of the dry soil (dried at 105°–110° C.) is often used as a base for expressing percentage moisture content in investigational work. Under this system the base is always the same regardless of moisture content. The following example will illustrate the difference in the two methods of calculation. If to 100 grams of oven-dry soil, 50 grams of water are added, 150 grams of moist soil are produced. Using moist soil as a base, the soil contains $50 \div 150 \times 100 = 33.3$ per cent moisture. Using dry soil as a base, the soil contains $50 \div 100 \times 100 = 50.0$ per cent moisture. It will be noted that at higher moisture contents there is a decided difference in the percentage moisture content calculated by the two methods.

There is a valid objection to expressing the moisture content of soil as a weight percentage for practical purposes. In farming, soil is dealt with on a basis of depth, for example: A farmer plows to a certain depth or lays tile lines at given distances below the surface; plant roots penetrate the soil to different depths depending on several conditions. Depth involves the concept of volume rather than of weight, as different soils have different volume weights. This is especially true of soils with high organic matter contents compared to those with little organic matter and also with well-granulated clays compared to sands. Frequently it is desirable to express moisture content on a volume basis as cubic inches of water per cubic foot of soil. In irrigation practice applied water is measured in inches of depth over an acre of surface, designated as *acre inches*. Rainfall is also expressed in terms of inches in depth. If the moisture content of a soil is expressed in percentage based on the weight of the dry soil, the percentage of water by volume may be obtained by multiplying by volume weight. Table 11 shows the moisture content of an imaginary soil expressed by different methods.

Determination of Moisture Content. The common method of determining the moisture content of soil is to weigh a sample of the moist soil, dry it at a temperature of 105°–110° C., and reweigh when cool.

Although this method is simple, time is required for the drying of the soil, samples of field soil must be taken to a laboratory where balances and an oven are available, and in studying moisture changes in field soils over a period of time repeated samplings are necessary which introduces an error due to lack of uniformity in the soil area. For years attempts have been made to devise a method for determining soil moisture content quickly and without removing a sample. The conductance of soil between two electrodes placed in the ground and measured by some form of Wheatstone bridge has been studied. The conductivity would vary with moisture content, between certain limits, if no interfering factors were present. Changes in content of soluble salts, however, affect the conductivity to a greater extent than changes in moisture content, and so the method has not yet proved satisfactory. Bouyoucos has suggested the use of a plaster of Paris block cast around two electrodes. The block can be buried at any desired depth and insulated wires from the electrodes brought to the surface. The moisture content of the block quickly comes to equilibrium with that of the soil in which it is embedded, and as the conductivity of the block is a function of its moisture content, readings may be quickly made by attaching the lead wires to a specially designed bridge.

TABLE 11

THE MOISTURE CONTENT OF A SOIL EXPRESSED BY DIFFERENT METHODS. THE SOIL IS HYPOTHETICAL AND ASSUMED TO HAVE 43.4% PORE SPACE WITH A VOLUME WEIGHT OF 1.50 AND SPECIFIC GRAVITY OF 2.65

Percentage of Moisture Based on			Cubic In. Water per Cubic Ft. of Soil	Pounds of Water per		Inches of Water for 3-Ft. Depth
Moist soil	Dry soil	Soil volume		Cubic ft.	Acre ft.	
20.0	25.0	37.5	648.00	23.43	1,020,697.9	13.49

Fletcher and Baver also have proposed new methods for measuring soil moisture content. Fletcher makes use of a specially constructed condenser which is placed in the soil and after equilibrium is attained with the soil moisture, readings are taken with a suitable capacitor. Baver suggests calculation of the moisture content from the heat conductivity of the soil.

Another device proposed by Richards and associates for measuring the movement of water in soils with moisture contents above the moisture equivalent is the tensiometer. This method is the same as that used for measuring capillary potential as described on p. 158.

Classes of Retained Water. The attraction between soil and water is so great that all the moisture is not removed by drying at a temperature of 110°C. although the quantity of moisture remaining is very small and a soil so dried is considered to be moisture free for general purposes. If a thin layer of oven-dry soil is exposed in a desiccator over 3.3 per cent sulphuric acid, which will create an atmosphere with 98.2 per cent relative humidity, the soil will adsorb appreciable quantities of moisture. The percentage of moisture so adsorbed is designated as the *hygroscopic coefficient* and will vary approximately with the colloidal content of the soil. In fact, this is one method of estimating the quantity of colloidal material in soils. Any amount of moisture a soil contains up to and including the hygroscopic coefficient is known as *hygroscopic moisture*. Water held in this condition does not move in the soil, is not used by plants, and in fact may exist in other than the liquid state.

Water held by a soil in excess of the hygroscopic coefficient is considered to exist in the form of thin films around soil particles and granules, in the larger pores of the granules, and in the angles or crevices made by the contact of particles. In general this water may be thought of as a series of small wedges occupying the angles formed by particle contact and connected by films. This water is defined as capillary from the old conception that the pore space in soil is similar to a series of capillary tubes containing moisture. Capillary moisture is susceptible to movement in the soil to a certain extent, may be absorbed by plant roots, and exists in the liquid state. The percentage of capillary water held by different soils varies with texture, structural conditions, organic matter content, etc. There has been much conjecture as to the effect of organic matter on the capacity of soils to hold water and particularly water which is available for crop use. Information on this question was obtained by collecting samples of soil from fence rows and from adjoining fields which had been under cultivation for a long period of years. This procedure gave pairs of samples which were of similar texture but of different organic matter content. An average of the data for 25 pairs of samples is presented in Table 12.

An empirical method of measuring the power of different soils to hold moisture has been devised by submitting a thin layer of the soil to a pull of one thousand times gravity in a porous cup placed in a specially constructed centrifuge. The percentage of water retained by

TABLE 12

THE EFFECT OF ORGANIC MATTER CONTENT ON THE MOISTURE-RETAINING CAPACITY OF SOILS

Sample	Organic Matter, %	Hygroscopic Moisture, %	Wilting Point, %	Moisture Equivalent, %	Available Moisture Capacity, %
Fence row	3.45	1.51	9.71	24.93	15.22
Cropped	2.09	1.29	9.08	19.94	10.85
Difference due to organic matter	1.36	0.22	0.63	4.99	4.37

a soil when submitted to this treatment is known as the *moisture equivalent.*

The percentage of moisture in a soil after drainage has ceased and after capillary adjustments have taken place is variously known as *normal field capacity, normal moisture capacity,* and *field-carrying capacity.* The value is also close to that of the moisture equivalent and is approximately two to three times the hygroscopic coefficient.

TABLE 13

THE INFLUENCE OF TEXTURE ON THE CAPACITY OF SOILS TO RETAIN MOISTURE AS INDICATED BY SEVERAL MOISTURE CONSTANTS *

Kind of Soils	Organic Carbon	Organic Matter †	Hygroscopic Coefficient	Moisture Equivalent	Maximum Water Capacity ‡
Loess, Lincoln, Neb.	2.86	4.93	10.2	27.8	60.9
Loess, McCook, Neb.	1.23	2.12	10.5	24.1	63.7
"Sandy land," Imperial, Neb.	0.71	1.22	3.3	7.9	34.2
Dune sand, Dunning, Neb.	0.05	0.08	0.6	1.5	25.8
Black Adobe, Douglas, Ariz.	1.29	2.22	12.9	25.8	60.3
Red Loam, Cuervo, N. M.	0.62	1.07	10.0	19.2	49.9
Red desert sand, Orogrande, N. M.	0.20	0.34	1.7	3.0	27.1

* "Relation of Movement of Water in a Soil to Its Hygroscopicity and Initial Moistness," F. J. Alway and G. R. McDole, Jr. Agr. Res., Vol. 10, No. 8, 1917, p. 399.

† Organic Carbon X 1.724. ‡ By Hilgard method.

Hilgard proposed that the capacity of different soils to hold moisture might be compared by placing a short column of soil (about ¾ inch) in a porous-bottomed cup. The cup is then set in a pan of water of such depth that the water surface touches the bottom of the soil column. When the soil becomes saturated, the cup of soil is removed from the pan and allowed to stand until drainage ceases. The percentage of water retained is called the *maximum moisture capacity*. Table 13 shows the hygroscopic coefficient, moisture equivalent, and maximum moisture capacity of several soils of different textures as reported by Alway. The variation in organic matter content probably influenced the quantity of water held to some extent.

Movement of Soil Moisture. Moisture may move in the soil in the liquid and in the vapor states. The movement of moisture in the soil in the liquid state takes place because of the unequal tensions developed under different degrees of curvature of the surface film, and is known as capillary movement. The cause of this movement may be explained as follows.

If a drop of water is placed on a glass plate covered with a film of grease, it will assume an almost spherical shape because the cohesion of the molecules tends to force the water into as small a volume as possible. The attraction between the grease and water offers small resistance to this tendency. A drop of water placed on soil particles has the same tendency to form a sphere, but in this case the cohesive force is largely overbalanced by the adhesion of the soil and water, and so the water is spread as a thin film over the soil particles. Now if water occurs as wedges in the angles of adjoining soil particles with connecting films, as shown in Fig. 38, there is a tendency for it to form spherical droplets at each surface, as *A* and *B*. Droplet formation is prevented, however, because of the attraction of the soil and because the two surfaces are acting against each other. Now if the curvature of one film is greatly increased through the removal of moisture by a

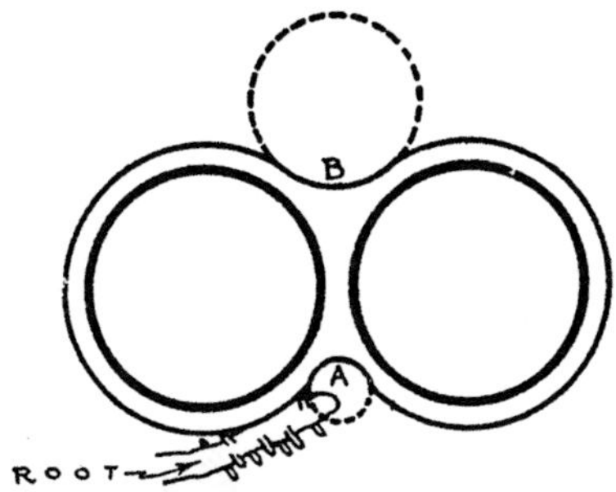

FIG. 38. As the root absorbs moisture from the accumulation between two soil particles, the film curvature increases, as is shown by the projected circles. As the force tending to draw water into a given portion of the film varies inversely with the radius of the curvature $\left(P = \frac{2T}{r}\right)$, it follows that moisture will move to the feeding point of the root.

root, as is shown at A, its pulling power will be proportionately increased since the pull exerted by such a curved film is inversely proportional to the radius of curvature. As a result, water will be drawn toward A until the curvature of films A and B is equal. The equation

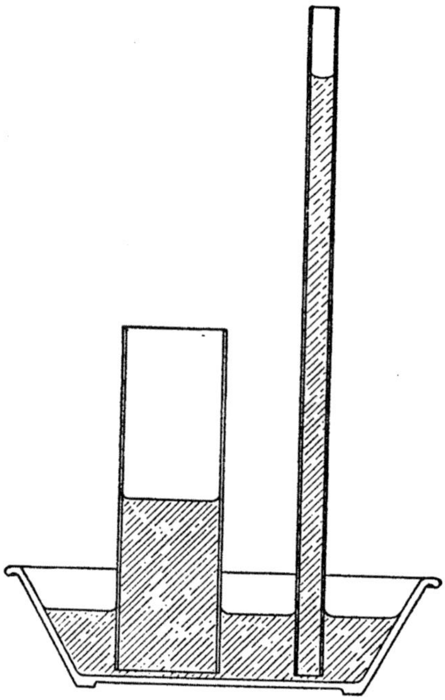

FIG. 39. Water rises to different heights in capillary tubes depending on the diameters of the tubes. In small tubes there is a maximum of contact between the water and glass, giving a strong adhesive force, while the diameter is so small that cohesive forces can support a high water column. In a large tube adhesion between water and glass can be exerted on only a small part of the water, and the diameter is so large that cohesion can support only a short column.

for capillary movement may be written: p (pull) $= \frac{2T}{r}$, in which T is surface tension and r the radius of curvature. Assuming, for example, that the r of A is 1 and of B is 2, then the pull exerted at the two points will be $p = \frac{2T}{1}$ compared to $p = \frac{2T}{2}$. As the liquid in each case is water, T will be equal, and so the pull exerted by film A will be twice as great as that exerted by film B.[1]

[1] Students particularly interested in capillary movement of soil moisture will find very intriguing the comparisons between this phenomenon and the flow of heat through a metal bar or the passage of an electric current through a wire.

The distance and the rate of movement of capillary water have been the subjects of much discussion and investigation. It can be seen that if the spaces between the soil particles are too large, the cohesion of the water molecules will be insufficient to raise it to any appreciable height; however, the movement will be rapid as friction will be small. This is the condition which exists in sandy soils. A somewhat analogous situation exists in the installation of telephone wires. If the posts are set too far apart, the wire will break because cohesion in the wire is insufficient to support the weight. On the other hand, in soils containing considerable silt and clay the spaces between particles will be small, and hence cohesion will support a relatively high column. Furthermore, in a fine-textured soil there is much more surface offering a greater opportunity for the action of the attractive force between soil and water which is the primary force holding the water in the soil. Water may then be expected to rise more slowly but to greater heights in soils of a fine texture.

The principle may be illustrated by Fig. 39, in which a small and a large capillary tube are set in a basin of water.

The rate of rise of capillary water in a sand, silt loam, and clay loam is shown by the data in Table 14 together with the heights to which the water has risen after different time periods.

TABLE 14

RATE AND HEIGHT OF CAPILLARY RISE OF WATER IN SOILS OF DIFFERENT TEXTURES *

Time	Height of Rise in Inches		
	Sand	Silt loam	Clay
½ hour	13.5	7.3	5.4
1 hour	14.3	11.2	8.0
6 hours	16.6	26.6	15.5
12 hours	17.2	35.3	18.5
1 day	18.4	46.4	21.0
3 days	20.3	65.4	24.7
6 days	21.8	78.5	27.3
9 days	23.0	86.3	28.8
18 days	25.3	99.2	33.2

* "Soils and Fertilizers," F. E. Bear, 3d Edition, 1942, p. 82. John Wiley and Sons. Used by permission.

On a theoretical basis it has been calculated that moisture may be lifted from a free-water surface to a height of 1½ feet by coarse sand

and to 150 feet by fine silt (particles from 0.01 to 0.002 mm. in diameter). However, these values have no significance from a practical viewpoint because the movement in the silt would be extremely slow, especially after the water had risen a short distance, and because under field conditions capillary movement of moisture into the root zone of most crops does not take place from a free-water surface except when the water table is at no great depth.

The role of capillarity in supplying moisture to crops has undoubtedly been misinterpreted. Under field conditions moisture moves from soil of somewhat greater moisture content to soil containing less moisture. Seldom is the difference in moisture content very great, and as a result movement is slow—so slow that a plant will wilt before an appreciable quantity of water can move within reach of its roots from any great distance. Furthermore, as the supply of capillary water becomes less, movement is correspondingly slower. Unquestionably, however, considerable amounts of moisture move through short distances to the roots of growing crops, and hence capillarity is of much significance in supplying moisture to plants.

Movement of Water Vapor. The relative humidity of the air in soils containing appreciable quantities of moisture is approximately 100 per cent. Also, the pressure of the water vapor in the soil atmosphere increases with temperature. If, therefore, there is a considerable difference in the temperature of the soil at different depths, water vapor may move from the warmer zone and condense in the cooler soil. It is conceivable that if a long period of hot weather which has warmed the soil to a considerable depth is followed by a cold spell which cools the surface soil rapidly, appreciable quantities of water vapor may move up from the warm soil below and condense in the upper soil layer. Such movement may also occur in the fall when the lower soil horizons have not yet cooled to the temperature of the surface soil. Likewise some moisture may condense in the soil surface from a warm atmosphere of high humidity.

WATER IN RELATION TO PLANT GROWTH

Although the farmer considers soil moisture in relation to the fitting of a seedbed and to the tillage of his crops, and precipitation in relation to the loss of fertile surface soil through erosion, his primary interest in moisture is as a factor in crop growth. Not only is moisture necessary for the absorption of nutrients by plant roots but also it is a nutrient itself and as such is as important as phosphate, potash, or nitrate. In fact, a growing plant contains a much larger proportion of water than of any other chemical compound. Plants are colloidal systems in which water serves as the liquid phase. The following questions emphasize pertinent points concerning the relation of water to plant growth.

Questions:

1. How do plant roots absorb moisture?
2. How does water function in the absorption of nutrients by plants?
3. Is water effective in bringing nutrients within the reach of plant roots?
4. Is transpiration of service to plants?
5. How much of the retained water can plants absorb?
6. What is the wilting coefficient?

Absorption of Moisture by Plants. Moisture enters plant roots by the process of osmosis, which may be roughly defined as the movement of a liquid through a semi-permeable membrane caused by unequal concentrations on the two sides. It is customary to think of the cell sap within the root as having a greater concentration of soluble material than the soil moisture, and hence water passes in to equalize the concentration. A more correct view is to consider the concentration of water molecules in the cell sap reduced because of the quantity of soluble substances present, and hence the number of water molecules in the soil solution is greater. As a result, more water molecules strike against a unit area of the exterior of the root membrane than against the interior, for the molecules in both cell sap and soil solution are in constant motion. As a result of the bombardment, water passes into the root from a zone of higher concentration of water to a zone of lower concentration.

If through any circumstance the concentration of soluble substances in the soil moisture exceeds that of the cell sap, the situation will be reversed and water will pass out of the root.

It should be recalled to mind that plant roots do not absorb the soil solution as such, as animals drink water containing soluble material. The water enters the roots as pure water without regard to the intake of any of the materials dissolved in it. The entrance of dissolved substances is entirely a separate process.

The Role of Water in Nutrient Absorption. It is generally believed that material must be in solution to pass through the plant-root membrane. Nutrients in the soil solution should therefore be in suitable condition for absorption. It must be pointed out, however, that the presence of a nutrient in the soil solution does not assure its use by the plant because the passage of ions and molecules through the absorbing membranes of the root is affected by various factors which cannot be discussed at this point. Furthermore, it cannot be assumed that nutrients must be in the soil solution to be used by plants. In the zone of intimate contact which exists between a root hair and a soil particle or a colloidal aggregate there is the possibility of the solution and absorption of nutrients. What proportion of the materials absorbed by plant roots comes from the soil solution and what directly from the particle surfaces is impossible to tell.

Movement of Nutrients in Soil Moisture. Nutrients dissolved in the soil solution move with it, and so when moisture moves by capillarity to replace that which has been taken up by plants, a supply of nutrients may be moved near the roots. Although this action takes place through short distances only, the net result in the course of a growing season may add materially to the food supply of the crop. Vertical movement of soluble salts in capillary water has been observed more extensively than horizontal movement. During periods of drought a considerable accumulation of salts may develop at the soil surface. In some soils this may be sufficient to be visible, especially in regions of neutral or alkaline soils. The appearance of small quantities of salt on the surface of clods is a familiar sight even in humid regions. The movement of salts to the surface is much more pronounced in soils not occupied by growing plants because the roots absorb the moisture and so reduce the quantity evaporated from the soil surface. Water from heavy rains has a tendency to displace the soil solution from the pore spaces and force it out of the soil into the drainage water. This process results in considerable loss of nutrients. There is also considerable mixing of the rain water with the soil solution, and a portion of the mixture drains away while the remainder constitutes the new soil solution which dissolves fresh supplies of nutrients from the soil complexes.

Transpiration. There has been much discussion concerning the value of transpiration to plants. Some cooling of the plant leaves may result from the evaporation of moisture from them, but there is a difference of opinion as to whether the plant is benefited by this effect. Some investigators think that transpiration is of assistance to plants in the movement of nutrients, particularly from the lower into the upper parts. Nutrients must be in solution for translocation to different parts of the plant, and since water serves as the dissolving medium it is of service in the growth processes aside from its role as a food material.

Available Moisture. As the quantity of moisture in the soil becomes less, the force by which it is retained increases. Finally, a point is reached at which the adhesion between soil and water equals the absorptive power of the plant roots, and hence no more moisture can be utilized by the plant. Theoretically this point represents the lower limit of available moisture. It varies greatly in soils according to the amount of surface exposed and therefore with the content of colloidal material. Since in the actual growth process it is impossible for plants to withdraw the last traces of available moisture from a soil, as is shown under the discussion of *wilting point,* the quantity of water a given soil can retain in an unavailable state can only be determined by some laboratory method.

The upper limit of available water is assumed to correspond roughly with the percentage of water retained. On the other hand, as the last traces of gravitational water leave the soil slowly and the division point between gravitational and retained water may vary somewhat with the temperature, there is reason to assume that plants use some gravitational moisture. The quantities of available moisture which three soils of different texture will hold are shown in Table 15.

TABLE 15

Available Water Held by Soils of Different Textures.* Water Table at 6-Foot Depth. Moisture Contents Are for 3-Foot Depth of Soil. Percentage Moisture Calculated on Dry Soil

	Dunkirk Fine Sand	Wooster Silt Loam	Brookston Clay
Water-retaining capacities	6.5	30.2	30.5
Wilting coefficients	3.5	7.7	17.5
Available moisture	3.0	22.5	13.0

* Data from "Soil Management," F. E. Bear, 1st Edition, 1924, p. 87. John Wiley and Sons. Used by permission.

Wilting Coefficient. It is pointed out (p. 171) that the rate of transpiration of plants varies with environmental conditions which affect evaporation. The quantity of water lost by a given plant in a given time, let us say 1 hour, will also vary with the extent of the transpiring surface. When the quantity of water absorbed by the roots is less than the quantity lost by transpiration, a negative moisture balance is set up in the plant and very shortly the cells cease to be turgid, and wilting sets in. Wilting will occur before a plant has taken all the moisture from a soil that it can, since the soil retains the last traces of available moisture with much force and the roots cannot take up moisture in the face of this resistance at a sufficiently rapid rate to equal transpiration. Furthermore, at this low moisture content capillary movement of moisture to the absorbing zone of the root is too slow to meet the demands of the plant. When a plant has wilted to such an extent that it will not revive when placed in a saturated atmosphere, it is said to be *permanently wilted,* and the percentage of moisture in the soil when permanent wilting occurs is designated the *wilting coefficient* or *wilting point* of that soil.

As is to be expected, the wilting point of soils varies greatly and is markedly influenced by the content of colloidal material and organic matter. Any given soil has about the same wilting coefficient for all common plants. The influence of soil texture on the percentage of moisture in soils when plants wilt is shown by the data in Table 16.

TABLE 16

INFLUENCE OF SOIL TEXTURE ON THE WILTING COEFFICIENT

Crop	Coarse Sand	Fine Sand	Sandy Loam	Loam	Clay Loam
Corn	1.07	3.1	6.5	9.9	15.5
Sorghum	0.94	3.6	5.9	10.0	14.1
Wheat	0.88	3.3	6.3	10.3	14.5
Pea	1.02	3.3	6.9	12.4	16.6
Tomato	1.11	3.3	6.9	11.7	15.3
Rice	0.96	2.7	5.6	10.1	13.0

LOSS OF MOISTURE FROM THE SOIL

On all but a very small fraction of the farm land of the world, crops must draw their moisture from the supply stored in the soil, supplemented by rains. As the farmer sees his crops suffering from lack of water during periods of drought, he gives considerable thought to methods of increasing the moisture-retaining power of his soil and especially to ways of decreasing the loss of stored water. The conservation of soil moisture is uppermost in his mind as he cultivates his crops. A study of processes by which moisture is lost from the soil and of possible means of decreasing these losses is of practical significance. Questions which arise in a consideration of moisture loss from the soil are presented.

Questions:

1. How much moisture is transpired by plants?
2. May the amount of transpired water be reduced?
3. What conditions affect evaporation of water from the soil?
4. How do mulches decrease evaporation?
5. What effect does soil cultivation have on evaporation of soil moisture?
6. Can losses of water by percolation be decreased?

Loss by Transpiration. As transpiration is simply an evaporation of moisture from plant surfaces, it is influenced by the same factors that affect the evaporation of water from any moist surface—exposure to direct sunlight, air temperature, humidity, wind movement, and atmospheric pressure are among the most important. To these must be added the characteristics of the plant itself because transpiration is considerably more rapid from some plants than from others. This is due largely to the number, size, and location of the stomata although the nature and thickness of the leaf covering are also concerned.

Under uniform conditions a plant will transpire about the same quantity of water for each pound of dry plant material that develops. The weight of water in pounds transpired per pound of dry plant tissue

TABLE 17

VARIATION IN WATER REQUIREMENTS OF DIFFERENT VARIETIES OF CROPS *

Water Requirement Based on Pounds Produced of							
Wheat	Grain	Total plant	Oats	Grain	Total plant	Corn	Total plant
Turkey	995	364	Canadian	1,416	399	Iowa Silvermine	302
Kharkov	1,064	365	Swedish Select	1,103	423	Hopi	285
Kubanka	1,111	394	Burt	1,224	449	Northwestern Dent	280
Bluestem	1,573	451	Sixty Day	1,172	491	China White	315

* "Relative Water Requirements of Plants," by L. J. Briggs and H. L. Shantz, Jr. Agr. Res., Vol. 3, No. 1, 1914, pp. 8, 11, and 17.

produced above ground is known as the *transpiration ratio*. This ratio varies for different kinds of plants and for the same kind of plant grown under different climatic conditions. In some cases there is also a difference in transpiration ratio for different varieties of the same plant species, as is shown in Table 17. The great variation in water requirement of the same variety of plant, due to different climatic conditions prevailing in different years, is illustrated by the data in Table 18.

TABLE 18

EFFECT OF SEASON ON WATER REQUIREMENT OF CROPS *

		Pounds of Water Transpired per Pound of Dry Plant Tissue Produced in				
Crop	Variety	1911	1912	1913	Greatest variation	Average
Wheat	Kubanka	468	394	496	102	452.7
Oats	Swedish Select	615	423	617	194	551.7
Corn	Northwestern Dent	368	280	399	119	349.0
Sorghum	Red Amber	298	237	296	61	277.0
Alfalfa	Grimm	1,068	657	834	411	853.0

* Briggs and Shantz, p. 56, *op. cit.*

In Table 19 are shown the transpiration ratios of several crops grown at Akron, Colorado. These results are much higher than would

TABLE 19

WATER REQUIREMENT OF DIFFERENT CROPS AT AKRON, COLORADO *

Average for 2 or 3 Years and Two or More Varieties			
Crop	Pounds water per pound plant tissue	Crop	Pounds water per pound plant tissue
Millet	279	Sugar beet †	397
Sorghum	311	Potato †	554
Corn	375	Sweet clover †	770
Wheat	532	Alfalfa ‡	904
Barley	534	Oats	597

* Briggs and Shantz, pp. 58–60, *op. cit.*
† One variety for 2 years. ‡ Two selections of Grimm for 2 years.

be expected for the same crops grown at lower altitudes and in regions of higher humidity with more cloudy days. Nevertheless, the data show that different crops vary greatly in their water requirement under similar climatic conditions.

Reducing the Transpiration Ratio. Alteration of environmental factors which affect evaporation will alter the transpiration ratio. Unfortunately little can be done along this line except in the production of special crops on a limited scale. The erection of shades to protect plants from direct sunlight is practiced in the growing of special types of tobacco and medicinal herbs, such as ginseng. Dense windbreaks have some influence on transpiration for a limited distance. A high air humidity, obtained by sprinkling walks and floors of greenhouses, reduces transpiration. On a general scale, however, man can do little along this line.

It has been demonstrated that plants receiving an abundant supply of moisture transpire more freely than plants supplied with a limited quantity of water. Also, plants growing in a medium containing relatively small quantities of nutrients appear to transpire more moisture per pound of plant tissue produced than those growing in a medium containing an abundance of plant-food materials. Whether or not the transpiration ratio of plants can be appreciably reduced under field conditions by applying liberal quantities of fertilizer has not yet been demonstrated.

Evaporation of Soil Moisture. Aside from the environmental factors which influence evaporation of moisture from any surface, previously mentioned, the following may affect loss of water from the soil surface. The shade afforded by growing crops and their protection of the soil from wind action are immeasurable but influential factors in reducing evaporation. Likewise the higher humidity in the air in contact with the soil, resulting from the protection afforded by the crop, particularly by a low-growing, leafy crop, must have an influence on evaporation, especially in the morning and evening.

The nature of the soil surface is also significant. If the soil is loose so that air moves freely through it, evaporation will be increased. Unevenness or roughness of the soil that exists after cultivation or plowing also exposes more surface for evaporation, as do cracks in the soil. Cracking is especially prevalent in heavy soils during dry periods and results in much loss of water. A hard, compact surface in soils very high in clay may permit more capillary movement of moisture to the surface and therefore greater evaporation loss.

The Function of Mulches. A mulch may be roughly defined as any protective covering resting on the soil. It may be of entirely foreign material, such as shavings, gravel, boards, straw, and paper, or it may be a layer of dry soil. Whatever it is, the chief function of the mulch is to protect the moist soil from the direct rays of the sun, and from direct contact with wind and warm dry air. Through reduced evaporation mulches sometimes reduce cracking of the soil, and in some instances they may limit capillary movement of moisture. An increased absorption of rainfall may result from limitation of surface runoff. Furthermore, a mulch prevents the "running together" of the surface soil under the impact of rain drops and the development of a structural condition which restricts the intake of water. It has also been shown that mulches of peat sometimes reduce the intake of moisture by absorbing the rain and later losing it through evaporation. A mulch of gravel, on the other hand, permits ready absorption of rain by the soil.

Cultivation and Loss of Water by Evaporation. The stirring of the moist surface soil with tillage implements results in increased aeration and exposure of more surface to wind and sun, all of which hastens drying. A layer of dry soil is thus produced which serves as a mulch over the moist soil beneath. This mulch functions as any other in reducing evaporation and in moisture conservation. Also, cultivation during dry periods reduces loss of moisture by filling cracks and keeps the soil in a suitable physical condition for the absorption of rainfall. In soils very rich in clay, cultivation may interrupt the movement of capillary moisture to the surface and so prevent excessive loss of soil water.

In the great majority of soils, however, the rate of evaporation from the surface is more rapid than the capillary movement of moisture, and hence a layer of dry soil (constituting a mulch) is developed without the aid of cultivation. The loss of moisture during the development of this natural mulch appears to be no greater than that during the formation of a mulch through tillage, and there seems little difference in the effectiveness of the two in preventing further moisture losses. In this connection it should be mentioned that Veihmeyer has shown that before the soil has become dry enough to cultivate into a fine mulch most of the moisture which will be lost has already evaporated. Numerous experiments have been conducted to determine the effect of cultivation on the yield of crops. Most of them have been carried out on soils which tend not to pack or cake but to retain a good state of tilth either because of a moderate clay content or of a relatively high organic matter content. In Table 20 are presented sample results from several of these experiments.

It is well recognized that utilization of water by weeds is one of the greatest sources of soil-moisture loss. Tillage is the most practical method of weed control, and hence the tendency of soils to mulch

TABLE 20

COMPARATIVE EFFECT OF REMOVAL OF WEEDS WITHOUT DISTURBING THE SOIL, NORMAL CULTIVATION, AND WEED GROWTH ON THE YIELD OF CORN

State	Soil Type	Duration of Experiment (Years)	Soil Treatment and Yield (Bushels)	
Ohio	Brookston silty clay loam	6	scraped, 72.3	cultivated 3 times, 69.8
Ohio	Miami silty clay loam	6	scraped, 56.3	cultivated 3 times, 59.7
Illinois	Brown silt loam	6	none—weeds allowed to grow, 7.0	cultivated with blades, 53.0
Illinois	Brown silt loam	6	scraped, 53.3	cultivated with shovels, 51.1

themselves does not eliminate the need for careful cultivation of the soil in the preparation of a good seedbed and of adequate tillage of row crops during the growing season. However, if the soil is comparatively free of weed seed and care is taken to stimulate the germination of as many weed seed as possible during seedbed preparation, the number of cultivations of the growing crop may be reduced.

In heavy soils which do not maintain a good state of tilth but easily become compact, cultivation may be of value from the standpoint of increased aeration. Baver cites an example from the results of Ohio experiments, as follows: Corn on Paulding clay yielded 27.7 bushels of corn in 1927 when weeds were removed by scraping without stirring the soil, and 41.4 bushels with three normal cultivations.

Percolation Losses. Traveling eastward from the arid areas, one passes through the western plains region where only occasionally is rainfall sufficient to wet the soil to the depth of the permanent water table. As the climate becomes more humid, losses of water by percolation assume proportions of practical significance. Distribution of rainfall is probably as important as total precipitation in inducing percolation, for a moderate amount of rain falling in the course of a few weeks may result in more drainage water than a much larger amount distributed throughout the year. Some other factors affecting percolation are (1) soil texture, (2) soil structure, (3) depth of soil and nature of underlying material, (4) loss by evaporation and transpiration, and (5) organic matter content of the soil. In some of the sandy soils in areas with long warm seasons and high rainfall the quantity of moisture percolating through the soil is unbelievably large.

Unfortunately, little can be done to reduce the loss of water through percolation. Increase in moisture-retaining power of the soil by inducing granulation of heavy soils and by compaction of sandy soil, coupled with an increase in organic matter content, is about all that can be suggested.

RUNOFF WATER

An appreciable proportion of the precipitation does not enter the soil but runs off over the surface in virtually all climatic zones regardless of how small the total precipitation may be. The damage done through this process of runoff of water is so widespread that the problem is one of universal concern, and as a result much attention has been given to methods of controlling runoff, particularly within the last decade. As this subject is covered in considerable detail by a special course in most colleges and universities, only a general discussion of it will be included here in the way of answers to the following questions.

Questions:

1. What factors influence the proportion of the precipitation which runs off?
2. Why is the runoff of water detrimental?
3. By what practices may runoff be reduced?
4. What agencies are concerned with the control of runoff?

Factors Influencing Runoff. Many conditions have a bearing on the proportion of the precipitation which runs off over the soil surface instead of soaking into the soil, and so only the most prominent conditions will be listed here.

Rate of rainfall is important, for even the most receptive soil cannot absorb moisture as rapidly as it falls during the torrential storms which occur at least occasionally in practically all parts of our country. In many sections where total precipitation is limited, a considerable number of the rain storms are torrential in nature, and hence floods in the Great Plains area and deserts are not uncommon. Likewise, there are few seasons in which heavy rains do not cause much damage in parts of the humid section of the United States.

Rolling topography or steepness in slope and the length of slope of the land also contributes to increased runoff. Rains in the mountains result in many floods in arid valleys, and swollen streams draining steep watersheds are familiar to all residents of hilly communities.

The type of vegetation covering the land has much to do with the speed and amount of water which finds its way down the slopes and into drainage channels. The mat of leaves on the forest floor and the cover of leaves and stems in pastures and hay fields slow down the movement of water over the land surface and keep the soil receptive of moisture by preventing destruction of the granular soil structure through beating rain drops. Small grains such as wheat and oats are less effective, whereas row crops such as corn and cotton are of still less value in preventing runoff. Bare soil, unless kept ridged across the slope and permeable to water by cultivation, offers little obstruction to runoff water. The data in Table 21 give an idea of the relative effectiveness of different kinds of crop cover and methods of tillage in preventing runoff of rainfall.

TABLE 21

TONS OF SOIL LOST THROUGH EROSION PER ACRE PER YEAR AND PERCENTAGE OF PRECIPITATION WHICH RAN OFF UNDER DIFFERENT CULTURAL OR CROPPING SYSTEMS. AVERAGE OF 14 YEARS. SLOPE 3.68%. AVERAGE PRECIPITATION 40.37 INCHES*

Cultural or Cropping System	Plowed 8 In. Uncropped	Bluegrass Sod	Wheat Each Year	Rotation Corn, Wheat, and Clover	Corn Each Year
Tons of soil	41.08	0.34	10.10	2.78	19.72
Number of years to erode 7 in. of soil	24	3,043	100	368	50
Percentage of runoff	30.3	12.0	23.3	13.8	29.4
Pounds of soil lost for each inch of runoff	6,734	142	2,149	993	3,314

* Prepared from "The Influence of Systems of Cropping and Methods of Culture on Surface Runoff and Soil Erosion," M. F. Miller and H. H. Krusekopf, Mo. Exp. Sta. Res. Bul. 177, 1932, p. 22.

Detriment Due to Runoff. One of the objections to having water run off over the soil surface is the loss of the moisture. Many of the torrential rains come during the hot season of the year when crops need much moisture, and, as previously mentioned, many of the summer storms in areas of limited rainfall are of a nature that produces runoff. If a larger proportion of this moisture could be induced to enter the soil, crop growth would be much benefited. In running over the surface, the water carries away the soil and thus depletes fertility. When the runoff water is not concentrated into channels, the whole soil surface suffers loss and *sheet erosion* results. This type of erosion takes a heavy toll of the rich plow soil which contains the major portion of the humus and the available plant nutrients. With concentration of the water into narrow runways comes the spectacular *gully* formation with its self-evident damage to the land.

The soil removed from slopes may be deposited in the depressions in the field or it may be carried into streams, depending on the local situation. In streams the sediment is deposited as mud or sand bars and so obstructs the flow of water, which increases floods and thus does further damage. The flood damage resulting from the accumulation of sediment in the Ohio, Mississippi, Missouri, and many smaller rivers is well known to engineers battling with the problem of flood control.

The decreased capacity of reservoirs, through silt accumulation, where water is impounded for the purpose of irrigation or development of electric power or for use by cities is another problem resulting from erosion caused by runoff water.

Practices Reducing Runoff. Methods developed to reduce the proportion of precipitation that runs off over the soil surface are based on two principles, namely, to reduce the rate of movement so that the water has more time to soak into the soil and to keep the surface of the soil in such a condition that water will enter it freely. Incidentally, these same practices tend to reduce erosion damage not only by reducing the amount of water running down the slopes and the velocity of movement but also through holding the soil in place by the dense-root systems of certain types of plants and by the cover over the soil surface.

The first provision, in a farm plan for the purpose of controlling runoff and erosion through a proper use of the different kinds of land, is to plant very rough areas to trees. Areas in which the slopes are not so steep and yet are sufficiently rough to make cropping to cultivated or small-grain crops difficult are seeded for permanent pasture. On the remaining crop land an effort is made to plant the different crops or crop rotations on the areas where they may be grown with the least detriment to the soil. For example, fields with appreciable slope should be in hay or rotation pasture as much of the rotation period as possible, and intertilled crops should be grown on the comparatively level fields as much as is feasible.

Crops grown in rotation on sloping fields should be planted in strips across the slope. The width of the strips should vary with the steepness of the slope, and fences may need to be moved and fields rearranged so as to make the strips as long as possible. Strips of cultivated crops are usually placed below strips of meadow and are followed by a small grain. This arrangement results in a greatly reduced runoff of water and in consequent erosion. Strip cropping seems inconvenient at first, but use of the system has convinced thousands of farm operators that it is not and that in reality crop yields are increased, less power is required for operation of farm implements, and work may be done more efficiently because of the long strips which require fewer turns.

On some slopes broad, low terraces are used which stop the flow of water and cause most of it to soak into the soil. The excess of water is conducted slowly across the slope, for the terraces are not placed exactly on the contour, and is allowed to flow down the grade over a waterway which is left in sod. Wherever it is necessary to allow water to accumulate and flow down a slope, the waterway should be left permanently in sod. The terraces are made low and broad so that farm implements may be operated over them with little inconvenience, although it is desirable to operate machinery parallel to the terrace.

Agencies Concerned with Runoff Control. For many years a number of the agricultural experiment stations have been concerned over the great soil damage resulting from erosion, and experiments have been inaugurated to measure these losses and to devise methods of controlling them. Also, in some sections of the United States farmers have been utilizing for years some of the practices for reducing soil erosion which have been brought recently to the general attention of farm operators. However, the enormity of the damages caused by water runoff did not receive any notable consideration by the public until 1933, when the Soil Erosion Service was established in the Department of the Interior. During 1935 the Service was made a bureau in the Department of Agriculture and hence became a permanent governmental agency. For convenience of administration the United States is divided into seven regions with a headquarters office located at a suitable point in each region.

The purposes of the service are twofold. Through research pertinent facts relative to erosion, water conservation, sedimentation, floods, irrigation, and farm drainage are gathered. And through educational methods, including extensive demonstrations, those concerned with land use are acquainted with the damages resulting from runoff water and with practical methods of reducing runoff and soil erosion. In many cases technical guidance is given in outlining soil-management programs to accomplish the desired conservation of soil and water.

In addition to various types of publications, newspaper articles, illustrated lectures, and personal conferences, the service carries on its educational program by means of three types of demonstrational enterprises. The projects are definite areas set off by physiographic features, within which farmers are encouraged to sign a five-year cooperative agreement for operation of a soil-conserving program on their farms. Farm plans are prepared by the service, based on a detailed soil map and on a study of the farm business as a whole. Considerable assistance is given the farmer in getting the plan under way. This type of enterprise rapidly is being discontinued, and services formerly used in projects are now being made available to legally organized soil-conservation districts.

Districts are definitely defined areas in which the farmers realize the need for a soil-conservation program and have organized for the purpose of sponsoring such a community activity. Operations are in charge of a board of directors composed of farmers. A committee composed of state officials, when requested by local farmers, conducts a hearing to determine if a district is needed, after which organization proceedings are consummated. The Soil Conservation Service and other agencies then supply technicians to make soil maps of farms, study the farm enterprise as a whole, and prepare a farm plan. Guidance is given in getting the farm plant in operation, but all labor and materials needed must be supplied by the farmer.

In order to introduce soil-conservation principles into areas in which farmers are not aware of the damage being done through improper soil-management practices, a series of demonstration farms in cooperation with the extension service of state agricultural colleges is established. Soil-management plans are supplied by trained technicians, and guidance is given in getting the plans under way. Farmers in the community are encouraged to study the methods being followed and to observe results so that they may adopt similar practices. Furthermore, the prevention of soil depletion on level lands is considered just as urgent as controlling erosion on the rolling and hilly lands.

The service may also purchase areas of submarginal land, control soil blowing, make flood-control surveys, and perform numerous other useful services. Close cooperation between the extension service and experiment stations of state institutions and the Soil Conservation Service is highly desirable.

CHAPTER VII

SOIL ORGANISMS—THEIR RELATION TO SOILS AND SOIL PRODUCTIVITY

The soil is the home of innumerable organisms that range in size from those too small to be seen with a powerful microscope to larger forms such as earthworms. Through a great variety of activities these organisms contribute to the productive capacity of the soil; for example, they are intimately concerned with the cycles or series of transformations of plant-food elements from one form to another. A better appreciation of the importance of organisms in soil development and productivity comes through a knowledge of the general nature of the soil population, of the specific functions of the various forms, and of the influence of environmental factors on their survival and activities.

A better control of soil microbial activities can be obtained where unfavorable soil conditions exist, by such practices as liming, drainage, adding organic matter, proper tillage operations, using good cropping systems, controlling parasites, and inoculation. The study of microorganisms living in the soil and the part they play in plant growth may be undertaken under four headings.

Objectives:

A. The general nature and extent of the soil population.
B. Activities of soil microbes in relation to the growth of higher plants.
C. The role of microorganisms in the development of soils.
D. Interrelationships between higher plants and soil microorganisms and among microorganisms themselves.

NATURE AND EXTENT OF THE SOIL POPULATION

The original concept of the soil population as being predominately bacterial has been modified through years of investigation to include also the numerous groups of Fungi, Actinomyces, Protozoa, Algae, and many small invertebrate animals. Carefully conducted studies have shown that soil characteristics such as structure, aeration, drainage, organic matter content, and reaction, as well as tillage practices and kinds of crops grown, exert pronounced effects on the abundance of the soil organisms as a whole and upon the relative abundance of the various species. A general knowledge of the nature of this population should be of considerable interest to all who are concerned with the use of the land because it is to these tiny living things we owe the continued development of higher plants. A few questions will serve as guides in the study of this subject.

Questions:

1. What are the kinds, numbers, and characteristics of organisms found in soil?
2. What are the nutritional requirements of soil organisms?
3. How do the environmental factors of soil temperature, moisture, acidity, aeration, salts, light, and organic matter affect the nature of the soil population, and can these factors, in any way, be controlled by man?
4. What are the causes for the distribution of microbes through the soil profile?

Kinds of Organisms Found in Soils. If we were to consider all life in or springing from the soil, it would be necessary to include all plants ranging from the large trees to the smallest microbes, with an almost indefinite number of life forms between the two extremes in size. In our discussion, however, we shall direct our attention to the bacteria, fungi, actinomyces, and algae belonging to the plant kingdom; and to the protozoans, nematodes, and a few of the larger animals that live in the soil. A summary of the more important groups of soil organisms is presented in Chart 6.

Bacteria. In cultivated soils bacteria usually exceed all other organisms in numbers and kinds. Generally they are regarded as the simplest and smallest forms of life. They are single-celled organisms whose larger individuals seldom exceed 0.005 mm. in diameter. Most of them fall within the colloidal-particle size range. The numbers of bacteria present in soils vary greatly since so many conditions affect their growth and since they can multiply in numbers with extreme rapidity by elongating and dividing. From each individual a new one may be formed in less than 20 minutes, thus causing sudden fluctuations in number with the ever-changing soil environment which occurs under field conditions.

The numbers of bacteria in soil, as determined by the plate method, vary widely, generally between 0.3 of a million and 95 millions per gram of soil. With the aid of microscopic methods, it has been found that the numbers may range from 1 billion to 4 billions per gram of soil. It is not unlikely that the total weight of bacterial substance

Chart 6. Summary of the More Important Groups of Soil Organisms

PLANTS

- I. Bacteria
  - *A*. Heterotrophic
    - 1. Nitrogen fixers
      - (*a*) Symbiotic
      - (*b*) Non-symbiotic
        - (1) Aerobic
        - (2) Anaerobic
    - 2. Those requiring fixed nitrogen
      - (*a*) Spore formers
        - (1) Aerobic
        - (2) Anaerobic
      - (*b*) Non-spore formers
        - (1) Aerobic
        - (2) Anaerobic
  - *B*. Autotrophic
    - 1. Nitrite formers
    - 2. Nitrate forms
    - 3. Sulphur oxidizers
    - 4. Iron oxidizers
    - 5. Those that act on hydrogen and other hydrogen compounds
- II. Fungi
  - *A*. Yeasts and yeastlike fungi
  - *B*. Molds
  - *C*. Mushrooms
- III. Actinomyces
- IV. Algae
  - *A*. Blue-green
  - *B*. Grass-green
  - *C*. Diatoms

ANIMALS

- I. Protozoa
  - *A*. Ciliates
  - *B*. Flagellates
  - *C*. Amoeba
- II. Nematoda
  - *A*. Those that feed on decaying organic matter
  - *B*. Those that feed on earthworms, protozoa, bacteria, etc.
  - *C*. Those that infest the roots of higher plants
- III. Earthworms *
- IV. Other larger animals (worms, insects, ants, snails, spiders, mites, rodents, millipedes, centipedes, etc.) *

* These are frequently referred to as macroorganisms, the others in this chart as microorganisms

per acre to a depth of 7 inches of soil exceeds 500 pounds (live weight) in good soils.

Soil bacteria may be divided broadly into two large groups, based on their energy requirements: (1) the autotrophic bacteria, which can obtain their energy from the oxidation of inorganic elements or compounds, their carbon from carbon dioxide, and their nitrogen and other minerals from inorganic compounds; and (2) the heterotrophic bacteria, which obtain their energy and carbon from complex organic substances. In the former group are found such organisms as the nitrite formers, the nitrate formers, the sulphur-oxidizing bacteria, the iron-oxidizers, and those that act on hydrogen and its compounds. The heterotrophic group may be subdivided into two groups: (1) the nitrogen-fixing bacteria and (2) those that require fixed nitrogen. There are two groups of nitrogen-fixing bacteria, the symbiotic and non-symbiotic, which are discussed in more detail elsewhere (pp. 198–204). The non-symbiotic group may be subdivided further into aerobes and anaerobes. The former require an abundant supply of oxygen, and the latter function in the absence of oxygen. There are two groups of bacteria requiring fixed nitrogen. They are the spore formers and non-spore formers, and each of these groups may be further divided into aerobic and anaerobic forms. These numerous soil bacteria differ considerably in their nutrition and in their response to environmental conditions. Consequently, the nature and abundance of the various types of bacteria depend both upon the available nutrients present and upon the soil environmental conditions.

Fungi. On the basis of their morphology, fungi may be classified into three groups: (1) yeasts and yeastlike fungi, (2) molds, including the various filamentous fungi, and (3) the mushroom fungi. These organisms vary greatly in structure and size from the simplest yeasts and molds to the more complex mushroom forms. They are all devoid of chlorophyll and must therefore obtain their energy and carbon from complex organic substances. The members of the first group are found in the soil only to a limited extent and are believed to be of no great importance in soil development or productivity. The other two groups normally are found in great numbers, with the mushroom fungi being particularly abundant in forest soils. The fungi, in group three, form a very extensive mycelium, sometimes producing fruiting bodies in the form of mushrooms, puffballs, and toadstools. Some of the fungi of this group produce an associative growth with the roots of higher plants, especially forest trees, often referred to as fungus rot or mycorrhiza. In this relationship it is believed that the fungi in some way aid the higher plants in obtaining their nutrients and that the fungi obtain food and energy from the roots of the trees. The soil fungi are usually outnumbered by bacteria except in acid soils, in soils that have been heavily manured, or in those that are naturally rich in organic matter. Fungi respond especially well to aeration and can make extensive growth only under aerobic conditions. They make a vigorous growth in acid, neutral, or alkaline soils, and many fungi are favored by the lower *p*H values.

The number of fungi ordinarily found per gram of soil has been usually from 8,000 to somewhat over 1,000,000. This would probably be the equivalent of 1,000 to 1,500 pounds per acre (7 inches deep) of living substance. It is observed, therefore, that fungi may be outnumbered by bacteria but that the fungi present a greater mass of growth per unit volume of soil. This is due, of course, to the much greater size of a fungus mycelium in comparison to a bacterial cell.

Actinomyces. The actinomyces may be looked upon as occupying a position, from the morphological point of view, between that of the bacteria and fungi. They are frequently spoken of as *ray fungi* or *thread bacteria.* The actinomyces resemble bacteria in that they are unicellular and of about the same size in cross section. They resemble the filamentous fungi in that they produce a very extensive and profusely branched filamentous network. In many cases these organisms reproduce by means of spores, and these spores appear very much like bacterial cells.

These organisms are present in great abundance in soil, making up as much as 60 per cent of the colonies that develop on plates containing artificial media inoculated with a soil extract. The numbers of actinomyces may vary between 0.1 million and 36 millions per gram of soil. In actual weight of live substance per acre, they may exceed bacteria but as a rule will not equal that of fungous tissue. Under optimum soil conditions the total live weight of actinomyces on the average may exceed 700 pounds per acre 7 inches of soil.

For the most part, actinomyces are aerobic and in comparison to most bacteria they are more sensitive to changes in reaction and are active over a narrower range of acidity and alkalinity. The optimum development of actinomyces occurs between *p*H 6.0 and *p*H 8.0, and practically no growth takes place below *p*H 5.0 with most species. They make more extensive development than bacteria in soils of low moisture content although the actinomyces develop well in fairly moist soil under aerobic conditions. These organisms are especially numerous in soils of high organic content where the acidity is not too high. The odor of freshly plowed land noticeable at times is believed due to the products of the activities of actinomyces.

Actinomyces perform a very important function in soils by breaking down organic matter and setting free the plant nutrients it contains. Apparently these organisms are able to attack humus which would be very slowly decomposed by bacteria, and so release nitrogen which might long remain unavailable to higher plants. The actinomyces thus perform a most important function, from the standpoint of soil fertility, in helping keep the soil nitrogen in circulation.

Algae. The chlorophyll-bearing microscopic plants are called algae, and they form a very extensive and important group of soil organisms. They are universally distributed in the surface layers of soil wherever moisture and light are available. The humidity of the soil appears to be a very important ecological factor affecting their distribution, for moist soils contain many more species than dry soils although they are able to retain their vitality for long periods even after prolonged drought.

Algae obtain nitrogen and minerals from the soil, carbon from the carbon dioxide of the atmosphere, and utilize the energy of the sun so long as they have free access to light. Below the soil surface, in the absence of light, they act in a manner similar to fungi in that they utilize the energy derived from the decomposition of organic materials. Algae may thus live and function much like higher plants or they may perform like the lower forms of plant life.

The algae vary greatly in size and shape; but the species commonly found in soils are largely microscopic in size and are either unicellular or filamentous in structure. They are usually classified into the following three groups: (1) blue-green, (2) grass-green, and (3) diatoms. Members of the first group frequently contain pigments other than chlorophyll and may thus vary in color from blue-green to violet or brown. The second group usually contains only chlorophyll, but occasionally other pigments may give them a grass-green or yellow-green color. Members of the last group have a golden brown color due to a blending of chlorophyll with other pigments. In the absence of light the chlorophyll in any species may or may not be retained.

As would be expected, algae are most abundant in the upper soil horizons where environmental conditions are most favorable for their development. Members of the blue-green and grass-green groups usually outnumber the diatoms. It is estimated that a soil under favorable conditions may contain 100,000 of each of either the blue-green or grass-green algae per gram. Diatoms usually are quite numerous in old garden soils.

The development of algae in the soil results in increasing the supply of organic matter and in transforming, temporarily, soluble forms of nitrogen and minerals into organic or insoluble forms. By so doing they aid in preventing the leaching losses of soluble nutrients, but during the cropping season this may result in competition with higher plants. Furthermore, algae aid in the decomposition of organic material and exert a solvent action on certain insoluble rocks and minerals and thus affect the formation of soils. They appear to encourage the fixation of nitrogen by living symbiotically with Azotobacter and perhaps the blue-green algae are able to acquire some free nitrogen themselves. It has been suggested that algae, by taking in carbon dioxide and giving off oxygen, aid rice plants growing in swamp soils. It is impossible to state specifically or with any degree of certainty the role played by algae in the various soil processes.

Protozoa. The Protozoa are generally accepted as being the simplest form of life belonging to the animal group. They are all microscopic in size and unicellular but larger than bacteria and more complex in their activities. In general, cells of protozoa exhibit a much higher development than bacteria. These organisms may move in the soil by means of cilia,[1] flagella, or pseudopodia, and this difference in the nature of their locomotion offers a convenient basis for classifying them. Thus they may be grouped as follows: (1) ciliates, (2) flagellates, and (3) amoeba. All three groups are abundantly represented in the soil, members of the last two groups being especially abundant. It is reported that soils may contain 500,000 to 1,000,000 flagellates, 100,000 to 500,000 amoeba, and 80 to 1,000 ciliates per gram. They might amount to a total weight of 200 or 300 pounds to the acre.

Because it is difficult to free protozoa from bacteria, the exact nature of the physiology of these small animals is still a matter of conjecture. Their nutritional habits are not well known although it is certain that in the soil they depend on the organic matter as a source of food. Physiologically, however, the protozoa vary considerably. Certain groups may ingest bacteria and thus indirectly affect the growth of higher plants. Several years ago a theory was advanced in explanation of soil fertility based on the interrelationships of protozoa and bacteria. The protozoans feeding upon bacteria were believed to decrease soil fertility because the bacteria at that time were looked upon as the only organisms capable of decomposing organic matter. Obviously this theory does not hold because many types of organisms are now known to have the capacity for breaking down organic matter. Granting that protozoa may feed on bacteria, then their effect will be either beneficial or harmful, depending upon the kind of bacteria consumed. By feeding on the nitrifying bacteria, the protozoa would prove detrimental, but decidedly beneficial if they aided in destroying disease-producing bacteria. Under certain special conditions, as in waterlogged soils or in the presence of an abundance of organic matter, the protozoa may keep the bacterial flora as a whole in check. It is believed, however, that the results of the relationship between protozoa and bacteria are largely beneficial through their feeding on

[1] Cilia are hairlike projections capable of vibratory motion; flagella are whiplike appendages; and pseudopodia are temporary protrusions of the cell and may be used for locomotion

pathogenic bacteria and by keeping important groups of bacteria in a youthful state for a longer period of time (by consuming the older bacterial cells), thus stimulating specific bacterial processes.

Nematoda. Among the various worms found in the soil, nematodes occupy a prominent position. Billions of them are found in each acre of soil, and on the basis of their food requirements three groups are distinguished: (1) those that feed on decaying organic matter; (2) those that feed on earthworms, other nematodes, plant parasites, bacteria, protozoa and the like, and (3) those that infest the roots of higher plants, passing a part of their life cycle imbedded therein. Nematodes are sometimes called eel worms; they are round or spindle-shaped, usually with a pointed posterior. Most nematodes are microscopic in size.

Members of the first group are in greater abundance in most soils than those of the other two groups, although the last group is the most important from the agricultural point of view. The roots of a great number of plants, such as the English pea, cowpea, tomato, and carrot, are entered by certain species of group three, and a great deal of damage is often done. They may also become serious pests in greenhouse soils unless special care is taken to avoid infestation. Not only do the nematodes injure the plant roots themselves but also by puncturing the plant they prepare an entrance for other parasites. All the activities of nematodes are not harmful to the growth of higher plants, for they aid in bringing about an intimate mixture of the mineral and organic matter in soil and in breaking down organic materials. They may also improve soil aeration.

Earthworms. Perhaps the most important group of the larger animals inhabiting the soil is the common earthworm, of which there are several species. These organisms prefer a moist environment with an abundance of organic matter and a plentiful supply of available calcium. Consequently, earthworms are found most abundantly, as a rule, in heavy soils which are high in organic matter and are not strongly acid and occur only sparingly in acid sandy soils low in organic matter.

Obviously, then, the number and activity of the earthworms vary greatly from one location to another and, as with other soil organisms, figures indicating numbers are merely suggestive. The number of earthworms in the plowed layer of an acre may range from a few hundred or even less to more than a million. It has been estimated that between 200 and 1,000 pounds of earthworms are present in an acre of soil.

It is believed that in some soils these organisms may pass several tons of soil through their bodies annually and in so doing bring about an increased availability of plant nutrients. Plant nutrients contained in both the mineral and organic portions of the soil are released, the nitrogen of the organic matter being affected particularly. Holes left in the soil by earthworms increase soil aeration and drainage. Furthermore, these organisms bring appreciable quantities of soil from the lower to the upper horizons, resulting in considerable soil mixing. Earthworms may drag undecomposed organic matter into the soil to serve as food and protection. This is particularly noticeable in uncultivated soils, especially in certain forest soils. The development of the so-called mull layer (A horizon) in certain forest soils is due to this type of earthworm activity.

Earthworms are often objectionable when present in great abundance in lawns and especially in the soil of golf greens. Frequently they are present when such soils have received heavy applications of organic manures.

Other Larger Animals. In addition to earthworms, other groups of the larger animals inhabit the soil, namely rodents, ants, snails, spiders, mites, millipedes, centipedes, and various other worms and insects. Some of these organisms may spend all and others only a part of their life cycle in the soil.

The effects of these animals on the character of the soil are beneficial for the most part. As a result of their activities (burrowing habits, as an example), considerable quantities of soil are transferred and some disintegration of soil particles occurs. Their activities tend to increase aeration and improve the drainage of soils. Although soils may be directly benefited by their activities, it is obvious that they may prove unfavorable to agriculture under certain conditions.

Some of these animals cause various chemical changes in the soil either directly by their digestive processes or indirectly by influencing the activities of the soil bacteria, fungi, and other microscopic organisms. They may feed also on certain other soil organisms like algae, fungi, and protozoa. Much damage is done to certain crops by representatives of these groups, especially by some of the insects. From these few remarks it is evident that the animal population contributes to the complex system of activities going on in the soil.

Nutrient Requirements of Soil Organisms. Soil organisms have, in general, the same nutrient-element requirements as higher forms of life. For their growth and development they all require supplies of energy in addition to the several essential elements, including carbon, hydrogen, oxygen, nitrogen, phosphorus, potassium, sulphur, and others. With the exception of algae all the important soil microbes are devoid of chlorophyll, and they must obtain their energy either from the oxidation of simple inorganic substances, as is true with the autotrophic bacteria, or from complex organic substances, as do most bacteria (heterotrophic organisms), all the fungi, and protozoa. Most soil organisms are thus dependent upon the organic matter of the soil for their supply of food and energy. The autotrophic bacteria and algae obtain their carbon from carbon dioxide. Oxygen and compounds containing hydrogen are usually present in soils in sufficient quantities, although the former may be deficient in poorly drained soils, and under such conditions certain microbes get their oxygen from inorganic compounds such as nitrates or sulphates. Nitrogen is one of the most important elements in the nutrition of microbes and may be used in the form of complex organic substances or in simple inorganic compounds, as ammonia or nitrate salts. Certain groups of bacteria have the ability to utilize the free nitrogen of the air and build it into their protoplasm, thereby increasing the soil's supply of combined nitrogen. Microbes obtain the other necessary mineral elements from the soluble salts in the soil and from the ash of decaying organic matter.

The Influence of Soil Environment on Microorganisms. As has been indicated, the environmental conditions determine the nature of the microbial population present at any given time in the soil. In general, the fertile, heavy soils rich in organic matter contain many more microbes than the light soils poor in organic matter. If we examine the nature of the changes which occur in the microbial population of soils, we find certain causes for these variations. Let us briefly consider some of these causes.

Temperature. The optimum temperatures for the growth of most soil microbes are considerably higher, as a rule, than those which prevail in the soil even in summer. Consequently, it may be assumed that microorganisms never reach their highest level of activity in soil, and thus they utilize only a part of the potential energy sources. Temperature regulates the reacting velocities of chemical and biological changes occurring in the soil. Within a rather narrow range the rate of biological reactions increases two to three times for each increase in temperature of 10° C. Roughly, the limits of microbiological functions are reached with a temperature of 80° C. For the majority of the soil organisms the optimum temperature is about 35° C., although they can grow at rather wide temperature ranges and may adapt themselves readily to gradual changes in temperature.

Moisture. Another major factor affecting the numbers and activities of soil microorganisms is soil moisture. The influence of moisture depends to a large extent upon the nature of the soil and the nature of the organisms concerned. The optimum amount of water for most soil organisms is between 50 and 70 per cent of the water-holding capacity of the soil, about the same as for most higher plants. The majority of soil organisms has the ability to withstand rather wide extremes in soil moisture content, thus insuring their wide distribution in soils in spite of periodic changes in the environment.

Acidity. The degree of acidity or alkalinity of the soil is of particular importance in influencing the activities and relative abundance of the different groups of soil organisms. It is frequently noted that the proportion of fungi to bacteria and actinomyces is greater in acid than in neutral soils; thus it appears that an acid soil favors the development of fungi but is unfavorable to the development of other forms. Usually the beneficial organisms function best in a soil that is approximately neutral in reaction. As a rule, the actinomyces prefer a reaction of 7.0 to 7.5, the bacteria and protozoa from 6.0 to 8.0, and the fungi from 4.0 to 5.0. Below *p*H 6.0 azotobacter in general, are inactive. In strongly acid soils legume bacteria fail to develop and function normally, and as a result poor inoculation is obtained and the organisms do not persist in the soil for any great length of time. The nitrifying organisms are also sensitive to a highly acid condition. Thus it becomes evident that the optimum in acidity for the majority of the soil population (especially the more desirable groups) is essentially the same as for most higher plants. The tolerance of soil organisms to acidity, as is also true for higher plants, is influenced considerably by other conditions like nutrient supply and favorable moisture content and temperature.

Aeration. The development and activities of soil organisms are greatly affected by the concentration and rate of supply of certain gases (particularly oxygen, carbon dioxide, and nitrogen) in the air. Oxygen is needed for oxidation processes, carbon dioxide as a source of carbon for autotrophic organisms, and nitrogen for the nitrogen-fixing organisms. Abundant oxygen favors the activities of the nitrite and nitrate formers, the nitrogen fixers, fungi, actinomyces, and other organisms which oxidize organic matter. Light soils frequently are excessively aerated, which restricts the accumulation of organic matter, whereas heavy soils frequently are insufficiently aerated. Poor aeration favors reduction processes. Soil aeration is governed largely by fluctuations in soil moisture. Aeration increases with a decrease in soil moisture, while an excess of water tends to encourage anaerobic conditions.

Salts. The addition of mineral elements to the soil influences the nature and activities of the soil population in several ways. They may stimulate the growth of higher plants, giving greater crop residues, thereby increasing the available energy supply, and resulting in increased microbial activities. With this increase in available energy, the mineral elements may then become limiting factors. This is especially true of nitrogen, phosphorus, potassium, calcium, and magnesium. The addition of mineral nutrients tends to produce a more desirable balance in the concentration of the soil solution for microbial activities, although an excess of soluble salts may prove injurious. The action of the various mineral salts depends on the nature of the salt, type of soil, and the nature of the organism concerned.

Light. Direct sunlight is highly injurious to most forms of soil microorganisms, and many are instantly killed when exposed to it. Diffused daylight appears to have an inhibiting effect on most bacteria although it has little effect on the growth of fungi. The growth of algae, however, is stimulated by diffused daylight, but direct sunlight appears to be detrimental.

Organic Matter. Among the various factors affecting soil organisms, the influence of organic matter is certainly one of the most important, especially under humid conditions. Since organic matter is the source of food and energy for the majority of the organisms, obviously those soils abundantly supplied with this material are capable of supporting a more dense microbial population than those low in organic matter. There is also an influence of organic matter on the structure, aeration, water-holding capacity, and temperature relations of soils.

Distribution of Microbes in the Soil Profile. The greatest number of microbes as a rule occurs in the surface layers (A horizon) of the soil, as shown in Fig. 40, although the number at the very surface of cultivated soils may be relatively low owing to a lack of moisture and to the germicidal action of the sunlight. In forest and meadow soils the greatest number of individuals and varieties of microbes is at or very near the surface of the soil. The number in all soils decreases with depth, below the A horizon, and the rapidity of this decrease in numbers varies with the soil conditions, especially the distribution of organic matter, aeration, and acidity.

The rate of decrease is relatively slow in most arid and semi-arid soils. The microbial population, therefore, is densest in that part of

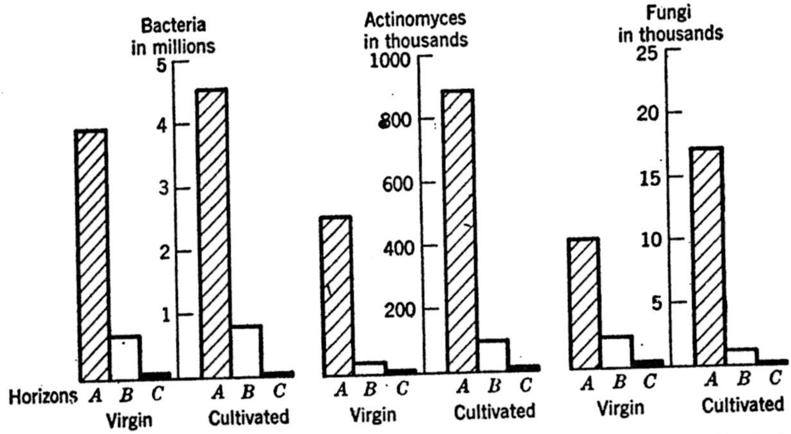

FIG. 40. Vertical distribution of microorganisms in a cultivated and virgin Carrington silt loam. [Data from Ia. Res. Bul. 132.] The values for the A horizons are the averages for the A_1, A_2, and A_3 horizons. The values for the C horizon of the cultivated soil are the averages for the C_1 and C_3 horizons. All values refer to the number of organisms per gram of air-dry soil.

the soil profile where the food supply is plentiful and where the environmental conditions are most favorable, and this condition usually is found at or near the top of the soil profile.

ACTIVITIES OF SOIL MICROBES IN RELATION TO THE GROWTH OF HIGHER PLANTS

The chemical processes that take place within the soil are, for the most part, dependent upon the activities of living organisms. The growth and activities of microbes are intimately concerned with the availability of nutrients, and hence the existence of higher plants depends upon the activities of soil microbes. After all, a productive soil is characterized not necessarily by the mere presence of large quantities of plant nutrients but by the rapidity with which these nutrients are liberated in an available form. A knowledge of some of the more fundamental microbial processes is of prime importance for the proper application of certain principles relative to good soil-management practices. Several questions will serve as an outline for this study.

Questions:

1. What are the most important beneficial effects of soil organisms?
2. What changes are brought about in the soil organic matter and its decomposition products by the soil organisms?
3. Is the fixation of atmospheric nitrogen a microbiological process?
4. In what ways are the inorganic soil constituents affected by the activities of the soil organisms?
5. What soil microbial processes are detrimental?
6. Is the soil a habitat for disease-producing organisms?
7. What is denitrification and under what soil conditions does it occur?
8. Do soil microorganisms compete with higher plants for available nutrients?
9. What toxic substances sometimes are encountered in soils?

Beneficial Functions of Soil Microbes. The beneficial effects of various microbes may be conveniently discussed under the following three general topics: (1) changes brought about in the soil organic matter and its decomposition products, (2) fixation of atmospheric nitrogen, and (3) changes brought about in the soil mineral constituents.

Changes in Organic Matter and Its Decomposition Products. One of the chief activities of microorganisms is the mineralization of organic matter. By mineralization of organic matter is meant the breaking down of plant and animal organic materials into simple inorganic elements or compounds through the activities of the various organisms. Plant or animal residues, as such, cannot be used by higher plants, but through decomposition by microorganisms they become a valuable source of nutrients. All forms of soil organisms are concerned with the processes of organic matter decomposition, but the actions of bacteria, fungi, and actinomyces are the most important. These organisms bring about various transformations such as hydrolysis and oxidation by means of enzymes, which are usually called organic catalysts. The physical and chemical changes which organic materials undergo when changed to simple compounds by soil organisms are termed *decomposition.* The processes of decomposition vary greatly, however, depending on the nature of the organic substance and on various other conditions, especially the degree of aeration.

When organic material is added to soils in the presence of sufficient oxygen (aerobic conditions), the non-nitrogenous compounds, such as starches, sugars, lignocellulose, organic acids, and oils, are decomposed by microorganisms and converted into various intermediate simpler compounds. But regardless of the kind of intermediate products formed from the non-nitrogenous organic compounds, the process leads to the formation of a succession of simpler compounds. The carbon, hydrogen, and oxygen contained in the plant and animal materials are finally reunited to form carbon dioxide and water; the phosphorus is combined with oxygen to form phosphoric acid; and the sulphur with oxygen to form sulphuric acid. The mineral constituents of the original plant or animal material are left as a residue, mainly as salts of calcium, magnesium, potassium, and sodium. Examples of such salts are calcium phosphate, magnesium sulphate, potassium carbonate, and sodium chloride.

Decomposition in the absence of sufficient oxygen (anaerobic conditions) is brought about by entirely different kinds of organisms from those that function under aerobic conditions, and the chemical changes produced are not the same. The rate of decomposition is much slower and the chemical changes are less complete. Although some carbon dioxide, water, and ammonia are formed from the complex organic materials, most of the nitrogen, carbon, hydrogen, and oxygen are found in the form of more or less complex intermediate compounds, many of which are quite resistant to further decomposition. Under anaerobic conditions many compounds having offensive odors and a few poisonous compounds are produced. In the list of such compounds may be found hydrogen sulphide, phosphine, skatole, indole, and methane.

Owing to the great diversity of conditions affecting microbial activity and to differences in the nature of the organic materials, the processes of decomposition of organic matter in soils are of extreme complexity. When fresh organic substances are added to soils, the numerous bacteria, fungi, protozoa, and other organisms attack the various constituents, changing them back to more simple compounds. During the processes of decomposition the chemical changes as a rule go through an orderly progression although many of the changes may occur simultaneously.

Ammonification. Ammonification may be defined as the formation of ammonia by soil organisms as a product of the decomposition of organic nitrogenous compounds. Ammonia is formed by a large number of soil organisms, including various fungi, actinomyces, and aerobic and anaerobic bacteria. Thus the soil organisms responsible for ammonification are widely distributed and are found in abundance in most soils. The rate of ammonification and the quantity of ammonia produced are influenced markedly by soil conditions, nature of the organic material, and nature of the microbial population. The soil-management practices that especially favor ammonification are liming acid soils, draining wet soils, adding easily decomposible organic matter high in nitrogen, and proper tillage. The process of ammonification is of extreme importance in crop growing not only because certain plants have the ability to use ammonia compounds directly but also because ammonification is a necessary preliminary step to nitrification.

Nitrification. Another beneficial effect of soil organisms is the conversion of ammonia nitrogen to nitrate nitrogen—the process of nitrification. The fact that the nitrates, produced in the soil, are the principal sources of available nitrogen for practically all agricultural plants makes this a most important biological process. The conversion is carried on largely by two specific types or groups of bacteria, of which there are several species, often referred to as the nitrifiers or nitrifying bacteria. Although seldom ever found in great abundance in any soil, fortunately they are widely distributed, being present in every cultivated soil. As would be expected, they are usually more numerous and especially more active in rich than in poor, unproductive soils.

The conversion biologically of ammonia into nitrate nitrogen is the result of two distinct processes. First, the ammonia is changed to nitrite nitrogen principally by the action of two groups of bacteria called Nitrosomonas and Nitrosococcus, and secondly, the nitrite nitrogen is changed to nitrate nitrogen by the action of a group of bacteria called Nitrobacter. These nitrifying organisms use the energy liberated (as heat) in the nitrification processes for their growth and development, and the carbon for the synthesis of their cells is taken from the carbon dioxide of the atmosphere.

The reactions indicating the two steps in nitrification may be represented as follows:

$$(1) \qquad NH_3 + 3O = HNO_2 + H_2O$$

$$(2) \qquad HNO_2 + O = HNO_3$$

These reactions show the change to be an oxidation process. As a rule, in soils of humid regions, there is no appreciable accumulation of either ammonia or nitrites because they are rapidly oxidized. It is apparent that the process of nitrification does not increase the total supply of nitrogen in the soil; merely the form of nitrogen is changed.

In good agricultural soils nitrous acid and nitric acid as such probably do not exist, but they are combined with basic elements, forming salts. For example, nitrate nitrogen in soils usually is found as the nitrates of calcium, potassium, magnesium, and sodium, depending on the chemical condition of the soil.

The soil conditions under which active nitrification occurs are much more restricted than those for ammonification because of the limited number of types of organisms concerned with nitrification and because the nitrifying organisms in general have similar growth characteristics. Consequently, the process is affected much more by changes in soil reaction, aeration, moisture, salt concentration, and other environmental conditions than is ammonification. Perhaps the most important factors affecting the activity of nitrifying bacteria in cultivated soils and which the farmer can more or less control are drainage, organic matter supply, fertilizers, and soil reaction. Drainage increases aeration and since nitrification is an oxidation process the need for oxygen is obvious. Crops frequently suffer for want of available nitrogen in soils poorly drained owing to the inactivity of the nitrifying bacteria. The optimum reaction for the nitrifiers is *p*H 7.0 to 8.0, although some strains will grow between the *p*H limits of 3.5 and 10.0, their activity is retarded in both strongly acid and highly alkaline soils. Since the nitrification process results in the production of an acid, and if good conditions for further nitrate production and for good crop growth are to be maintained some neutralizing material must be present or must be added to combine with the acid. Lime is usually the most desirable material to use for this purpose.

Sulphofication. Organic materials commonly added to soils contain compounds of sulphur, and when the organic matter decomposes, the sulphur is released from its organic combination and eventually appears in an inorganic form, usually as H_2S and perhaps as some free sulphur. These inorganic products are oxidized by certain specific groups of bacteria into sulphurous acid (H_2SO_3) and finally into sulphuric acid (H_2SO_4). Some fungi are also believed capable of bringing about the latter reaction. The transformation of sulphur in organic and inorganic compounds to available sulphate forms by soil microbes is called sulphofication. It is the same type of chemical change (oxidation) that takes place in nitrification. When sulphurous acid is formed, it usually combines with some of the soil mineral elements to form sulphites of calcium, magnesium, sodium, and potassium. They are then oxidized to sulphates. The importance of the sulphofication process is realized when it is remembered that the higher plants take in most of their sulphur in the sulphate form.

Assimilation and Synthesis of Organic Matter Decomposition Products. When soil microbes have access to an abundance of food and energy materials, and if environmental conditions are favorable, they build up a considerable amount of organic matter of their own in the form of synthesized cell substance. In so doing they assimilate certain simple soluble products formed as a result of their own activities. In other words, large quantities of nitrogen, phosphorus, and other nutrient elements may be temporarily stored in the bodies of soil microorganisms in their processes of synthesizing cell substance. This assimilation of soluble plant nutrients may or may not be harmful, depending on whether or not it occurs during a time when they are needed by higher plants. The process would be highly desirable under conditions where there is danger of leaching before the nutrients can be used by crops. The addition of carbonaceous material, such as straw, to soil may greatly stimulate microbial growth and materially decrease leaching losses of soluble plant-food materials. It is obvious that conditions frequently may arise where soil microbes offer serious competition with higher plants for soluble nutrients. This is especially true with nitrate nitrogen.

Another important function of soil microbes, related to the assimilation and synthesis of organic matter decomposition products, is associated with the formation of humus, which is discussed in the chapter on soil organic matter.

Fixation of Atmospheric Nitrogen. Nitrogen in its elemental form is a colorless, odorless gas and very inert. There is an inexhaustible supply of it in the air, but it is in the free state and does not easily combine with other elements. It is estimated that there is the equivalent of 148,000 tons of nitrogen in the atmosphere for every acre of land area; the atmosphere being approximately 80 per cent nitrogen. Regardless of the fact that nitrogen is so inert in the free state, certain groups of soil organisms have the ability to take that element out of the air and utilize it in the synthesis of their cells. The nitrogen of the air is thereby changed to a "fixed" form in which it can be of subsequent use to higher plants. This changing of atmospheric nitrogen into fixed-nitrogen compounds in the soil by microorganisms is known as *nitrogen fixation.* The process is accomplished largely by two groups of bacteria, that is, symbiotic and non-symbiotic, of which there are several species of each.

Non-Symbiotic Nitrogen Fixation. There are certain groups of heterotrophic bacteria living in the soil independently of higher plants that have the ability to use atmospheric nitrogen in the synthesis of their body tissues. Since these bacteria do not grow in association (mutual relationship) with higher plants, they are termed non-symbiotic. These bacteria may be separated into two groups, based on their oxygen requirements, that is: (1) the aerobic group and (2) the anaerobic organisms. The former are known as the Azotobacter group and the latter as the Clostridium group.

Members of the Azotobacter group are widely distributed in nature. They have been found in soils (of *p*H 6.0 or above) in practically every locality where examinations have been made. The greatest limiting factor affecting their distribution in soils appears to be the soil reaction. These organisms may exist in soils below *p*H 6.0, but as a rule they are not active, as far as nitrogen fixation is concerned, under such conditions. Azotobacter are favored by good aeration, abundant organic matter (particularly of a carbonaceous nature), the presence of calcium carbonate and sufficient quantities of available nutrient elements, especially phosphorus, and proper moisture and temperature relations. So responsive are azotobacter to phosphorus supply that they have been used as an agent in testing soils for the quantity of available phosphorus. The individual farmer, by applying lime and fertilizers and by returning all crop residues as far as possible, may make the soil conditions more favorable for these nitrogen-fixing bacteria. Attempts have been made from time to time to increase the crop-producing power of soils by inoculating them with azotobacter. For the most part, this practice has not been justified for the reason that unless soil conditions are favorable, especially with respect to lime, the bacteria will not survive or function and, if soil conditions are made favorable, the organisms will naturally become established without the aid of man.

Members of the anaerobic group, Clostridium, are much more acid-tolerant than members of the aerobic group and perhaps for that reason are more widespread. It is believed that these organisms can be found in every soil and that under suitable conditions they fix some nitrogen. It is not necessary that soils be waterlogged in order for anaerobic bacteria to function. A soil in good tilth may contain considerable areas within the granules favorable for the activities of these anaerobic nitrogen-fixing bacteria.

It is not to be inferred by this brief discussion of non-symbiotic nitrogen fixation in soils (frequently called azofication), that Azotobacter and Clostridium, although responsible for most of the non-symbiotic fixation, are the only soil organisms that have the ability to take nitrogen from the atmosphere and incorporate it in their bodies. It is possible that several different groups of bacteria and perhaps fungi have that capacity, and there is definite evidence to show that certain species of blue-green algae have the ability to gather nitrogen from the atmosphere.

A question that naturally comes to mind is, how much nitrogen is fixed per acre per year by these non-symbiotic microorganisms under favorable field conditions? This question cannot be answered definitely because of the many difficulties encountered in making such a measurement under field conditions. However, in fallow soils increases in nitrogen have been secured which can be accounted for only through the activities of non-symbiotic organisms. These organisms can be isolated from soils, and they will fix nitrogen under laboratory conditions; it is therefore assumed that they will do likewise under field conditions, and thus they must be responsible for observed increases in soil nitrogen. Although accurate measurement of the quantity of nitrogen fixed cannot be made, the best estimates range from 10 pounds to 50 pounds per acre per year. The importance of this biological process to agriculture is realized when one considers that the amount so fixed, assuming an average fixation of 20 pounds per acre per year, is equivalent to the nitrogen in a 100-pound application of ammonium sulphate.

Nitrogen fixation is fundamentally a different process from nitrification. The former process increases the soil's supply of nitrogen, whereas the latter process merely changes the compounds in which the nitrogen is held. The quantities of this element taken from the air and synthesized into proteins and other organic compounds by nitrogen-fixing bacteria are finally broken down by various heterotrophic organisms to ammonia, which in turn undergoes nitrification. Thus, in general, nitrogen of the air goes through the processes of fixation, synthesis into proteins, decomposition, and ultimately nitrification before higher plants can use it. It should be mentioned, however, that certain soluble nitrogen compounds may be secreted from the bacterial cells during their active growth period and in turn utilized directly by higher plants. In other words, it may not always be necessary for the decomposition of the cells of nitrogen-fixing bacteria to occur in order for higher plants to derive benefit from the nitrogen-fixation process.

Symbiotic Nitrogen Fixation. The most important bacteria, from the agricultural point of view, capable of utilizing the free nitrogen of the air are those that cause the formation of nodules on the roots of legumes. These organisms, when growing in the nodules of legume plants, derive their food, energy, and minerals from the legume, and in turn they supply the legume with some of its nitrogen. This growing together for a mutual benefit is called *symbiosis,* and hence the organisms are designated symbiotic nitrogen-fixing bacteria. The economic importance of symbiotic fixation is realized when it is said that the extent of the growth of legumes is a major factor in the attainment and maintenance of a high level of agriculture. It has been estimated that nearly 2,000,000 tons of nitrogen are fixed annually by legume bacteria in the United States.

When these bacteria come in contact with the roots of the legume plant, some of them enter the single-celled root hairs. A rapid increase in the growth rate and in the numbers of bacteria then takes place because of the abundance of easily accessible food. These bacteria form an infection thread toward the base of the root hair that eventually penetrates the cortex of the root. This infection injures the legume plant and, in response to this stimulus, numerous plant cells in the meristematic tissue are produced in the immediate vicinity of the infection, eventually forming the nodule. The nodules, then, are essentially nothing more than masses of root tissue in which the bacteria live. The nodules are connected with the vascular system of the root, which permits transporting materials to and from them. Each nodule may contain millions of bacteria and the number of nodules that may develop on a single plant may vary from a few to a thousand or more.

Nodules are usually found in bunches, mainly on the younger roots; or in annual legumes they are frequently distributed about the tap root or the first-formed lateral roots.

The individual nodules may vary greatly in size and shape on different kinds of legumes. For example, the cultivated annual legumes generally have large spherelike nodules, while those on the biennial and perennial legumes tend to be smaller, elongated, and in clusters.

Occasionally, nodule-like growths may be produced by nematodes or by crown-gall bacteria on the roots of legumes. These false nodules should not be confused with true nodules when determining the degree of infection by legume bacteria. Furthermore, certain non-leguminous plants frequently are found to possess nodule-like growths, which are produced by mycorrhiza, crown-gall organisms, and certain nematodes; but on examination they are not difficult to distinguish from the leguminous nodules.

In order to obtain the full benefits from the growing of legumes, it is necessary that they possess true nodules. If active nodule-forming bacteria are not present in soils, they should be supplied—a process which is known as inoculation. This may be accomplished by the use of soil that contains the desired bacteria or by the use of specially prepared pure cultures of the desired organism. As a rule the most effective although not necessarily the most convenient method of inoculating soil with legume bacteria is to apply 200 to 300 pounds per acre of soil, taken from a field on which a good crop of the kind desired has been growing and is known to have produced an abundance of nodules. The application may be made in any convenient manner, but it should be worked into the soil within a short time in order to avoid undue exposure to direct sunlight. In obtaining the inoculating soil, the immediate surface (about 1 inch) should be removed and the soil taken from beneath. If it is to be scattered with a drill, it will be necessary to put the soil through a screen in order to remove the coarser material. A more convenient and generally satisfactory method of inoculating a soil is to use pure cultures of legume bacteria, usually prepared in liquid and dust form or on agar. The liquid and dust cultures are ready for immediate use, but the agar cultures first must be shaken with water to obtain a suspension of the bacteria. The liquid or the suspension is applied to the seed, which are allowed to dry in the shade before being sown.

No one particular kind of bacteria can be used to inoculate all legumes successfully. Different plant species as a rule possess their own particular bacteria although sometimes one kind of organism may inoculate a half dozen or more plant species. Legume bacteria have been classed in the genus Rhizobium, of which there are several species and strains. Some of the more common groups, each group representing a separate species of bacteria, are as follows:

1. Alfalfa group (*Rhizobium meliloti*), including bacteria that will produce nodules upon white sweet clover and yellow sweet clover as well as alfalfa.
2. Clover group (*R. trifolii*), which will produce nodules upon clovers such as red, alsike, crimson, mammoth and white Dutch.
3. Pea group (*R. leguminosarum*), containing bacteria that produce nodules upon garden and field peas, sweet peas, vetches, etc.
4. Bean group (*R. phaseoli*), including bacteria capable of producing nodules upon garden, kidney, and navy beans and scarlet runner.
5. Soy bean group (*R. japonicum*), those bacteria that produce nodules upon soy beans.
6. Cowpea groups (*R.* ———), including bacteria producing nodules upon cowpeas, peanuts, lespedeza, velvet beans, and lima beans.
7. There are several other small and less important groups.

In planting a legume crop on a piece of land for the first time it is essential that the soil be inoculated with the proper bacteria. The fact that red clover may have been grown on the land and produced an abundance of nodules is no indication that, for example, alfalfa bacteria will also be present. When in doubt always inoculate because the expense is small compared to the cost for seed and preparation of the seedbed. If a legume crop is grown without proper nodule production, the full benefit from that crop has not been realized. If there is no nodule production, the effect of legume crops on the soil would be similar to that of non-legumes, for the plants must get all their nitrogen from the soil. Good inoculation, when needed, results in an increase in yields and nitrogen content of the crop grown.

The question is frequently asked, how often should I inoculate my soil for the growing of certain legumes? That depends largely on

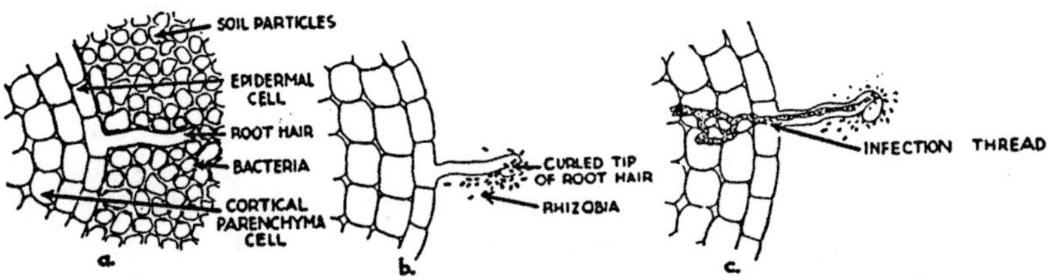

FIG. 41. Early stages in the formation of a nodule. *a.* Response of the bacteria to a product of the host plant; organism moves toward root hair. *b.* Curling of the root hair. *c.* Early penetration of the infection thread. [Courtesy of P. W. Wilson and the University of Wisconsin Press, "The Biochemistry of Symbiotic Nitrogen Fixation," 1940.]

the nature of the soil conditions. Some of the more important factors which influence the growth and longevity of nodule bacteria are the air supply, temperature, sunlight, reaction associated with the calcium content, supply of mineral elements, and the antagonistic action of other soil organisms. It can be said that unless killed by unfavorable soil conditions, the rhizobia may live and maintain their vitality in a soil for a number of years, possibly 10 to 20 years, even in the absence of the particular host legume. Generally, however, the legume bacteria in a soil seem to be rejuvenated by association with their host plant every few years or the effectiveness of the bacteria can be increased, as evidenced by increased nodule production and nitrogen fixation, by reinoculation. It is not to be inferred that there is always a correlation between numbers or quantities of nodules and the amount of nitrogen fixed, because certain strains apparently have the ability to produce nodules but do not gather nitrogen. Of the various factors affecting the longevity of legume bacteria in normal soils, their ability to produce nodules and their effectiveness in fixing nitrogen, calcium supply seems to be the most important. A good supply of active calcium not only in the soil but also in the plant is necessary for proper nodulation and active nitrogen fixation. In most soils available calcium seems to be more of a controlling factor in this respect than does *p*H.

The nitrogen gathered by legume bacteria may be utilized in three ways. It may be absorbed by the legume plant itself, it may be excreted from the nodule into the soil and be utilized by some crop growing in association with the legume, or when the legume crop is plowed down or dies the nitrogen may be released after the decomposition of the nodules.

The quantity of nitrogen fixed by a legume varies greatly according to many conditions, such as the kind of legume, the nature of the soil conditions, the effectiveness of the bacteria present, and seasonal conditions. In general, it can be said that the poorer the soil in available nitrogen and the richer in lime and available phosphorus and potash, the greater will be the gain in nitrogen by growing legumes. It appears that the intimate relations existing between nodule bacteria and their host plants are determined largely by the carbohydrate supply in the host plants. Any environmental condition affecting the production of carbohydrates in the plant would automatically affect the quantity of nitrogen fixed by good strains of legume bacteria. Well-inoculated legumes growing under favorable conditions will usually gather 50 to 100 pounds of nitrogen per acre per year. With good crops of such legumes as alfalfa and sweet clover the quantities fixed may be even greater than 100 pounds per acre per year.

In actual farm practice the amount of nitrogen added to a soil by legume bacteria is determined by the methods of disposing of the legume crop. If the crop is turned under as a green manure, then the total quantity of nitrogen taken from the air is added. If the crop is cut for hay and sold off the farm, little or no gain is realized; with some legumes there may even be a net loss of nitrogen. And if the crop is cut for hay and fed on the farm, about one-half the nitrogen that was taken from the air by the legume bacteria can be returned to the soil if special care is exercised in handling the manure to prevent loss. It is generally assumed (although not necessarily true for all legumes) that the amount of nitrogen in the roots and stubble equals the amount of nitrogen taken from the soil, which would mean that the quantity of nitrogen removed by the harvested crop is equal to the nitrogen obtained from the air.

Changes in Inorganic Soil Constituents by Microorganisms. It becomes evident that in the decomposition of organic matter and in the synthesis of microbial cells there are numerous reactions involving the solubility of various soil mineral elements. Among those mineral elements which are affected directly or indirectly by the action of microorganisms, the following should be mentioned: phosphorus, sulphur, calcium, magnesium, iron, manganese, aluminum, and silicon. In some cases soluble compounds of the elements are formed, whereas in others insoluble compounds result.

The most common acids produced in the soil as a result of the action of microorganisms are nitrous, nitric, phosphoric, sulphurous, sulphuric, carbonic, and numerous other organic acids. The following changes result from the action of these acids: (1) Insoluble phosphates are changed to more soluble compounds; (2) calcium, magnesium, and other slightly soluble carbonates are changed to more soluble forms; and (3) insoluble potassium compounds are also changed to more soluble combinations. The reactions for some of the above changes may occur according to the following equations:

$$CaCO_3 + H_2SO_4 = CaSO_4 + H_2O + CO_2 \tag{1}$$

$$Ca_3(PO_4)_2 + 4HNO_3 = Ca(H_2PO_4)_2 + 2Ca(NO_3)_2 \tag{2}$$

$$3KAlSi_3O_8 + H_2CO_3 = KH_2Al_3Si_3O_{12} + K_2CO_3 + 6SiO_2 \tag{3}$$

Practically all mineral elements related to the growth of higher plants either as nutrients or "stimulants" are subject to the action, in one way or another, of microorganisms in the soil.

These few statements serve to show that extensive and far-reaching chemical changes are produced in the inorganic soil constituents through the various activities of soil microorganisms. They are the primary agents which convert the elements of the insoluble mineral and organic compounds into soluble forms available as plant nutrients.

Soil Microbial Processes of a Detrimental Nature. For the most part microbial processes are highly beneficial, but certain of the detrimental effects should not go unmentioned. The harmful activities of soil organisms may be grouped as follows: (1) the production of plant and animal disease, (2) denitrification, (3) competition with higher plants for available nutrients, and (4) the production of toxic compounds.

Production of Disease. The soil frequently contains a rather large number of organisms that cause diseases either in plants or animals. Some of these organisms live in the soil only temporarily and others use it as a permanent habitat. The soil may harbor organisms that cause such bacterial diseases as wilt of tomatoes and potatoes, soft rots of a number of vegetables, leaf spots, and galls. Some of the most destructive parasites are the disease-forming fungi such as those that cause damping-off of seedlings, cabbage yellows, mildews, blights, certain rusts, wilt diseases, scab, dry rot of potatoes, and many others. Certain species of actinomyces may cause diseases like scab in potatoes and sugar beets and pox in sweet potatoes. Nematodes of a parasitic nature commonly infest soils, except possibly in the extreme northern states, and cause injury to many plants, especially to the roots. Abnormalities or swellings on the roots, usually referred to as root knot or root gall, are produced.

So-called "soil sickness," frequently encountered in greenhouses, is often due to the presence of parasitic soil organisms of one kind or another. The ability of soil-borne microbes to infect a plant and cause disease is often a function of the temperature, moisture, nutrient, or acidity relationships. Partial or complete control may be exercised by regulating or modifying the soil conditions accordingly, the theory of control then being that the parasitic organisms have a range through which they function narrower than that of the higher plants which they infect.

Denitrification. Denitrification may be considered the reverse of nitrification. It is a reduction process in which oxygen is taken away from nitrates to form first nitrites, then either ammonia or gaseous nitrogen. In the ammonia or gaseous nitrogen form, the nitrogen may escape into the atmosphere. This destructive or unfavorable process takes place in the absence of oxygen and in the presence of an abundance of soluble organic matter, which serves as a good source of energy for the organisms concerned. The fact that denitrification takes place in the absence of air explains the decomposition of nitrates in the lower layers of soils as well as in surface soils that are deficient in oxygen. Furthermore, that is why heavy, compact soils tend to suffer greater loss from denitrification than do the more open, sandy soils. Any soil-management practice that promotes the circulation of air in soils lessens the danger of denitrification. In ordinary cultivated soils denitrification is of little economic importance as a rule, although if a cropped field becomes flooded for a considerable length of time, a considerable loss in nitrogen may occur owing to denitrification.

Microbial Competition with Higher Plants for Available Nutrients. Attention has already been called (p. 198) to the beneficial effect of the assimilation of nutrient elements by microorganisms in decreasing leaching losses, but a pertinent question now arises as to whether microorganisms seriously compete with higher plants for nutrients. The soil organisms require nitrogen, phosphorus, potassium, calcium, and other elements, as do higher plants, and obviously even a slight increase in the microbial population might appreciably reduce the available nutrients of the soil to a point at which the growth of higher plants would be impaired. Aside from carbon, hydrogen, and oxygen, nitrogen and phosphorus are the elements usually consumed in largest quantities by microorganisms and for which the keenest competition between microbes and higher plants exists. Of the two elements, nitrogen is the one most vigorously contested.

If straw or any easily decomposable carbonaceous material is added to a soil containing a liberal supply of nitrates, biological activities are greatly accelerated. The soil organisms, having an abundant supply of food, will multiply rapidly and will need a large supply of nitrogen for the synthesis of their protoplasm. Since the carbonaceous food material will not supply the necessary amounts of nitrogen, the organisms will utilize the soluble nitrates of the soil, thereby decreasing the supply available to higher plants. Conditions may thus exist whereby higher plants suffer seriously for available nutrients, owing to competition of the soil microorganisms. For this reason, crops following immediately the turning under of straw or similar material may suffer for the want of available plant nutrients. The remedy in this case would be the liberal application of a fertilizer supplying these nutrients. Soluble nutrients taken up by microorganisms are "tied up" in an unavailable form only temporarily. When the organisms die and their bodies decompose, the plant nutrients are again released in an available form.

The Production of Toxic Substances. Substances toxic to plant growth as well as to certain microorganisms accumulate, under anaerobic conditions, in some soils as a result of the incomplete decomposition of organic matter and some of its products. For example, methane, hydrogen sulphide, phosphine, skatole, indole, and numerous organic acids are toxic compounds produced, but they are generally considered merely products of improper soil conditions and when these soil conditions are corrected the toxic materials will disappear. If good drainage and tillage are provided, together with the proper use of fertilizers and lime, these toxins will not be a problem in soils.

THE ROLE OF MICROORGANISMS IN THE DEVELOPMENT OF SOILS

Without living organisms there could be no natural soils, and hence no one will question the contribution of biological processes in soil formation. The earliest students of soils were fundamentally students of geology and chemistry, and consequently they did not pay much attention to the roles played by organisms in rock weathering and soil development. Recently more consideration has been given to the part played by microorganisms, and it is suggested that, possibly, biological processes are of equal if not of more importance in soil formation than are the non-biological processes. It is obvious that the biological and non-biological forces act together. Answers to the questions below should be of interest to the student of soils.

Questions:

1. How do the microorganisms affect the decomposition of rocks and minerals?
2. Is the decomposition of soil organic matter related to soil development?
3. Do microorganisms influence soil-leaching losses?
4. Is the movement of soil material within the profile affected by microorganisms?
5. Do organisms aid in the development of a good soil structure?
6. In what ways may the animal population affect soil development?
7. What is the main driving force of the biological factors of soil formation?

Decomposition of Rocks and Minerals. Mention has already been made that microbes are capable of bringing about extensive changes in a number of minerals and compounds of mineral elements. They assimilate such elements as phosphorus, calcium, magnesium, iron, and potassium from slightly soluble mineral compounds. They produce various organic and inorganic acids which exert a solvent action on most minerals. Soil organisms such as the algae and lichens have a corroding effect on rocks and minerals because of the production of acids. Other organisms increase the solubility of feldspars and other silicates (p. 205), and many of them are active in promoting the mechanical weathering of minerals.

Organic Matter Decomposition. The amount and nature of the organic matter is of major concern in the formation of a soil, and the control of organic matter decomposition is exercised to a large extent by the microorganisms. In the cycle of mineral elements which are taken up by plants from the lower soil horizons and then returned to the surface horizons, there exists a close relationship to the microbial population. Decomposition and mineralization, as has been pointed out, are important functions of the soil organisms whose activities are largely controlled by climatic conditions. In cool, humid climates biological activity is slowed down and decomposition of organic matter is slight, thus permitting considerable accumulation which in extreme cases may result in the formation of peat or tundra soils. In warm, humid climates, on the other hand, rapid decomposition does not permit a great accumulation of organic matter in well-drained soils. The role of humus in soil formation is discussed elsewhere (p. 59).

Leaching of Soil Materials. The function of microbes in affecting leaching losses is closely associated with the decomposition of minerals and organic matter. Leaching losses are greatly enhanced by the activities of microorganisms, owing to their effect in increasing the solubility of both mineral and organic constituents. The leaching process is of major importance in determining the nature of the soil profile that is developed. It is to be pointed out in this connection, however, that certain quantities of the nutrient elements which are made available are assimilated by the microorganisms and are not immediately lost by leaching.

Movement of Soil Material. Microorganisms play a part in the translocation of soil material through the production of substances which exert a solvent effect on certain soil constituents. The dissolving effect of organic and inorganic acids on more or less insoluble compounds is obviously a contributing factor in the translocation of some products by soil-moisture movements. Furthermore, some of the mineral elements that are released combine with the organic colloidal constituents and move in the colloidal state. Products of microbial activity may appreciably affect the dispersion of soil colloids and thus influence their movement in the soil profile. Also, modification of the soil reaction not only will affect the solubility of certain soil constituents but also will have a pronounced influence on soil structure, both of which actions are concerned in the movement of soil material.

Structure of Soils. In the development of soil structure, microorganisms have an important but indirect effect. They are responsible mainly for the formation of organic colloidal material which promotes the formation of a granular structure which is so desirable in soils.

Effects of the Animal Population. The work of the various animal species as soil formers is largely mechanical. Ants, worms, termites, and rodents, for example, are instrumental in mixing the soil, thereby distributing the microbial population and incorporating organic matter with the mineral material. This action is of especial significance in timbered soils. Burrows made by these organisms also increase aeration and drainage. Chemical effects come about largely through the decomposition and mineralization of substances composing the bodies. Some chemical action may also take place when soil materials pass through the digestive system or when some chemical substance, such as an organic acid, is produced by the animal.

General Remarks. In relation to soil formation and development enough has been said to indicate definitely that microbes through their diverse activities play parts of major importance. In some instances they aid the purely chemical, physical, and mechanical changes; in some instances they are the primary agents in the transformation processes; and in others they play a prominent role, as in the formation of forest soils and of peat and muck soils. The main driving force behind the various biological factors is the *climate.*

INTERRELATIONSHIPS BETWEEN HIGHER PLANTS AND SOIL MICROORGANISMS AND AMONG SOIL MICROORGANISMS THEMSELVES

Neither microorganisms nor higher plants can develop normally for any length of time under natural conditions in the absence of the other. Even a very superficial examination of the relationships of microorganisms to higher plants brings to our minds a large number of interrelated factors, the most important of which are suggested by the questions below.

Questions:

1. Does an increase in microbial activities accompany the growth of higher plants?
2. What is meant by the antibiotic effects of microorganisms?
3. Are microorganisms concerned with the production of growth-regulating substances in soils?

Effect of Higher Plants on Microbial Activities. The growth of higher plants in soils increases many microbial activities, and it is noteworthy that their activities become especially intense in the vicinity of the roots. The enhancement in microbial action begins when the higher plants are young and continues throughout the active growth period of the plants. This modification in the microbial flora is due in part to physical changes produced in the soil environment by the plant roots, but the most pronounced influence is believed exerted by the organic substances, arising from the roots themselves, which serve the heterotrophic organisms as a source of food and energy. Evidently it is the soil of the root zone that plays the most important part in the nutrition of higher plants, and it is here that microorganisms are most active and exert their most pronounced effects. It is in the root zone (rhizosphere) that microbes are most active in increasing the availability of nutrients for plants, and it is here also that they may exert their most injurious effects under certain conditions.

Antibiotic Effects. The effects of one organism upon the development of another may assume particular significance because certain disease-producing organisms may be destroyed by other organisms. These effects may be of considerable importance although such effects are extremely difficult to isolate and demonstrate. Bacteria, for example, sometimes appear to be antagonistic to the actinomyces causing potato scab. It has been demonstrated that some fungi and bacteria suppress the activities of the wheat root rot fungus. The antagonistic effects among the soil organisms may be very common.

It has frequently been observed that plants affected by disease are greatly influenced by soil conditions, perhaps reflected through the general activity of the microbial population. A certain disease, for example, may be suppressed by the addition of fertilizers; the treatment increases the activities of the organisms other than those responsible for the disease. Such a practice has been used to decrease the effectiveness of certain root rot diseases.

Growth-Regulating Substances. Recently considerable attention has been given to the subject of the effect of vitamins particularly of vitamin B_1 (thiamine chloride) on plant growth. The beneficial effects of organic matter in some instances have been ascribed to growth-regulating substances of a vitamin-like nature which are believed produced by microorganisms in the decomposition of the organic materials. The power to synthesize vitamin B_1, for example, is not limited to green plants but is also possessed by certain fungi and perhaps by other microorganisms. Favorable effects have been observed from application of some of these materials from time to time, but differences of opinion exist concerning the real nature of these phenomena or their importance in agricultural practice. It is established, however, that vitamin B_1 is essential for plant growth; but since it is produced in the decomposition of organic matter, perhaps synthesized by certain microorganisms and higher plants, it is not likely to be a limiting factor in plant growth under good soil conditions.

CHAPTER VIII

SOIL ORGANIC MATTER

In the study of soils centuries ago, it was observed that their capacity to produce crops was more or less directly related to the amount of organic matter which they contained. Likewise the farmer of today, in designating soils which he considers highly fertile, will usually select the dark-colored ones. He makes this choice because experience has taught him that such soils are usually more productive than the light-colored ones, and hence in selecting dark-colored soils the farmer is involuntarily paying tribute to the value of organic matter in soils.

In general, it can be said that organic matter exerts a controlling influence on soil properties, including productivity, and without it the surface layer of the earth could hardly be designated correctly as soil. The organic matter content of soil is one of our most important and also easily exhausted resources. This subject may be studied under the following headings.

Objectives:

A. Organic matter accumulation in soils.
B. Effects of organic matter on soil productivity.
C. The decomposition of organic matter and humus formation.
D. Loss and restoration of soil organic matter.

ORGANIC MATTER ACCUMULATION IN SOILS

The stock of organic matter in our virgin (uncultivated) soils as taken over by the pioneers was an accumulation during many generations. Its collection in the surface layers of the earth began several thousand years ago. As the rocks and minerals of the earth's crust decomposed, mineral elements were made available to plants; and as supplies of nitrogen in usable chemical combinations were produced from the store of nitrogen in the air, plants grew, died, and contributed their remains to the soil. Thus organic matter began to accumulate. As the supply of available plant nutrients in the soil increased, the accumulation of soil organic matter increased accordingly. This condition continued until an equilibrium point was reached, at which the accumulated organic matter held in combination most of the mineral nutrients that the soil could supply in readily soluble forms. Thereafter, the quantity of organic matter remained more or less constant because the supply of minerals became a limiting factor in the growth of plants. This so-called equilibrium level is determined to a large

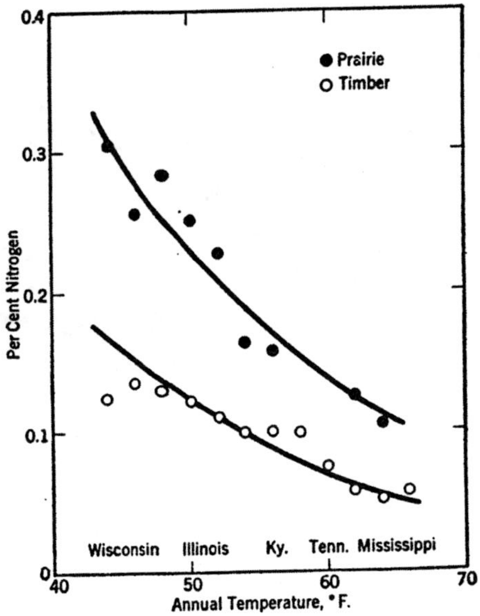

FIG. 42. Nitrogen—temperature relation in humid prairie (upper curve) and humid timber soils (lower curve) for silt loams. [Mo. Res. Bul. 152.] These curves show that nitrogen and organic matter contents of soils under similar moisture conditions decrease from north to south, and that under similar conditions the prairie soils are considerably higher in nitrogen and organic matter than the timber soils. In these studies the average nitrogen or humus content of a county area was chosen as a comparable unit. All nitrogen analyses for a county were arranged according to topography, vegetation, and texture of the soils, and their respective averages were tabulated with the mean annual temperature and annual humidity factor of the county.

extent by (1) the climatic conditions (temperature and rainfall), (2) the nature of the vegetation, (3) the topography of the soil (which influences erosion losses), (4) texture of the soil, and (5) drainage conditions.

The source of most of the soil organic matter is plant tissue. It is true that animals contribute to the organic supply of the soil, but the contribution is so small, relative to that made by plants, that only passing mention need be made of it here. (See Chapter VII for information concerning the quantity of organic matter supplied by the animal population.) Almost all virgin soils, where there is sufficient rainfall, are kept covered with a mat of dead grass or leaves. Under grass vegetation the entire top growth is left annually on the soil surface to decompose while much of the root system decays within the soil. In forested areas a certain amount of leaf litter accumulates on

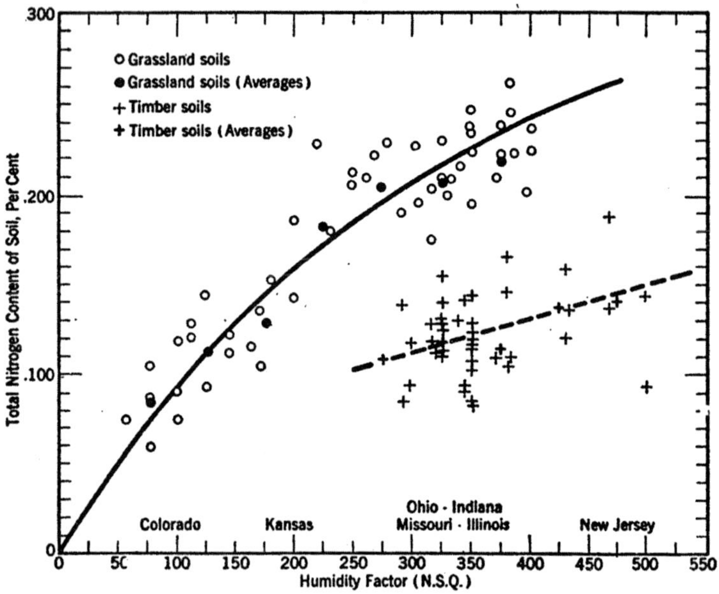

FIG. 43. Soil nitrogen—humidity factor relationship along the annual isotherm of 11° C. [Mo. Res. Bul. 152.] It is observed that with an increase in humidity, with all other factors constant, including the mean annual temperature, there is an increase in content of soil nitrogen in both the grassland and timber soils. N.S.Q. is the ratio of precipitation to the absolute saturation deficit of the air. These values are used instead of precipitation values as such since rainfall alone is not a satisfactory index of soil-moisture conditions because of the great variations in evaporation. The N. S. quotients include therefore the effect not only of temperature but also of air humidity on evaporation.

the surface of the soil each year together with fallen twigs, branches, and some logs. A large part of the year's growth, however, does not return to the soil. A small part of the forest litter is carried into the soil by worms and insects, and some of the more highly decomposed portions are carried downward in suspension in water.

The conditions that favor the accumulation of organic matter in soils are, in general, the reverse of those favoring its decomposition. The accumulation of organic matter represents the difference between the addition through plant growth on the one hand and the decay activities of microorganisms on the other. Under natural soil conditions, an equilibrium exists between the formation of organic matter and its decomposition. The balance is determined largely by soil and climatic conditions. Cultivation disturbs this natural equilibrium because less organic matter is usually returned to the soil and decomposition processes are speeded up by the farming operations. The consideration of the accumulation of organic matter may be taken up in answer to the following questions.

Questions:

1. What soil factors are conducive to the accumulation of organic matter?
2. Is it feasible to attempt to maintain a high organic content in soils?
3. What fundamental principles are involved in the accumulation of soil organic matter?
4. What materials can be used to add to the soil organic supply?
5. How much organic matter do mineral soils contain and how is it distributed throughout the profile?

Soil Conditions Favoring the Accumulation of Organic Matter. The quantity of organic matter in soils and its rate of accumulation are determined by numerous factors, the most important of which are (1) the nature and amount of organic materials returned, (2) soil moisture, (3) soil temperature, (4) the degree of soil aeration (which is determined largely by the soil texture, tillage practices, and drainage conditions), (5) topography, and (6) the nature and supply of plant nutrients. All these factors except the first may be classed as soil factors. Of course, soil moisture and soil temperature are governed largely by the corresponding climatic factors.

Organic matter decay is essentially a burning process (carried out by microorganisms) and, like any burning process, the greater the supply of air, the more rapid will be the burning and the more completely will the organic matter disappear. This explains why cultivation tends to increase the rate of organic matter decomposition in the top soil and emphasizes the importance of a minimum of cultivation on sandy soils.

Soil moisture and soil temperature have an important influence on organic matter accumulation. With ample moisture and high temperatures organic matter disappears very rapidly, especially from light-textured soils. With lower temperatures and heavier soils the disappearance is less rapid. Poor soil drainage aids in the accumulation of organic matter, but it is to be remembered that the activity of the organic matter is perhaps of more importance than its mere presence.

In general, organic matter accumulation is greater in comparatively level soils than in soils of rougher topography, particularly if water erosion is in evidence. Since most of the organic matter is found in the upper soil horizons, any water erosion that may occur will result in significant organic matter losses. Frequently soils of level topography exhibit poor profile drainage or have a high water table, either of which conditions may aid in the accumulation of organic matter.

The humus level in mineral soils is very closely associated with the supply of the nutrient elements such as calcium, nitrogen, phosphorus, and potassium. Many other elements, of course, are needed by plants, but these are the ones that commonly need to be considered in agricultural practices. Since plant growth is the source of nearly all the soil humus, the quantity of plant growth and hence the rate of humus formation will depend to a large extent on the supply of available plant nutrients in the soil. For the maximum accumulation of organic matter, the soil must be well supplied with the essential plant nutrient elements. Crops used for adding organic matter to the soil will not only make more top growth but will also make denser and more extensive root development. Many soils are too low in fertility to produce either a good sod, cover, or green-manure crop. With such soil conditions, it is utterly impossible to build up the humus content by attempting to grow soil-conserving or soil-improving crops. If the organic matter is to be restored in soils that have lost their organic matter and their soluble minerals as well, provision must be made for restoring these mineral deficiencies before attempting to grow crops for the sake of adding organic matter.

The Feasibility of Attempting To Maintain a High Soil Organic Content. The maintenance of an adequate supply of organic matter in soil is rendered more difficult because of the large quantity that is dissipated each year. Since the rate of organic matter loss from soils increases rapidly as the organic content is raised, the maintenance of a high organic level is not only difficult but also expensive. Therefore, it is wise to maintain the organic matter at a level that will result in good crop yields. Obviously, this level will vary with the type of soil, with the rotation followed, and especially with climatic conditions. It is unwise to attempt to maintain as high an organic level in soils of the southern states as in the soils of the northern states. The higher temperatures in the South speed up decomposition processes and almost prohibit the maintenance of a high organic level. Under conditions which prevail over most of the eastern half of the United States, the most feasible practice is to make frequent additions of fresh organic materials with no special attempt to maintain a high organic matter level.

Fundamental Principles Involved in the Accumulation of Soil Organic Matter. The best practices for maintaining or increasing the organic supply in soils will vary from one section of the country to another or from one farm to another, and will frequently vary from one field to another on any particular farm. It is logical, then, that this problem be discussed from the standpoint of the fundamental principles involved, and certain of them may be applied as they pertain to the particular conditions and circumstances concerned.

The accumulation of organic matter in soils is largely a nitrogen problem. Theoretically, there can be no increase in effective soil organic matter without first a proportionate increase in soil nitrogen. This implies that there is a very constant and close relationship between the nitrogen and organic matter contents in soils. This close relationship does actually exist (p. 252). Since the ratio of C to N in humus is roughly 10 or 12 : 1, it must be concluded that neither carbon nor nitrogen and hence soil organic matter can be permanently or appreciably increased or decreased without a corresponding change in the other. Permanent soil productivity requires practices that maintain the soil nitrogen, and such practices will then also automatically maintain the supply of soil organic matter, provided the mineral constituents are present in adequate amounts.

The quantity of humus produced in soil from added organic materials is governed largely by the amount of nitrogen present in the soil or in the organic materials added. In general, the increases in humus will be only as large as the quantity of nitrogen will permit; this explains why many tons of certain organic materials, low in nitrogen content, are needed per acre to produce any appreciable increase in the amount of soil humus.

If the nitrogen content of plant residues is low, added nitrogen will be required to meet the demands of the soil organisms which produce the soil humus. We must therefore come to the conclusion that the accumulation or the restoration of soil organic matter is a problem of utilizing nitrogen as a means of holding carbon and other materials that constitute humus. Since nitrogen is an essential constituent of humus, the value of organic materials added to soils as a source of humus is in a general way directly proportional to their content of nitrogen. The carbon-nitrogen ratio of several organic materials is presented in Table 22.

TABLE 22

THE CARBON-NITROGEN RATIO OF SOME ORGANIC MATERIALS *

Material	C : N Ratio
Sweet clover (young)	12
Barnyard manure (rotted)	20
Clover residues	23
Green rye	36
Cane trash	50
Corn stover	60
Straw	80
Timothy	80
Sawdust	400

* Data taken from several sources. The values are approximate only, and the ratio in any particular material may vary considerably from the values given.

Materials Used To Increase the Supply of Soil Organic Matter. The conservation and restoration of humus is a problem of soil management that must include, first, methods which eliminate or reduce to a minimum needless losses of humus and, second, methods that provide more or less regular additions to the soil's supply of organic matter. We are here concerned with the latter problem.

Green Plants. They include the various plants used as cover or green-manure crops. These crops aid in conserving soil humus by virtue of the fact that (1) they decrease erosion losses; (2) they conserve plant-food elements by decreasing leaching losses; (3) when turned under, they increase the supply of soil organic matter; and (4) with leguminous crops, the total supply of nitrogen is increased, provided the legumes are well inoculated with effective strains of bacteria. The ability of leguminous crops to increase the nitrogen content of soils is one of the primary reasons why these crops generally are considered more efficient as sources of humus material. The difference between legumes and non-legumes for this purpose is one not only of nitrogen-gathering ability but also of a narrower C : N ratio in the leguminous materials.

Green plants are an excellent source of active organic matter; that is, they decompose rapidly. This active decomposition is accompanied by a rapid release in available forms of nitrogen and of the mineral elements contained in them.

Crop Residues. Leaves, plant stubble, plant roots, and mature plant materials such as straw and corn stalks are included in this group.

They decompose and form humus slowly. No nitrogen or mineral elements are released until after a considerable period of time has elapsed. These materials have a relatively wide C : N ratio, and under many circumstances it is advisable to supplement these materials with nitrogen fertilizer to hasten decomposition. If this is not done, the microorganisms may rob the soil of available nitrogen and thus offer serious competition with the growing plants for available nitrogen. This temporarily harmful effect can be overcome by incorporating these materials into the soil well in advance of the time a crop is planted.

Farm Manures. The importance of farm manures as a source of humus and plant nutrients is well recognized. Farm manures are ideal sources of humus. A large portion of the nitrogen and a part of the minerals are liberated rapidly in forms available for plant growth, and a rather large amount of humus is formed. On the average, a ton of farm manure contains about 500 pounds of dry matter and about 25 pounds of plant nutrients (10 pounds N, 5 pounds P_2O_5, and 10 pounds K_2O).

Peat, Muck, and Forest Litter. These materials (with the exception of fresh forest litter and raw peat) have already undergone extensive decomposition. They are largely in the humus state, and when added to soils they decompose very slowly. These materials must be considered only as sources of humus and not as fertilizers. They have a very low content of phosphorus and potassium, and the nitrogen is only slowly available. From the general farming point of view, these materials have limited use. They are frequently used in the preparation of lawns, golf courses, and greenhouse soils.

Organic Fertilizers. This group of materials includes all fertilizers of plant or animal origin, such as dried blood, tankages, cottonseed meal, and sewage sludge. Generally speaking, they decompose rapidly and leave a rather small quantity of humus. The nitrogen in most of these materials is made available rather rapidly. Only in special cases can these materials be considered important sources of humus because the quantity usually added to soils is so small.

Artificial Manures. Composts of plant residues are excellent sources of humus if they have been well prepared and if sufficient nitrogen, phosphorus, and lime have been used to bring about rapid decomposition. When compost materials are added to soils, they produce or leave considerable humus in the soil and some of the nutrients are readily liberated in a form available for plant growth.

Artificial manures are produced by composting plant residues, such as leaves, straw, corn stover, grass, and weeds, under favorable conditions of moisture, temperature, and aeration. These residues are low in nitrogen, phosphorus, and lime, which are necessary for rapid decomposition of the organic materials by microorganisms. These three elements (N, P, and Ca), in readily available substances, must be added to the compost. Several formulas have been suggested for chemical mixtures that are satisfactory for composting purposes. The following formula has been used with considerable success:

	(By Weight)
Ammonium sulfate	45 parts
Superphosphate (20%)	15 parts
Limestone (finely ground)	40 parts

This mixture is used at the rate of 150 pounds per ton of dry straw. A ready-mixed fertilizer such as a 10–6–4, in addition to lime, can be substituted for the chemical mixture given above. From 100 to 125 pounds of the 10–6–4 fertilizer and 50 or 60 pounds of finely ground limestone should be used per ton of straw. As the compost pile is built up, the fertilizer can be mixed with each layer of organic material. The time required to produce a good grade of manure is determined largely by the conditions of temperature, moisture, and aeration in the compost pile.

The Quantity and Distribution of Organic Matter in Representative Soils. Although highly important in soils, organic matter makes up only a very small fraction of the total weight of mineral soils. The actual amount of humus found in any soil depends upon texture of surface and subsoil horizons, topography, drainage, climatic factors, native vegetation, and treatment. The quantity ranges from a fraction of 1 per cent for the very poor sandy soils and desert soils to 12 or more per cent for prairie soils. The humus content is much higher in organic soils. Generally speaking, lighter soils (sands and sandy loams) are lower in organic matter than the heavier soils (clays, clay loams, etc.) because the sandy soils did not originally support as dense a vegetative growth, because they have better aeration which promotes more rapid decomposition, and because they are more subject to leaching losses.

In humid-region soils the nitrogen is carried or held almost entirely by the organic matter and, as the content of organic matter varies, so does the nitrogen. The nitrogen content of soils in the United States, according to the results of analyses of a large number of samples carefully collected from selected areas under natural conditions, is given in Table 23. These samples are from six natural soil groups or regions. Each region contains soils that are extremely high and soils that are extremely low in nitrogen, and obviously such figures can only approximate an average content, yet they are of interest and they present the best available picture of the quantity of nitrogen in the soils of the United States.

TABLE 23

AVERAGE NITROGEN CONTENT IN VARIOUS SOIL REGIONS OF THE UNITED STATES *

Soil Region	Percentage Nitrogen		Pounds of N/Acre to Depth of 40 Inches
	Surface 6 inches	Average to depth of 40 inches	
Brown forest	0.05–0.20	0.05	6,700
Red and yellow	.05– .15	.03	4,000
Prairie	.10– .25	.12	16,000
Chernozem	.15– .30	.12	16,000
Chestnut	.10– .20	.08	10,700
Brown	.10– .15	.06	8,000

* "Soil Nitrogen," O. Schreiner and B. E. Brown, 1938 Yearbook, U.S.D.A., p. 366.

It is interesting to note that the nitrogen supply is highest in the soils of the prairie and chernozem regions and that the quantity declines both east and west of these groups. The approximate quantities of organic matter may be obtained for the different soil regions by multiplying the values in Table 23 by the factor 20.

The data presented illustrate, furthermore, the fact that the humus is usually concentrated in the upper soil layer and diminishes rapidly in the subsoil, although it should be noted that in the prairie and chernozem soils the organic matter extends to a much greater depth than in the timbered soils. The general distribution of organic matter in representatives of several of the great soil groups is shown in Fig. 44. Since most of the organic matter is found in the top soil, any erosion that occurs will result in significant organic matter losses. Furthermore, it is the topsoil that is disturbed by various tillage operations which greatly increase the loss in organic matter and nitrogen.

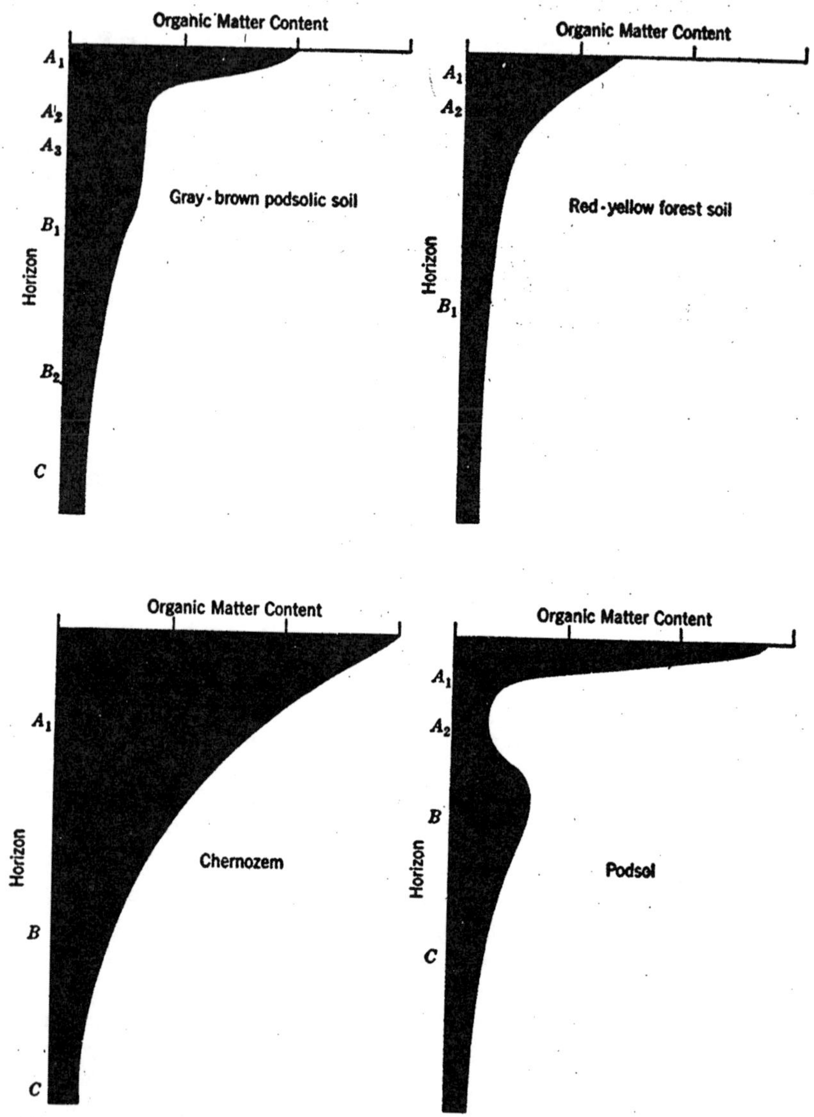

FIG. 44. The distribution of organic matter in the profiles of heavy soils from four of the great soil groups (generalized presentation).

EFFECTS OF ORGANIC MATTER ON SOIL PRODUCTIVITY

All organic materials after they have been partially digested by the numerous soil organisms are gradually changed into a dark-colored, structureless mass having colloidal properties and called humus. The humus exerts a most important influence on soil productivity, and its value can be fully appreciated only after the various physical, chemical, and biological effects produced by it have been considered. Some questions which suggest themselves in considering the effects of organic matter on soil productivity follow.

Questions:

1. What effects does organic matter have on the physical properties of a soil and how are they related to soil productivity?
2. Does organic matter perform chemical functions in the soil of particular significance?
3. From the standpoint of soil productivity, what biological processes are favorably influenced by the presence and additions of organic matter?

Physical Effects of Organic Matter in Soils. Organic matter is of extreme importance in regard to crop growing because of the many useful effects it produces in soils, and many of these effects are purely physical. Organic matter increases the water-retaining power of soils, decreases water-runoff losses, improves aeration especially on the heavier soils, and produces a better soil structure or tilth. Owing to the fact that organic matter improves the tilth of soils by promoting granulation and to the fact that organic matter decreases water-runoff losses, the damage done by either water or wind erosion is greatly reduced. That portion of the soil organic matter which has undergone considerable decomposition assumes colloidal properties for the most part and as such has a very high absorptive capacity for water. On the dry-weight basis, humus has a water-holding capacity of several hundred percent. The humus may be considered in this case to act as a sponge, and it becomes obvious why it has the ability to hold several times its weight of water. Soil water is also retained in the small pores or air spaces between soil particles. In the more sandy soils these spaces frequently are too large for maximum water retention, and humus tends partially to fill these large spaces and make them a more effective size for holding water. The humus also tends to pull the sand particles together, thereby increasing water retention. In heavy clay soils the pore spaces between the mineral particles frequently are too small for the greatest moisture storage. Organic matter improves this condition by forcing the soil particles apart, thus increasing the ability of those soils to retain water. The water-holding power of all mineral soils can be increased by raising the humus level.

Soil tilth refers to the physical properties of a soil as related to crop growth. The term tilth is used frequently to denote the structure of a soil. A soil in good tilth is easily cultivated, is loose and mellow, and is characterized by being well granulated. The degree of granulation in all mineral soils is determined to a large extent by the quantity of soil organic matter. Sandy soils are not well granulated because they lack sufficient binding material to hold the particles together. Organic matter, added to these soils, tends to promote granulation by binding the sand particles into clusters, thereby improving the tilth of these soils. Heavy clay soils, low in organic matter, often are in poor tilth because of clodding. The thorough mixing of organic matter in these soils promotes granulation, and clodding may be greatly decreased. A heavy clay soil that contains ample organic matter is usually mellow, friable, and easily tilled. Organic matter makes heavy soils less sticky, enables them to receive, distribute, and hold water more effectively, and makes them easier for roots to penetrate.

The degree of aeration in heavy soils is determined to a large extent by the structure or tilth of the soil, and so indirectly aeration is improved by the presence of organic matter. Sandy soils frequently are excessively aerated, and the addition of organic matter to these soils tends to diminish excessive ventilation.

The dark color of soils is due largely to organic matter. It is known that dark-colored surfaces are better absorbers of radiant energy of the sun than light-colored surfaces; therefore, dark-colored soils would be expected to warm up faster than light-colored soils, which are similar in all other respects, including the moisture content. This would lead one to expect dark soils, with all other things being equal, to have a somewhat higher average temperature, which would tend to promote a more rapid seed germination. It is obvious that under certain circumstances, as in early market gardening, a hastening of seed germination and a more rapid early growth of plants may be highly desirable. Attention, however, should be called to the fact that, if the higher organic content is associated with a higher moisture content, the effect of the dark color on soil temperature may be more than offset by the increased quantity of heat required to warm up the additional water.

One of the most important physical functions of organic matter in soils is its assistance in controlling both wind and water erosion. A soil containing ample organic matter is usually in good tilth, it will retain more water, water-runoff losses will be lessened, and the net result will be less destruction of the soil by water erosion. Sandy soils often are subject to wind erosion, and this can be controlled to some extent by increasing the soil organic content. Organic matter serves to bind sandy soils and lessens their movement by wind.

Chemical Functions of Organic Matter in Soils. Organic matter improves the soil chemically by serving as a storehouse or supply of plant-nutrient elements. As the organic materials are decomposed, the plant-food elements contained therein are gradually released. Most of the soil's nitrogen supply exists in organic form. Nitrogen cannot be held in the soils in humid regions to any great extent except as it is held in organic combination. This organic nitrogen gradually undergoes conversion into nitrate under normal soil conditions, and in the absence of growing plants most of the nitrates may leach out. Decomposition of organic matter favors the release of plant-food elements from the soil minerals. Various organic and inorganic acids are produced in soils when organic matter decays (p. 96), and they have a very pronounced dissolving effect on soil minerals. One of the important end products of organic matter decay is carbon dioxide gas. This gas dissolves in the soil water and forms carbonic acid, which is an effective dissolving agent for the soil minerals; the dissolving effect of this carbonated water is several times that of pure water.

After the organic matter has undergone considerable decay, it largely assumes the colloidal state. The colloidal properties thus exhibited exert important physical and chemical effects which are directly concerned with soil productivity. The organic colloidal materials have a much greater base-exchange capacity per unit weight than do the mineral colloidal materials, and hence they may act as buffers in the soil, thereby retarding the processes by which changes in soil reaction (acidity or alkalinity) are produced. Furthermore, these colloidal substances have a strong ability to adsorb or hold on to the constituents of fertilizers and nutrients released from soil minerals, thus decreasing their rate of loss by leaching.

Biological Effects of Organic Matter in Soils. Organic matter improves the soil for the growth of microorganisms which, after all, are the agents whereby the plant-food elements of the soil are kept in circulation. It serves as a source of food and energy for the majority of the soil microorganisms.

The soil may be considered a factory operating to produce plant nutrients. Soil microorganisms may be considered the power or driving force in this factory, and the soil organic matter the fuel or energy for this power. The organic matter is burned to carbon dioxide, water, ash, and various other products, the nature of which is determined largely by the degree of soil aeration. That the organic constituents constantly are being burned in the soil is shown by the continuous evolution or output of carbon dioxide.

The complex constituents of organic matter, upon being decomposed, are simplified, and nitrogen in the ammonia form is released and changed to the nitrate form. The energy stored in compounds of growing plants for the most part eventually is either used or released by soil microorganisms whose activities within the soil make food elements available for a new generation of crop plants. Thus it can be said that, without the presence of organic matter to supply food and energy for the soil microbes, the plant-food elements of the soil could not be changed to usable forms.

Only the more important physical, chemical, and biological effects of organic matter in soils have been briefly discussed, but enough has been said to indicate definitely that organic matter exerts a controlling influence upon the soil and its productivity. The presence of organic matter in soils is of value, but the activity provoked by organic matter decomposition is probably of more concern. Attention, therefore, should be given to the factors affecting the rate of organic matter decomposition and to the nature of the products formed.

THE DECOMPOSITION OF ORGANIC MATTER AND HUMUS FORMATION

As has already been pointed out, soil organic matter has its origin for the most part in plants. Its accumulation in any soil at any particular time represents the difference between the quantity of plant residues and the decay activities of microorganisms. This organic material, regardless of its origin, is found in soils in all stages of decomposition. It may vary from the fresh material to that which has undergone extensive chemical changes; it may occur as leaf or vegetable mold, peat, muck, or humus, depending on the nature and extent of changes it has undergone.

The decomposition of organic matter is a biochemical process and is brought about primarily by microorganisms, the most important of which in most soils are bacteria, fungi, and actinomyces.

The composition of organic matter in soils, whether fresh or in various stages of decomposition, is extremely variable; it is a complex mixture of numerous compounds, the kinds and quantities of which constantly vary. Some pertinent questions relative to the composition and decomposition of organic matter in soils are listed below, and without going into too great detail a few fundamental facts regarding these points will be given.

Questions:

1. What is the chemical nature of plant residues commonly added to soils?
2. What factors largely determine the rate of organic matter decomposition?
3. Are all soil organisms concerned in the decomposition of organic matter?
4. How is humus formed?
5. What are the outstanding characteristics and properties of soil humus?
6. In connection with the decomposition of organic matter, what changes are brought about in the nitrogen compounds?
7. What happens to the mineral constituents in the organic matter when it decomposes?
8. Do organic compounds serve directly as plant food?

Composition of Plant Materials Commonly Added to Soils. Numerous substances synthesized by higher plants are continuously added in the form of plant roots, stubble, stems, leaves, twigs, etc. Various residues are added to the soil in the form of manures and as by-products of industry, such as dried blood, meat scrap, bonemeal, and cottonseed and linseed meals. There are large numbers of small animals which find the soil either a temporary or permanent habitat, and there is a considerable quantity of microbial cell substances being continuously added to the soil. Thus, we may find in the soil any chemical compounds contained in higher plants, microorganisms, or animals. Needless to say, the composition of fresh organic materials added to soils is of extreme variability; this would be true even if the plant materials alone were considered. A knowledge of the chemical composition of plant residues undergoing decomposition is important in aiding us to understand the nature and rapidity of the liberation of the nutrient elements in forms available for plant growth and of the formation and nature of the residual organic matter.

Green plant tissue contains on the average about 75 per cent water. The dry matter, making up about 25 per cent of the plant, is composed mostly of carbon, hydrogen, and oxygen. Over 90 per cent of the dry-matter weight is furnished by these three elements, the remainder consists of nitrogen, sulphur, calcium, phosphorus, potassium, and other elements. Although the elements other than carbon, hydrogen, and oxygen make up only a small portion of the plant, they play an important role in plant nutrition as the plant residues are decomposed in the soil.

As a matter of convenience, the chemical compounds may be grouped into three general classes: (1) nitrogen-free organic compounds, (2) nitrogen-containing compounds, and (3) inorganic or mineral constituents. The compounds of the first group are composed of carbon, hydrogen, and oxygen and include the carbohydrates such as sugars and starches, the celluloses (lignocellulose or lignin), fats or oils and waxes, and organic acids. The nitrogen-containing compounds are mostly proteins, and in addition to carbon, hydrogen, oxygen, and nitrogen they may contain sulphur, phosphorus, iron, and other mineral elements. The inorganic or mineral constituents consist primarily of compounds of phosphorus, calcium, magnesium, potassium, silicon, sulphur, aluminum, iron, manganese, etc. More than 30 different elements have been detected in the chemical analyses of plants. Some of the mineral elements may exist in the plant as combinations with organic compounds; some may occur as organic salts and others as inorganic salts.

Factors Affecting the Rate of Organic Matter Decomposition. Since the decomposition of organic matter in the soil is a biochemical process, any factor that affects the activities of the soil organisms will necessarily affect the rate of organic matter decay. Several influential factors which have a bearing on the rate of organic matter decomposition may be placed in the three following groups: (1) factors concerned with the nature of the plant material (including such points as the kind of plant, age of plant, and chemical composition); (2) soil factors (including aeration, temperature, moisture, acidity, and fertility level); and (3) climatic factors (the effects of moisture and temperature are particularly influential).

In general, the younger the plant the more rapid will be its rate of decomposition when incorporated into the soil. This is largely due to the higher content of water-soluble constituents, a higher nitrogen content, a more narrow carbon-nitrogen ratio, and a smaller percentage of lignin and other constituents resistant to decomposition. Of the various plant and animal residues used to increase the organic content of the soil, as a rule those having the higher nitrogen content decompose most rapidly.

The influence of several soil factors on the rate of microbiological activities and on the accompanying rate of organic matter decomposition is discussed in considerable detail elsewhere (p. 190). Soil conditions which are favorable for the growth of the common crop plants are also favorable for organic matter decomposition. One of the most influential factors in this respect is the degree of soil aeration. Since organic matter decay is largely a burning or oxidation process, the greater the degree of aeration the greater in general will be the rate of decay. Soil moisture and especially soil temperature are controlled largely by climatic influences. With other conditions favorable, the rate of soil organic matter decay increases with increasing soil moisture to a point where aeration is inadequate. The rate of decay increases with increasing temperature to a certain limit, assuming all other conditions constant and favorable.

Organisms Concerned in the Decomposition of Soil Organic Matter. When plant and animal residues are added to the soil under favorable conditions, they are subject to various changes involving chemical agencies and the activities of the various groups of soil organisms. The animal population aids in the breaking down of plant residues; it brings about a reduction in quantity and a change in the chemical composition of the residues. Furthermore, it causes a mechanical mixing of the organic with the inorganic soil constituents. It is the fungi, bacteria, actinomyces, and protozoa, however, which usually accomplish the most pronounced changes in the organic residues. All the organisms active in the soil are associated in some way, either directly or indirectly, with the decomposition of organic matter, for it is on this material that most of them depend for food and energy.

The organic residues are not decomposed as a whole, but the chemical constituents may be attacked independently of one another. A particular group of organisms is usually associated with the active decomposition of one special group of compounds; some organisms perform definite or specific activities. Attack by soil organisms on organic materials seldom results directly in complete decomposition; various intermediate products usually are formed. Other members of the soil population which have the ability decompose the intermediate products. Sooner or later the original material loses its identity and is transformed to a variety of end products and *synthesized substances.* Soil organisms differ greatly in their food requirements; in other words, what may be a source of readily available food for one group may not be available to others. This makes possible the decomposition of whatever organic material may be added or formed in the soil.

For additional information on the soil organisms concerned in the decomposition of soil organic matter, see Chapter VII.

Humus Formation. Humus denotes the soil organic matter which has undergone extensive decomposition. It is not a homogeneous compound; it has no definite chemical composition. It is a dark-colored, heterogeneous mass, consisting of the residues of plant and animal materials together with the synthesized cell substances of soil organisms. Humus is not static but dynamic in soils; it is continually undergoing change.

It has already been pointed out that, during the decomposition of plant and animal residues in soils, some organic constituents are more readily attacked than others and that some are extremely resistant to decomposition. The starches, sugars, proteins, and amino acids are rapidly attacked by a great variety of organisms and, associated with these changes, is a considerable synthesis of microbial cell substance. The celluloses and especially the hemicelluloses are decomposed rapidly by a rather large variety of microorganisms. Of the various plant constituents that exist in considerable quantity, the lignins are particularly resistant to decomposition under anaerobic conditions, but under aerobic conditions they undergo some change although not so extensively as the cellulose and hemicelluloses.

Thus in the formation of humus from plant residues there is a rapid reduction of the water-soluble constituents, of the celluloses, and of the hemicelluloses; a relative increase in the percentage of lignin and lignin complexes; and an increase in the protein content unless the plant residues are unusually high in protein. The new protein is believed to be formed for the most part through the synthesizing activities of the microorganisms. The C : N ratio in fresh plant materials that are commonly added to soils is usually from about 80 : 1 in mature straw to 12 to 20 : 1 in leguminous green manure crops (Table 22). In any case, the ratio will narrow or drop to about 10 : 1 in a relatively short time after these materials have been incorporated into soils. In other words, the C : N ratio of humus in mineral soils is roughly 10 : 1. This means that for each pound of nitrogen in the soil there exists about 10 pounds of carbon.

The increase in the relative amount of nitrogen (as shown by the more narrow C : N ratio) in the organic residues following decomposition can only be explained by the fact that the nitrogenous complexes are rendered resistant to further rapid decomposition. This indicates, therefore, that either the organic nitrogenous complexes in humus are not of a true protein nature or they are not in a free state, otherwise they would decompose readily. This slow rate of decomposition of humus is obviously of considerable practical importance. It offers a means whereby nitrogen can be stored in the soil and released gradually.

Humus can be considered as consisting largely of two chemical complexes—lignin and protein. Perhaps these two groups of substances combine, forming a lignin-protein complex. Lignin makes up 40 to 45 per cent of the total humus, 30 to 35 per cent of the humus consists of protein, and the remainder is composed largely of fats, waxes, and other residual materials. The lignin-protein combination renders the protein resistant to microbial attack. It is thus observed that humus has a high protein content; yet the nitrogen is only slowly available for the growth of higher plants. The lignin in humus originates largely from plant residues with perhaps certain chemical modifications, whereas the proteins in humus are largely synthesized through the activities of microorganisms. This combination of lignins and proteins has been called a ligno-proteinate complex. Although humus is considered to be organic, it probably contains various inorganic elements which are an integral part of the complex. Phosphorus, sulphur, calcium, magnesium, potassium, iron, aluminum, and possibly other elements may be bound chemically with the humus. In acid soils the humus complex is likely to contain large quantities of hydrogen, iron, and aluminum; whereas in soils close to the neutral point the humus usually is nearly saturated with calcium and magnesium, and in strongly alkaline soils it may contain considerable sodium.

Since the lignin and protein account for 70 to 80 per cent of the total humus, the formation of a lignin-protein complex suggests evidence to explain the more or less constant C : N ratio which exists in mineral soils. These complexes, however, are not absolutely resistant to decomposition; therefore the nitrogen content of humus may not always be constant. Furthermore, the chemical nature of humus is determined to a certain extent by the chemical composition of the plant residues and by soil and climatic factors.

A general schematic summary of the processes leading to humus formation is presented in Chart 7. The material commonly referred to as humus includes the mass of plant residues undergoing decomposition, together with the synthesized cell substance and certain intermediary and end products. It is constantly changing in composition. It is better, therefore, to speak of humus not as a single group of substances but rather as a state of matter, which is different under varying conditions of formation.

Characteristics and Properties of Humus. In the preceding section, in discussing the formation of humus in mineral soils, it was pointed out that some plant constituents become completely decomposed, others are modified to a greater or less extent, and others are only slightly attacked, and that these processes are accompanied by the synthesis of new complexes (fats, carbohydrates, and proteins) by the microorganisms. It is in order next to point out that humus as a whole is characterized by certain specific physical, chemical, and biological properties that distinguish it from other forms of organic matter in nature.

Humus is practically insoluble in water although some of it may go into colloidal solution in pure water. To a large extent it is soluble in dilute alkali, and certain of the humus constituents may dissolve in acid solutions.

One of the most important and characteristic properties of humus is its nitrogen content, which usually varies from 3 to 6 per cent, although the nitrogen concentration may be frequently lower or higher than these figures. The carbon content is usually 55 to 58 per cent. According to Waksman,[1] the average theoretical carbon content of soil humus is 56.24 per cent, and the average nitrogen content is 5.6 per cent. This gives a theoretical ratio of carbon to nitrogen of 56.24 : 5.6, or 10.04, which is very close to the one commonly found in the humus of soils. This ratio varies with the nature of the humus, the stage of its decomposition, the nature and depth of the soil, and climatic and other environmental conditions under which it is formed.

Chart 7. Schematic Representation of the Mechanism of the Formation of Humus in the Decomposition of Plant Residues in Soil

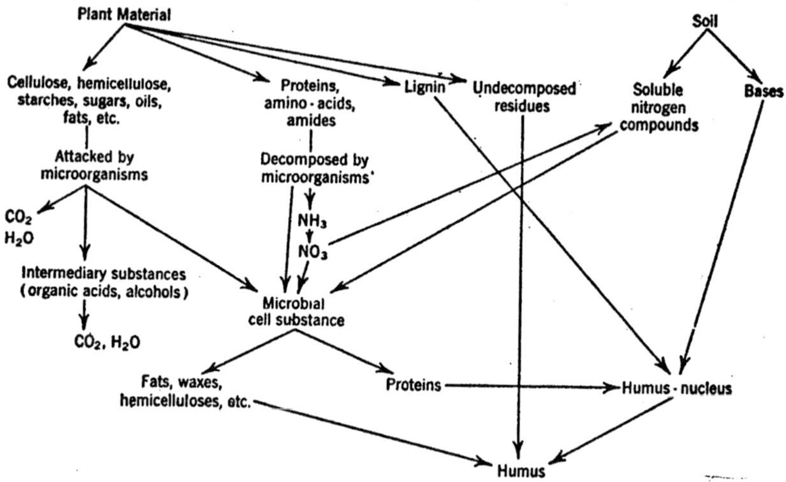

[After Waksman. From "Humus." By permission of the William and Wilkins Co.]

Humus is usually in a highly dynamic condition, for it is constantly being formed from plant and animal residues and is continuously being decomposed by microorganisms. It serves as a source of energy for

[1] "Humus," S. A. Waksman, 2d. Edition, 1938, p. 182. Williams and Wilkins Co.

various groups of microorganisms and with decomposition gives off carbon dioxide, ammonia, and other simple end products of decay.

Humus is highly colloidal and, as clay, it functions as an acidoid or micelle and usually carries a large number of negative charges. It is composed principally of C, H, O, N, S, and P, in contrast to the mineral acidoids which are composed chiefly of Si, O, Al, and Fe.

Humus is characterized by a high base-exchange capacity; it combines with various inorganic soil constituents; it absorbs large quantities of water and exhibits the properties of swelling and shrinking. It does not exhibit so pronounced properties of adhesion and cohesion as do the mineral colloids and is less stable because it is subject to microbial decomposition. It has already been shown that soil humus is an important factor in the control of aeration, water-holding capacity, and granulation of field soils. Humus possesses other physical and physicochemical properties which make it a highly valuable soil constituent.

Changes Brought about in the Nitrogen Compounds. Most of the nitrogen added to soils by the plowing under of sod, crop residues, stable and green manures, and organic nitrogenous fertilizers is in the form of proteins and their derivatives. These forms of nitrogen are relatively insoluble. Since most of the nitrogen in the soil occurs as a constituent of the organic matter, the rate of its depletion from soils is decreased. For the most part, these organic nitrogenous substances cannot be used directly by higher plants but first have to be broken down into more simple compounds. The breaking down of any organic constituent added to soils is carried out by the action of microorganisms, such as bacteria, fungi, and actinomyces, and is in general in the direction of the formation of simpler forms from more complex compounds.

Plant and animal proteins and their derivatives in the soil break down into acid amides and amino acids [2] of various kinds, which in turn may be readily broken down to carbon dioxide, ammonium compounds, and certain other simple end products.

The final nitrogen product of the breakdown of nitrogenous organic materials in soil is, chiefly, ammonia. This ammonia may be used by plants or microorganisms, reduced by certain bacteria, oxidized to nitrates, or leached out in the drainage waters. In the utilization of

[2] Protein decomposition probably proceeds as follows:

$$\text{Proteins} \rightarrow \text{proteoses} \rightarrow \text{peptones} \begin{cases} \rightarrow \text{acid amides} \\ \rightarrow \text{amino acids} \end{cases} \rightarrow CO_2,\ H_2O,\ NH_3,\ \text{etc.}$$

these simple inorganic nitrogenous compounds by soil organisms, complex nitrogen compounds are again formed which in turn must be broken down by other organisms before the nitrogen becomes available to higher plants.

Changes in the Mineral Constituents of Organic Material as It Decomposes. As the soil organic matter decays with the gradual formation of humus, simple mineral products begin to be released. Most of these simple end products are readily lost from the soil in the drainage waters, unless taken up by growing plants; and if an adequate supply for normal plant growth is to be maintained in the soil, frequent additions of fresh organic matter must be made.

The principal mineral elements contained in organic matter are sulphur, phosphorus, potassium, magnesium, and calcium. Some are held in organic combinations, as are sulphur and phosphorus, and others are present in fresh organic materials as inorganic compounds.

As the decay of organic matter progresses, sulphur is left behind as a by-product of the general-decay types in the form of elemental sulphur or hydrogen sulphide or of both. Some sulphur may be released as carbon disulphide. Elemental sulphur and hydrogen sulphide, upon oxidation, serve certain autotrophic bacteria as excellent sources of energy (p. 183). The biochemical oxidation of sulphur or hydrogen sulphide to sulphites and in turn to sulphates is called sulphofication. The oxidation of sulphites to sulphates occurs so rapidly that there is little chance for the accumulation of any great quantity of sulphite. It is thus observed that sulphur which was originally present in complex organic combinations is eventually released as a simple by-product and is made available to higher plants as the SO_4 ion. This is the form of sulphur generally used by plants, and large amounts are lost in the drainage waters.

The soil organic matter carries considerable quantities of phosphorus which are rather readily liberated as the organic matter undergoes decay. The phosphorus is probably released as the PO_4 ion, which in turn quickly reacts with other soil constituents and usually is readily fixed or held in the soil. As a result of this rapid fixation, which occurs in most soils, very little phosphorus is lost in the drainage waters. Since the quantity of inorganic phosphorus in soils is generally low and only slowly available, phosphorus added in organic matter is of great importance.

Other simple mineral products of organic matter decomposition that should be mentioned are potassium, calcium, and magnesium. They are released in ionic form and, under favorable conditions, react with water and in turn with CO_3^{--} and HCO_3^- ions. Thus a certain portion of these mineral elements is again made available for a new generation of plants. This explains how it is possible for deep-rooted crops such as alfalfa and sweet clover to bring up considerable quantities of mineral elements to the plow soil from the lower soil horizons, especially when these crops are used for green-manuring purposes.

In connection with the discussion of the changes in the mineral constituents accompanying organic matter decomposition, it should be recalled that considerable quantities of mineral elements, as they are released, may be assimilated by microorganisms. Thus there is always competition between microorganisms and higher plants for available plant nutrients, and the needs for the microorganisms must be satisfied first. However, under normal soil conditions the competition offered by the microorganisms for mineral elements is in general of no particular consequence. Where large amounts of carbonaceous materials are added to soils, a nitrate deficiency frequently occurs owing to the assimilation of the nitrogen by microorganisms active in the decaying of the organic matter.

Organic Compounds as Direct Sources of Plant Food. Over a hundred years ago the *humus theory* of plant nutrition was advanced by Thaer, of Germany. According to this theory, humus was absorbed directly and assimilated by higher plants. This theory was generally accepted until it was ridiculed by Liebig and replaced by his mineral theory. The mineral theory indicated the mineral elements to be the principal nourishment of plants, and Liebig believed that the main value of humus in soils was due to its release of ammonia into the atmosphere where the ammonia was absorbed directly by plants. The belief that organic matter is not used directly by higher plants has been held since Liebig's time until rather recently. It is now generally believed that some organic substances do serve as suitable nutrient material and are used directly by higher plants without undergoing decomposition. These organic compounds are absorbed in a complex molecular condition, but apparently only a very small portion of the soil organic matter is absorbed as such.

A field of soil and plant research that recently has received considerable impetus is concerned with stimulants to plant growth. It is believed that they develop within the soil through microbial action or are produced by the plant itself. Many investigators believe these stimulants (frequently referred to as auximones, hormones, phytamine, or growth-promoting or growth-regulating substances) function directly in the nutrition of plants in a manner analogous to that of vitamins in animal nutrition. These substances affect rates of cell growth, root formation, and flower formation of certain plants, and they also favor the development of some groups of microorganisms. It must be added that comparatively little is known concerning the formation or liberation in the soil of organic substances through the direct or indirect activities of microorganisms, which modify the growth of higher plants.

It must not be inferred that all organic substances released or produced in the soil are beneficial to higher plants. Under certain conditions several organic compounds are found in soils that are known to be harmful. It is generally believed that such substances are the result of improper soil conditions and are not the direct cause of decreased crop yields. When such unfavorable soil conditions are corrected, the organic toxins usually will disappear rapidly as a result of microbial attack. Rational soil-management practices, such as providing good drainage and tillage and supplying lime and fertilizers when needed, will usually prevent the accumulation of organic soil toxins.

LOSS AND RESTORATION OF SOIL ORGANIC MATTER

Throughout the ages man has associated high productivity of mineral soils with high content of organic matter. The quantity of organic matter in soils at the time they were brought under cultivation represented an accumulation over an extended period of years. The organic matter level attained was largely determined by the environmental factors concerned with both the soil and climate. As these lands were brought under cultivation, losses of organic matter began to take place at a rate far greater than its annual accumulation. These losses serve as direct measures of the losses in soil fertility. The losses and additions of soil organic matter may be considered in answering the following questions.

Questions:

1. What can be said in regard to the rate of loss of organic matter from virgin soils after they are brought under cultivation?
2. Is it possible to restore soil organic matter?

Rate of Loss of Soil Organic Matter. Even on non-erosive land that is brought under cultivation, rapid losses of organic matter usually occur. It has been observed that the losses are most rapid immediately after farming is started, and thereafter the rate of disappearance is decreased; ultimately the organic content of the soil will reach a comparatively fixed level. Under all comparable conditions, this rather stable organic matter level is much higher for good systems of soil management than for poor ones

It has been illustrated at the Missouri Agricultural Experiment Station [3] that, as a result of cultivation over a period of 60 years, soils in a non-eroded condition lost over one-third of their organic matter, the losses being much greater during the earlier than the later periods. The organic matter losses amounted to about 25 per cent the first

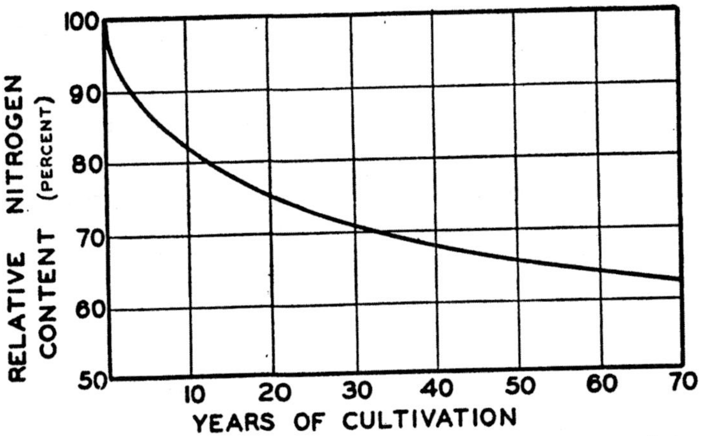

FIG. 45. Decline of soil fertility with length of cultivation period under average farming practices in the Middle West. [After Jenny, Mo. Agr. Exp. Sta. Bul. 324, 1933.]

20 years, about 10 per cent the second 20 years, and only about 7 per cent the third 20 years. In other words, a somewhat fixed equilibrium was attained after about 30 years (Fig. 45). In dry-land farming, the losses may be even greater.

That the organic matter of the soil which receives little or no added material becomes depleted is evident from the fact that there is a continuous evolution of carbon dioxide from soils under favorable moisture and temperature conditions. It has been estimated that a soil of average organic content under humid temperate conditions may lose 2.5 tons of organic matter per acre each year.

If a ton of fresh organic matter (dry weight), in the form of green manure, farm manure, or plant stubble, is plowed under or worked into the soil, decomposition begins immediately. Within a period of two or three weeks, under favorable conditions of moisture, tempera-

[3] "Soil Fertility Losses under Missouri Conditions," H. Jenny, Mo. Agr. Exp. Sta. Bul. 324, 1933.

ture, and aeration, only about one-half of the original ton will be left. After a month or six weeks have passed, only about 500 pounds of the original ton may remain. Of course, the quantity remaining will vary greatly from one set of conditions to another, but the point of emphasis is that there is a continuous dissipation of the soil organic matter.

The Maintenance and Restoration of Soil Organic Matter. Although there is a rapid depletion of the soil organic matter immediately following the bringing into cultivation of virgin lands, there is some consolation in the fact that this high rate will not continue indefinitely. It has been emphasized that after a period of heavy loss, a fairly constant level is attained during a long period of continued cultivation; the level is determined by the environment associated with a particular soil. Once the organic content has reached a low level, restoring the organic matter to its original level usually proves to be rather expensive and an extremely slow process, owing to the fact that the producing power of these soils has been permitted to decline to a degree that makes it difficult to grow green-manuring or soil-improving crops without special treatment or fertilization.

In other words, many soils are too low in fertility to produce a good sod, cover crop, or green-manure crop. With such soil conditions it is utterly impossible to build up the humus content unless material from the outside is brought in and that is usually impractical. If the organic matter is to be restored in the soils that have lost most of their organic matter and their soluble minerals as well, provision must be made to replace these mineral deficiencies before attempting to grow crops for the sake of adding organic matter.

The humus level in mineral soils is very closely associated with the supply of the nutrient elements, such as calcium, phosphorus, potassium, and nitrogen. Since plant residues are the source of nearly all the soil humus, the quantity of plant growth and hence the rate of humus formation will depend to a large extent upon the supply of available plant food in the soil. For maximum speed of organic matter restoration, the soil must be well supplied with the essential plant-food elements in order that greater crop residus will be produced.

The rate of organic matter loss increases rapidly as the organic content is raised and, as has been emphasized, its maintenance at a high level is not only difficult but also expensive. It is, therefore, usually unwise and uneconomical to maintain the organic matter above a level consistent with good crop yields. Attention, then, should be directed toward the frequent additions of small quantities of fresh organic materials rather than to practices of maintaining the organic matter content at any particularly high level.

Since it is the activity of the organic matter that gives it real value rather than the mere presence of it, the promotion of proper soil conditions in order that the organic matter may perform its normal functions is of great importance. Soil conditions should be the best. Drainage should be good in order to promote aeration and granulation. Lime should be added as needed, not only to give increased crop yields and permit the growing of legumes but also to increase the rate of organic matter decomposition by promoting biological activity and thus keep the organic matter cycle in more rapid progress.

In considering the maintenance and activity of soil organic matter, three phases demand special attention. First, make frequent additions of sufficient crop residues; second, supply enough nitrogen to permit a high organic matter level; and, third, provide for an adequate supply of calcium. Without doubt, one of the greatest limiting factors in crop production is the low humus supply or the low activity of the humus. The importance of nitrogen in this connection cannot be overemphasized. The organic matter content in soils can be maintained or restored and its activity promoted with good systems of soil management, involving the proper use of cover crops, green-manure crops, manure, fertilizer, lime, and by providing adequate drainage. The problem of regulating the humus supply of soils is only one, though a very important one, of the many phases of the scientific management of soils.

CHAPTER IX

COVER AND GREEN-MANURE CROPS

Green-manure crops are those which are turned under for the purpose of improving the soil. Usually such crops are turned under when they are still green, thus the name green manure. Occasionally these crops may be allowed to mature or even to lie over winter before being plowed under and still may be classed as green manures. A cover crop is one used to cover and protect the soil, particularly over the winter. A cover crop aids in preventing erosion and leaching, protects the soil from the impact of rain, shades the soil and aids in protecting the soil from excessive freezing and heaving. The term cover crop is usually applied only to crops planted for cover and soil-improvement purposes. Although the terms cover crop and green-manure crop are not identical, practically they are the same thing, and will be so considered in this discussion.

The use of green-manure crops is most general in the Atlantic and Gulf states. The most commonly used cover and green-manure crops in the southern states are Austrian winter peas and hairy vetch. In the potato-growing areas of the humid regions of the United States, cover and green-manure cropping is frequently practiced, primarily to increase the organic matter supply of the soil. Rye, red clover, crimson clover, and alfalfa are the crops most generally used. In the corn belt, sweet clover is the most popular green-manure crop before corn, the main object being to supply the corn crop with sufficient nitrogen throughout the growing season. Cover cropping in orchards is practiced in nearly all the fruit-growing sections of the United States.

It is generally conceded that there is ample opportunity for much improvement in soil conditions through the wider and more intensive use of cover and green-manure crops. In a discussion of this subject the following points are treated.

Objectives:

A. The effects of cover and green-manure crops on soil properties.

B. Principal cover and green-manure crops and their regional distribution.

C. Utilization of cover and green-manure crops.
D. Influence of turning under green manure on crop yields.

THE EFFECTS OF COVER AND GREEN-MANURE CROPS

As a whole the soils of the United States are not so fertile as they once were. The excessive production of cultivated crops, leaching losses, and the devastating effects of soil erosion on many farms have materially reduced the fertility of the soils. It is of great importance, therefore, that more attention be given to methods of maintaining and improving the productivity of our farm lands. Commercial fertilizers have an important place in this program, but many soils are badly in need of organic matter. Farm manure is an excellent source of organic matter, but in many instances it is not sufficiently abundant to go very far in soil improvement. The use of green manures will aid in supplying this need. Green-manure crops are used in the expectation that the yields of subsequent crops will be increased. In explaining how or why the turning under of crops has this effect, the following questions are answered.

Questions:

1. Can the organic matter and nitrogen content of soils be maintained or increased through the use of green-manure crops?
2. Do these crops aid in conserving soluble plant nutrients?
3. Does the use of green-manure crops tend to concentrate plant nutrients in the surface layer of soil?
4. Is the availability of plant nutrients in the soil affected by turning under green-manure crops?
5. How may green-manure crops improve the subsoil?
6. In what ways do green-manure crops protect the surface soil?
7. How are the activities of soil organisms influenced by turning under green crops?
8. What effect does green manuring have on soil acidity?
9. What are the effects of green manuring on the growth of fruit trees?

Addition of Organic Matter and Nitrogen. The amount of organic matter that may accumulate through the addition of green manures will vary greatly with the nature of the crop turned under, degree of soil aeration, and climatic conditions (Chapter VIII). A large part of the plant material may decay and disappear as carbon dioxide and water. In well-drained soils of hot climates this loss may be so rapid that no permanent increase in the soil organic matter is made even by turning under a heavy green-manure crop. Certainly no very large addition to the soil organic matter can be expected from turning under one or two green-manure crops. Because only a part of the plant materials added becomes a constituent of the soil humus on account of decomposition losses, it is obvious that a great many years would be

TABLE 24

THE AMOUNT OF NITROGEN TAKEN FROM THE AIR BY VARIOUS LEGUME CROPS (POUNDS PER ACRE PER YEAR) *

Name of Investigator	Place	Crop	Total Nitrogen (Pounds)
Arny and Thatcher	Minnesota	alfalfa	113 †
Arny and Thatcher	Minnesota	sweet clover	117
Whiting	Illinois	alfalfa (4 crops)	132
Fred and Graul	Wisconsin	alfalfa	64 †
Shutt	Canada	clover, red	75–150
Shutt	Canada	clover, red (average, 10 yr.)	51
Whiting	Illinois	clover, red (4 tons)	106
Graul and Fred	Wisconsin	clover, red	100
Duggar	Alabama	vetch, hairy	79
Whiting	Illinois	cowpea	86
Breal	France	bean	87
Fred	Wisconsin	soy bean (25 bu.)	57
Fred and Graul	Wisconsin	soy bean	108
Albrecht	Illinois	soy bean and cowpea	107

* Taken from Table 21, "Root Nodule Bacteria and Leguminous Plants," Fred, Baldwin, and McCoy, p. 218, 1932, University of Wisconsin Press.

† Average figure.

In several cases the high figures represent the gain in nitrogen from the growth of two or three crops.

required to raise appreciably the organic content of soils by green manuring. It is a very slow process. By adding 2,000 pounds of dry matter per acre each year in the form of green manure and assuming that it is all permanently retained, 10 years would be required to raise the organic matter content of the surface soil by 1 per cent. This rather crude illustration brings out the fact that the main object of green manuring in most instances must be to maintain rather than to increase the quantity of organic matter in soils and to supply continually organic matter in a readily decayable form.

Since legume crops have the ability to utilize atmospheric nitrogen if they are properly inoculated, they will increase the nitrogen content of the soil when used as green manure. The amount of nitrogen so added depends on the kind of legume (Table 24), soil conditions, thickness and height of the stand, and stage of growth at which the legume is turned under (Figs. 46 and 47). The quantity of nitrogen in a legume crop includes what it has taken from both the soil and

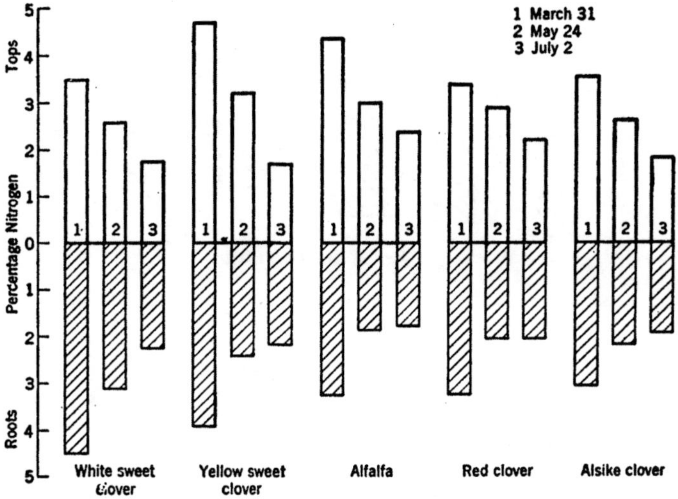

FIG. 46. Percentage of nitrogen in roots and tops of legumes at different stages of maturity. [Ohio Agr. Exp. Sta. Bul. 405.]

air, but only that nitrogen taken from the air represents a true addition to the soil when the crop is turned under. It is generally assumed that on the average about two-thirds of the nitrogen in a legume is taken from the air and one-third from the soil.

A generalized comparison of the distribution of dry matter and nitrogen in the tops and roots of several leguminous crops is shown in Figs. 47 and 48. It is observed that the annual legumes have root systems comparatively small in proportion to the mass of tops and that the percentage of the total nitrogen of the plant which is contained in the roots is less. The proportion of the plant nitrogen which is in the roots of annual legumes ranges from 5 per cent to 20 per cent and that in the roots of the biennial or perennial plants from 24 per cent to 35 per cent. An acre of well-inoculated legumes, growing under favorable conditions, will perhaps gather 50 to 100 pounds of

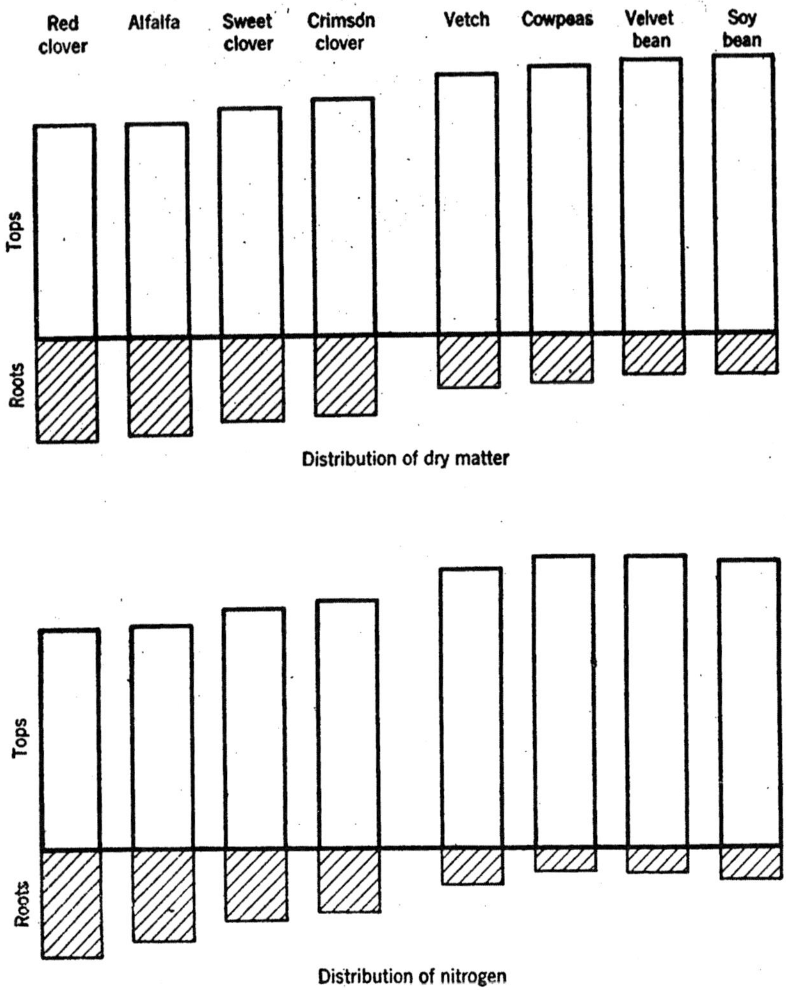

FIG. 47. The distribution of dry matter and nitrogen contents of several legumes between the tops and roots. [Calculated from data in "Root Nodule Bacteria and Leguminous Plants," Fred, Baldwin, and McCoy, p. 221, 1932, University of Wisconsin Press.]

nitrogen from the air per year. If the amount of nitrogen in the soil is incrèased 50 pounds per acre, it is equivalent to over 300 pounds of sodium nitrate. The nitrogen so added is important not only as a nutrient but also as a means of maintaining or increasing the humus level. The nitrogen-carbon relationships to humus formation are discussed elsewhere.

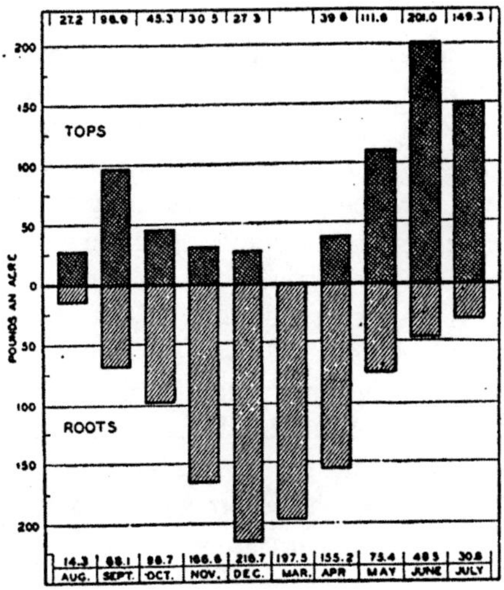

FIG. 48. Total nitrogen distribution in tops and roots of sweet clover at different dates during the year. "Although the distribution of nitrogen between roots and tops varies greatly with the season, the total nitrogen of the sweet-clover plant increases little after growth ceases in the fall of the first year." [From Ill. Exp. Sta. Bul. 394.]

Conserving Plant Nutrients. Leaching losses of plant nutrients from fallow soils are much greater than from similar soils occupied with crops. Land lying idle after the removal of a crop during summer or early fall is also often subjected to severe leaching losses. The nitrates are of particular concern in this connection, for they are highly soluble and are not adsorbed to any extent by the soil colloids. In the presence of a cover crop the soluble plant nutrients are taken into the plants and so are in little danger of being lost. Crops having extensive root systems are especially efficient in collecting plant nutrients, thus decreasing leaching losses. When these crops are turned into the soil, decomposition takes place and the nutrients are released for a new generation of plants.

It is to be emphasized that the mineral content of a soil will not be increased by green manuring although it may be conserved by decreasing wastage. The only elements actually increased by green manuring are carbon, hydrogen, oxygen, and (in legume crops) nitrogen.

Concentration of Plant Nutrients in Soil Surface. Green-manure crops, especially those which are deep rooted such as alfalfa, gather considerable quantities of plant nutrients from the subsoil. When the crop is turned under and decomposes in the upper layers of the soil, the nutrients as liberated are concentrated within a more limited area than previously. This permits the use of nutrients by succeeding shallower-rooted crops which otherwise would not have been available.

Increased Availability of Plant Nutrients. The addition of fresh organic matter to soils results in increased availability of certain inorganic constituents of the soil particles. The solubility of calcium, phosphorus, potassium, magnesium, and other elements is increased through the effect of the organic and inorganic acids produced as a result of the decomposing organic matter. These increases are obtained besides the mineral constituents contained in the organic matter itself.

A greater availability of certain nutrients may be brought about in another way. The action of growing roots of certain plants, used for green-manuring purposes, may be more effective in utilizing a particular element in the soil than the roots of a following crop grown for harvest. This plant-food element is taken into the plant and returned to the soil in the green manure in forms more available than formerly to succeeding crops. Sweet clover, for example, may be able to take in phosphorus which is not available to some other crops.

Improvement of the Subsoil. Plants having long tap roots penetrate the subsoil deeply where possible and, when the roots decay, numerous channels are formed which facilitate the circulation of air and the percolation of water. Furthermore, the character of the subsoil will be modified somewhat through the increased quantities of organic matter resulting from the green-manure crop roots, although no pronounced effect can be expected because of the limited extent of the root system in the subsoil.

Protection of the Surface Soil. Green-manure crops may serve as cover crops in that they may be used in many instances to protect the surface of the soil. On sloping land the soil is protected from water erosion, and light soils (both level and rolling) are protected from wind erosion. Furthermore, cover crops decrease the impact of beating rains which destroy the structure of the soil surface. In occupying the land over winter they hold snow, which keeps the soil warmer and decreases root injury to perennial plants. Root injury may result from either alternate freezing and thawing or excessive cold. This may be very important in certain fruit-growing areas. Green-manure crops shade the soil to some extent and thereby decrease the loss of moisture by evaporation.

Effect on Soil Microorganisms. Organic matter supplied by green manures serves as food for many soil organisms, and tends to stimulate certain biological changes to a marked degree. Such biochemical processes as the production of carbon dioxide, ammonia, nitrites, nitrates, other end products of decay, and the non-symbiotic fixation of nitrogen are all influenced by the incorporation of green-manure crops in the soil. As indicated above (p. 246), the increased availability of the inorganic constituents of the soil is a point that should be considered in this connection.

The effects produced by green manure on the microbial population will depend not only on the kind of green-manure crop and on its stage of maturity but also on soil conditions such as drainage, lime content, and soluble nutrients. Young, succulent green manures are relatively high in nitrogen and when turned into the soil result in an immediate and marked increase in growth and activity of the microorganisms. Such green manures decay rapidly, and usually there is little delay in the accumulation of nitrates and the release of other simple end products of organic matter decomposition. On the other hand, when more mature crops having a low nitrogen content are turned into the soil, the organisms multiply rapidly and compete with higher plants for the available nitrogen. The nitrates may all disappear from the soil and none will appear until after the rapid period of decomposition is past. Other things being equal, the lower the nitrogen content of the green-manure crop, the greater will be the delay in the accumulation of nitrates.

Effect on Soil Acidity. When a green-manure crop low in nitrogen is added to the soil, the first stage of decomposition may result in the formation of various organic acids, especially in the absence of free calcium carbonate. If soil conditions are favorable for further decomposition, the organic acids are broken down and the acidity disappears with the formation of carbon dioxide, water, and carbonates. On the whole, experimental results show that green manures do not increase soil acidity or, if so, such a condition is only temporary. With good soil-management practices green manures will have no lasting effect on increasing the lime requirement of soils.

Effect on Growth of Fruit Trees. The influence of cover and green-manure crops on the growth of fruit trees depends among other factors on the nature of the cover or green-manure crop, time of planting, locality, kind of trees, and soil conditions. These crops compete with trees not only for plant nutrients but also for water. In soils low in nitrates the competition for nitrates may become serious. Deep-rooted crops such as alfalfa and sweet clover draw heavily on the soil-moisture supply; and at a time when the demands of the trees are great and when the moisture supply is limited, this effect is of great concern. However, in soils that contain nitrate nitrogen in such quantities as to produce a prolonged, succulent growth of new wood, susceptible to winter injury, cover and green-manure crops serve to use the surplus nitrogen, slow up the growth of trees late in summer, and induce earlier maturity in the wood. Furthermore, crops that cover the ground during the winter hold snow, protect the tree roots from freezing, and on some soils help dry the soil in early spring.

THE PRINCIPAL COVER AND GREEN-MANURE CROPS AND THEIR REGIONAL DISTRIBUTION

Green-manure crops may be classified or grouped in various ways. On the basis of their time of seeding and their occupation of the land, four types are recognized: (1) those occupying the ground exclusively throughout the summer; (2) those seeded with a nurse crop; (3) those seeded on grain stubble after harvest; and (4) those serving as winter cover crops. Certain legume and non-legume crops fit into and may be used in each of the four categories given. For convenience, these crops will be discussed under the two groups—legumes and non-legumes—and the following questions answered.

Questions:

1. What legume crops are commonly used for green-manuring purposes and in what regions of the United States are they particularly adaptable?
2. Which of the non-legume crops are frequently used for cover and green-manuring purposes?

Legume Green-Manure Crops. Plants belonging to this group are commonly known as nitrogen gatherers. These plants, if properly inoculated and if grown under favorable soil conditions, contain nitrogen obtained from the air. When these crops are incorporated in the soil, they add nitrogen to the soil's previous supply. There are several legume crops grown in the United States for partial or complete use as green manures, among which the following are probably of most practical interest: hairy vetch, Austrian winter peas, crimson clover, crotalaria, sweet clover, alfalfa, cowpeas, soy beans, red clover, alsike clover, white clover, lespedeza, peanuts, Canada field peas, kudzu, mungbeans, and velvet beans. The most important ones, as determined by current use, are the first six listed above, and most of the discussion in this section will be devoted to these crops.

Two distinct forms or varieties of hairy vetch are commonly sold. One is distinctly hairy, whereas the other appears to be almost smooth. The hairy form is more winter hardy but will not grow at as low a temperature as will the smooth form. Therefore the former is to be preferred for planting in the more northern latitudes because of its winter hardiness, while the smooth variety is to be preferred for planting in the South as a winter cover crop.

Austrian winter pea is the most winter hardy of the field peas. It grows at quite a low temperature, and in this respect is superior to hairy vetch. It makes a good growth under the winter conditions of the South and is one of the best legumes for general use. In the southern states Austrian winter pea and hairy vetch are the crops most commonly used for winter cover and green manure.

Crimson clover makes an excellent winter cover crop for the northern part of the cotton belt, although stands are somewhat difficult to obtain. This has greatly limited its use. The seeds germinate easily, but the sprouts are readily injured by drying. A light rain may germinate the seed, but the young shoots may be subsequently killed by dry weather that often prevails at seeding time in the fall.

Crotalaria is one of the most recent and important soil-improving crops in the South. It is well adapted to sandy soil and is better suited to a soil of low fertility than many other crops. The fact that the seed can be produced locally and at a low cost makes it an especially useful crop in the cotton belt.

Alfalfa is seldom used exclusively for a cover or a green-manure crop; yet when used in rotations it is recognized as one of the most important soil-improving crops. It is a deep-rooted perennial and improves drainage and aeration in heavy soils. It furnishes large quantities of organic matter. It makes an effective permanent cover and is recommended for orchards, where moisture is not a limiting factor, or for other permanent plantings in regions where it succeeds. It is grown as a cover and a green-manure crop in apple orchards in parts of Washington and Oregon and in certain irrigated sections of the West. In many parts of the north central states, alfalfa is commonly used in the crop rotation. The crop is either pastured or used for hay for two or three years, sometimes longer, and the aftergrowth is turned under for green-manuring purposes.

There are both annual and biennial forms of sweet clover although the annual type is grown to only a limited extent in certain portions of the corn belt. The biennial type is used only in rotations and is perhaps the most satisfactory legume in the corn belt for use as a cover crop and as a green manure. It fits in well with a corn-wheat rotation. Sweet clover, like alfalfa, has a deep root system and produces a large tonnage of organic matter. These crops can be grown successfully only where the soil is well supplied with lime and phosphorus.

The various legume green-manure crops mentioned can be grown to best advantage only in definitely limited areas. It is to be pointed out, however, that the best crop to use in a given district must be determined finally by local conditions and by the objectives for which the green manure is grown.

In the northern parts of the United States, red clover, mammoth clover, alsike clover, field peas, common vetch, sweet clover, and alfalfa are commonly used. In regions having a moderate to warm temperate climate, cowpeas, crimson clover, red clover, sweet clover, and soy beans largely furnish green manures; and in the South crotalaria, Austrian winter peas, lespedeza, vetch, soy beans, cowpeas, crimson clover, red clover, peanuts, mungbeans, and velvet beans are commonly recommended for use as green manure.

Non-Legume Green-Manure Crops. The chief difference between legume and non-legume green-manure crops is that legumes add both organic matter and nitrogen to the soil, whereas non-legumes add organic matter only. From the standpoint of maintaining the organic matter content of the soil, quantity or bulk is of great importance, and this can often be supplied more efficiently by such a crop as rye, sorghum, or oats than by a legume. It is to be remembered, however, that while a non-legume will add organic matter to the soil, the soil microorganisms may require additional nitrogen to effect rapid decay. The turning under of mature non-legume crops frequently results in a decrease in the yields of the subsequent crops because of the exhaustion of available soil nitrogen. If the nitrogen contained in the green-manure crop is not sufficient, the decay organisms will draw on the available soil nitrogen.

The principal non-legume green-manure crops are rye, oats, millet, buckwheat, rape, rye grass, sudan grass, corn, sorghum, and weeds. Rye and oats are perhaps more extensively used than any of the other crops listed. They make a rapid, abundant, and succulent growth, and they easily fit into almost any rotation. These crops are hardy and will grow under rather adverse weather and soil conditions. They are particularly useful on poor soils, but their beneficial influence is greatly increased by seeding a legume crop with them. Combinations of rye and vetch and of oats and peas are very satisfactory for green-manuring purposes.

For the most part these non-legume crops which are used frequently as green manures are adapted to a wide range of soil and climatic conditions. They are not so exacting in their requirements as are many of the legumes. Most of these non-legume crops may be grown successfully throughout the humid regions of the United States. Rye is extensively used as a green-manure or a cover crop along the Atlantic seaboard; rye and winter oats are popular in the South; rye or rye and hairy vetch are successfully used in the North; and buckwheat is grown more often on the poorer soils of the North for green-manuring purposes than in any other section.

Weeds often serve as green-manure plants and are sometimes encouraged to grow for that purpose. Many orchardists allow weeds to grow in late summer, and they often make an effective cover crop. In certain tobacco sections weeds sometimes are encouraged in order that they may be turned under for green manure. In most instances it is believed that weeds are better than no cover or green-manure crop at all.

THE UTILIZATION OF COVER AND GREEN-MANURE CROPS

The extensive use of cover and green-manure crops in the United States is a rather recent development and has coincided with the realization that land under continuous cultivation often rapidly loses its productive capacity. Green manuring is only one of several practices used in the attempt to maintain soil productivity and is usually one of the last methods to be resorted to. To be most effective and to avoid detrimental results to soils and crops, the use of green manures should be well planned and the practice judiciously managed. In the practical utilization of cover and green-manure crops a number of problems, as indicated by the following questions, present themselves.

Questions:

1. Under what conditions should green manuring be practiced?
2. What points should be considered in the selection of green-manure crops?
3. What precautions should be observed in preparing the seedbed, in seeding and fertilizing green-manure crops?
4. When should green crops be plowed under?
5. What crops are usually best to follow green-manure crops?

Conditions under Which Green-Manuring Should Be Practiced. When green manures are used properly, they are highly effective, but when used injudiciously, they may be either useless or in some instances harmful. They can generally be used profitably in soils that are low in nitrogen and organic matter because such soils usually possess unfavorable physical properties. Green manures are particularly beneficial on many light, sandy soils and on heavy clay soils. They are quite generally used to advantage in orchards. On highly fertile soils in good physical condition it is necessary to use green manures only at intervals sufficient to maintain the organic matter content, and this can usually be done by using a legume hay crop as a part of the regular rotation. On livestock farms usually it is most profitable to grow those crops which, like alfalfa and clover, can be utilized largely for forage, and to depend upon the roots and stubble of these crops together with manure for the maintenance of the organic matter and nitrogen supply.

In general, green manuring in some form should be practiced on most farms of humid regions and on irrigated lands elsewhere. This does not mean a curtailment in the production of regular crops but means that cropping systems should be arranged so that a green manure may be worked in between regular crops. In other words, it usually is not advisable to sacrifice a summer crop for the purpose of growing a green crop to plow under because the value of the increased yields of following crops seldom equals the value of the crop plowed under. Summer green manuring, however, usually can be recommended on very poor soils where drastic measures are necessary to begin the soil-improvement program. The same principle may apply to extremely light or heavy soils to improve their physical properties. A summer green-manure crop may also be used in orchards without interfering with the main crops.

The use of green manures in dry-farming areas in arid and semiarid regions is as a rule impracticable. Not only is it difficult to produce such crops but also they decay very slowly because of insufficient moisture. Not only do the green-manure crops require large quantities of water for their production but also, when they are turned into the soil, evaporation losses are enhanced, owing to the increase in the amount of air space within the soil.

The Selection of Green-Manure Crops. In choosing crops which are to be used for green-manuring purposes it is obvious crops should be used which are best adapted to the climate, soil, type of farming, and special purpose for which they are being used. Legume crops should be given preference when they can be grown satisfactorily, because of their ability, when properly inoculated, to add nitrogen to the soil. The choice of the legume is determined primarily by the soil and climatic conditions prevailing and by the cropping system used. It is often difficult to obtain a catch of certain legumes and the seeds of many legumes are expensive.

Some of the legumes are annuals, some are biennials, and others are perennials. Some of them are warm-weather crops and others grow well in cool weather; some are best adapted to fall planting and others to spring planting. The biennials and perennials as a rule make a slow growth at the beginning, which is a serious objection where quick results are needed. Alfalfa is a perennial, whereas sweet clover (the variety commonly grown) is a biennial. Red clover, alsike, and mammoth clover, as usually handled, may also be considered biennials. Cowpeas, soy beans, common vetch, field peas, peanuts, velvet beans, lespedeza, and crotalaria are summer annuals; while hairy vetch is a winter annual in the North and crimson clover, common vetch, and field peas are winter annuals in the South.

Where there is a choice between two or more crops, the question arises as to which is to be preferred. In arriving at a decision, the following points should be considered: (1) its value as forage; (2) its ability to add nitrogen; (3) cost of seed; (4) ease of plowing under and of killing the crop; (5) the rooting habit of the plants; (6) rate of growth; (7) abundance and succulence of growth; and (8) soil requirements.

As a rule it is advisable to choose a green-manure crop that grows rapidly, one that produces abundant and succulent tops, and one that has the ability to grow well on poor soils. Cover and green-manure crops that least interfere with the regular cropping system should be used. In the North, catch crops following grain crops or planted with them are often used to advantage. On suitable land, sweet clover may be seeded with grain in the spring and turned under for corn the following spring. This is a common practice in the corn belt area. In the South, winter cover crops fit into most farm programs effectively. In truck farming, catch crops can be used, and legume crops in the regular rotation may serve as green manure. In general farming, no green-manure crop which requires an entire season for growing is likely to become widely used or generally profitable. Therefore, as a rule, it is necessary that crops be chosen which can be produced without the loss of the use of the land for an entire season. Such crops are often called "fill in" or "catch crops." For example, a crop such as rye may be seeded in the corn in the fall and plowed under for a second corn crop the next spring. In suitable climatic areas, soy beans may be seeded after oats or wheat and the crop turned under that fall. The second crop of red clover may be plowed under when the setting of seed is poor and when the clover is not needed for pasture. However, one is perhaps justified in giving over the entire crop season to a green-manure crop when the succeeding crop is more or less permanent, as in orchard plantings.

When green-manure or cover crops are used in orchards the season of growth of the trees must be considered. If an annual crop is used it is necessary to plant it at a time that will offer as little interference as possible with tree growth, fruit development, and harvesting operations.

Preparation for Green-Manure Crops. Most of the green-manure crops are easy to grow under proper climatic conditions, even on relatively poor soils; nevertheless in order to be grown most successfully certain precautions should be observed. Consideration should be given to the preparation of the seedbed, the need for fertilizers, the need for lime, and the matter of inoculation for legumes.

As a general rule more attention should be given to the seedbed preparation for legumes than for non-legumes. In any case, it is best that the soil be in good tilth and that a firm seedbed be established. Obviously, little or no seedbed preparation is necessary when cover or green-manure crops are seeded immediately after the removal of some cultivated crop, and if seeding is made in a nurse crop that has already been planted, little can be done to modify the existing seedbed.

Many soils are too low in fertility to produce a good cover or green-manure crop. With such soil conditions, it is utterly impossible to build the humus content by attempting to grow soil-conserving or soil-improving crops. If the organic matter is to be restored in these soils by green manuring, provision must be made to supply these mineral deficiencies before attempting to grow green-manure crops. In most instances where soil fertility has declined to a rather low level, a complete fertilizer—one carrying nitrogen, phosphorus, and potassium—is needed for non-legume crops. Legumes, if inoculated, usually will not require nitrogen fertilization. If fertilizer is needed in order to grow a good crop of green manure, it should be used; not only the crops will make more top growth but also they will make a more dense and extensive root development. The plant nutrients in the fertilizer that are taken up by the green manure are not wasted because they will be available to the following crops when the green-manure crop decays. The kind and amount of fertilizer to use will depend primarily on the crop and soil conditions.

An abundant supply of lime is necessary for the maximum growth of legumes. The majority of soils contains sufficient lime for most non-legume green-manure crops, but the supply frequently is inadequate for legume crops. Lime not only gives greater top and root growth but also will increase the amount of nitrogen fixed by the legume bacteria. The relationship of the lime supply in soils to humus formation is discussed in Chapter VIII.

It is perhaps well to emphasize again that, if legume green-manure crops are to be effective in increasing the nitrogen supply of soil, they must be well inoculated with an active strain of the proper bacteria. Most of the legumes require a particular kind of bacteria. For example, bacteria which produce nodules on soy beans will not do so on the alfalfa plant (p. 202). Inoculation may be accomplished by using soil from a field which is known to contain the organisms or by using artificial cultures. When one is doubtful as to whether inoculation is needed, it is advisable to inoculate because the cost is relatively insignificant.

Plowing under Green-Manure Crops. The time of turning under green-manure crops is an important consideration and is dependent, among other conditions, on the seasonal growth of the crop, the nature of the soil, the maturity of the crop, the weather conditions, and the crop which is to follow. Precautions in turning under green manures are of more concern where a crop is to follow immediately. Generally speaking, they should be plowed under when they are still green and succulent so that they will decay rapidly. This is particularly important in light soils.

A green-manure crop may remove much water from the soil and, when turned under, the soil may be left too dry for the crop which is to follow. The decay processes which break down this organic matter are hindered, and this material will lie in the soil for some time without decomposing. When a large amount of organic matter is turned under, it may leave the soil so loose that the entire furrow slice will dry out and thus injure the following crop. Thus when a crop is to follow immediately, it is better to plow the green-manure crop under some time in advance, when it is green and succulent and before the soil has become dry. If possible, green manure should be incorporated in the soil at a season when there is a plentiful supply of rain, for this will usually insure a rapid and effective decomposition. When large quantities of green organic matter are incorporated with the soil, decomposition products may cause considerable injury to the seedlings of the following crop. In order to avoid this type of injury, some time, perhaps two or three weeks, must be allowed to elapse before planting.

A deficiency of soluble nitrates is likely to occur where a green-manure crop is permitted to reach an advanced state of maturity before it is turned under, especially non-legume crops. The carbon-nitrogen ratio is wide and, in breaking down this carbonaceous material, the decay organisms may utilize most of the soluble nitrates in the soil in their life processes. The result is that the next crop may lack not only moisture but also available nitrogen. The younger the green-manure crop and the more water it contains, the shorter is the period of nitrate depression.

The effect which the time of turning under a green crop may have on leaching losses should also be considered. There may be considerable leaching of nitrates when a green crop is turned under in the fall for a spring-seeded crop. This point should be given due consideration when a decision is to be made as to fall versus spring plowing. As a rule, in the northern states, plowing under a green crop in the spring generally gives best results on light soils, whereas on heavy soils fall plowing is often preferred. Where erosion or leaching is a problem, it is better to leave a green crop on the soil over winter and to plow in the spring. In the South, from the standpoint of the leaching loss of nitrates, green-manure crops should be turned under in the spring.

In plowing under green crops, it is best to distribute the organic material throughout the plow soil instead of having it in a more or less continuous layer at the bottom of the furrows. This may be accomplished, to some extent, by turning the furrow slice partly over instead of turning it over flat. When the amount of material to turn under is large, it is sometimes advisable to double-disk the land before plowing in order to cut up the material and to facilitate working it into the soil.

Obviously, no definite dates can be given for the turning under of a green-manure crop. There are certain precautions to be taken into consideration in relation to turning under these crops, but on the whole there is little difficulty encountered and the farmer merely needs to use good judgment in order to secure good results.

Crops To Follow Green-Manure Crops. In general, it is a good plan to follow green manures with cultivated crops such as corn, tobacco, cotton, and potatoes. The cultural practices used in growing such crops produce conditions that favor decomposition of the green material and the consequent increase in available plant nutrients. Of the small-grain crops, rye and oats make a good response immediately after green manures; whereas wheat and barley usually do not respond so well, especially the first season after a green-manure crop. This brings us to a consideration of the effects of green-manure crops on the yields of the succeeding crops.

EFFECT OF GREEN MANURE ON YIELD OF CROPS

Many experiments with various green-manure crops under a great variety of soil and climatic conditions have shown that, when properly used, green manures increase the yields of succeeding crops and that the effects are often noticed for several years. Under certain conditions, however, some crops immediately following legume green manures may be unfavorably affected. For example, when large quantities of highly nitrogenous material decompose rapidly in soil, the nitrate content may be increased so that crops do not grow normally; they frequently develop too luxuriantly in stems and leaves. The ripening of fruit may be delayed and the yield decreased; with small grains, lodging may result and the yield and quality of grain be diminished. The probable detrimental effect of turning under mature or carbonaceous materials has been discussed in a preceding section.

In regard to the effect of green manures on the yield of subsequent crops, two questions appear pertinent.

Questions:

1. How do green manures affect the yield of crops?
2. Do green manures have a pronounced residual effect on crop yields?

Effect of Green Manures on Crops Immediately Following. The effect of green manures on the yield of subsequent crops has been studied by many of the state agricultural experiment stations although a large share of the experimental work has been confined to the South and to the Atlantic seaboard. The work of nine Southern states, dealing with this subject, has been recently summarized.[1] The increased yields resulting from turning under winter legume crops have ranged from 6 to 60 per cent over the yields of the check plots. In some of the Southern states cotton yields have been increased from 22 to 100 per cent in various experiments by turning under legumes. The turning under of winter legumes has generally increased the yields of corn in the South from 24 to about 78 per cent. A summary of the effects of previous crops of summer and winter legumes on the yields of seed cotton and corn in the southern region and nearby states is presented in Table 25. In the corn belt, where sweet clover is used extensively as a green-manure crop, substantial increases in corn yields usually are obtained. The turning under of summer legumes has also brought about marked increase in crop yields as has the turning under of the

[1] "The Use of Cover and Green-Manure Crops," A. J. Pieters and R. McKee, 1938 Yearbook, U.S.D.A., p. 436.

second growth of red clover. The great majority of records indicates that the turning under of a legume crop consistently increases subsequent crop yields.

TABLE 25

EFFECTS OF PREVIOUS CROPS OF SUMMER OR WINTER LEGUMES ON THE YIELDS OF SEED COTTON AND CORN IN THE SOUTHERN REGION AND NEARBY STATES *

Preceding Crop †	Without Legumes		With Legumes		
	Crop acre ‡	Yield per acre	Crop acre ‡	Yield per acre	Difference in yield
SEED COTTON					
	Number	*Pounds*	*Number*	*Pounds*	*Percentage*
Summer legumes	50	756	62	1,163	+53.9
Winter legumes	106	808	247	1,276	+58.0
CORN					
	Number	*Bushels*	*Number*	*Bushels*	*Percentage*
Summer legumes	51	25	103	36	+42.8
Winter legumes	53	25	175	35	+41.2

* Taken from "Summary of Effects of Legumes on Yields of Cotton and Corn in the Southern Region and Nearby States," Southern Region Agricultural Conservation, No. 3, 1937, p. 3 U.S.D.A., Agr. Adjustment Administration.

† The crops of both summer and winter legumes were used in various ways—sometimes for hay, sometimes for seed; for many crops only the st ıbble was turned under, and for others the vines were turned under after the seed was harvested.

‡ Data in these columns represent the number of crops on an acre basis. This number was arrived at by the number of years the tests were conducted. For example, the number 50 in the table means that the yields of 50 crops of cotton, following legumes, from several experiments have been considered.

Non-legume green-manure crops will add organic matter to the soil but will not in themselves increase the nitrogen content of the soil above that existing at the time the green-manure crop was seeded. It is largely on this account that the increases in yields of the crops following non-legume green-manure crops are not so great as those following legumes. As previously explained, the turning under of non-legumes frequently has resulted in a decrease in the following crop because of a deficiency of nitrates. It is evident, therefore, that non-legume green-manure crops generally should be turned under when young or soluble nitrogen must be added to the soil to facilitate decomposition of the organic matter and to offset or prevent a nitrate deficiency.

Residual Effect of Green Manures on Crop Yields. In determining the returns from the practice of green manuring, it is not sufficient to consider only the increase in the next crop. The green manure may have a marked effect on yields of subsequent crops for several seasons. The degree and the extent of the residual effect will be determined largely by the amount and kind of green manure turned under, the condition of the soil, and the climatic factors. Under otherwise similar conditions the residual effect likely would be extended over a longer period of years in cool, humid regions than in warm, humid regions. In some of the midwestern states a favorable residual effect of green manure on yields of succeeding crops has been observed for a period of 8 to 10 years.

The residual effect of green manures may be due in part to the improved mechanical condition of the soil or to any one factor or a combination of the other factors mentioned in the first part of this chapter (p. 240). At any rate, green manuring is considered a part of a well-established system of soil management, in maintaining or increasing the crop-producing ability of the soil.

CHAPTER X

FARM MANURES

Farm manure offers the farmer's chief opportunity for returning to the soil a portion of the plant nutrients removed in crops. Manure is one of the most important agricultural by-products; it is a perishable product and is frequently subjected to very severe losses. When calculated in terms of dollars and cents, the losses occurring with manure on American farms is astonishing. Only a fraction of the value of farm manure, perhaps not more than one-third, is actually realized. Farmers are acquainted with increased yields after an application of manure, but in general they do not understand the true nature of manure, the perishable nature of some of its constituents, or the losses occurring from its improper handling. In order to understand and appreciate better the value of manure, a few facts are presented regarding the production, losses, care, and field management of manure, from a practical point of view. The study may be pursued under the following divisions.

Objectives:

A. The production of manure.
B. The decomposition of manure.
C. Losses occurring with manure.
D. Methods of handling manure.
E. Field management of manure.
F. Fertilizing properties of manure.
G. Effects of manure upon the soil.

THE PRODUCTION OF MANURE

It is the common knowledge among farmers that the maintenance of soil fertility is much more easily accomplished with a livestock system of farming than with a grain or cash system. This is true because a large share of the nitrogen and minerals in the crops fed and, to a less extent, of the organic matter is excreted by the animals. However, it is to be remembered that not all the plant-food elements or the organic matter in the feed is excreted in the manure. Since the animal must be heated, have energy for performing work, and have nutrients for growth and body maintenance, it is obvious that a portion of the plant nutrients and organic matter is taken from the feed that is consumed by the animal. In order to obtain a better understanding as to the real nature of manure, it is necessary to consider the chemical changes certain constituents of feed undergo in the digestive tract of animals, the components of manure, the quantity and composition of excrements from different kinds of livestock, and the proportion of fertilizing constituents recovered from the feed eaten by the animals. Several questions present themselves in connection with a study of the production of manure.

Questions:

1. Which animals supply the largest proportion of farm manure?
2. Do pronounced chemical changes occur in the fertilizing constituents of feed in the digestive tract of animals?
3. What are the principal components of manure and how do they differ in quantity and composition?
4. How do the quantity and composition of excrements from different kinds of livestock compare?
5. What percentages of the fertilizing constituents are recovered from feed consumed by animals?

Quantity of Manure from Different Types of Animals. The term farm manure has reference to the excrements from all animals of the farm. The total quantity produced in the United States from different kinds of livestock is in the following descending order: cattle, horses and mules, hogs, sheep, and chickens. Most of the manure that is eventually returned to the land is excreted by cattle and horses because, on the average farm, these animals consume a much greater portion of the grain and roughage and because the methods usually employed in handling such livestock make it easier to collect the manure.

Chemical Changes in Feed in the Digestive Tract of Animals. Complex chemical changes occur in the food in the digestive tract of animals. These changes are brought about partly by digestive enzymes and partly by the numerous bacteria which live in the intestinal tract. Twenty to 30 per cent of the dry weight of the solid excrement consists of living and dead cells of bacteria.

The various constituents of feeds undergo different rates and different degrees of decomposition. Sugars and starches are easily broken down; celluloses and hemicelluloses are less easily decomposed; and lignins are very resistant. Proteins vary considerably in their ease of breakdown. Proteins in "concentrates" are usually much more highly digestible than proteins in "roughages."

Most of the potassium in feed is absorbed from the digestive tract and excreted in the urine. Only a small fraction of the phosphorus, except in hogs, is so absorbed. Consequently, most of the phosphorus in manure is carried in the solid fraction.

Components of Manure. Farm manure consists of two original components, the solid and the liquid. The solid material represents for the most part the undigested material and the liquid portion represents the digested material that has been absorbed by the animal and then excreted.

The solid excrement, on the average, contains about one-half of the nitrogen, about one-third of the potassium, and nearly all the phosphorus that is excreted by the animal. The nitrogen in the feces exists largely in two forms: first, the residual proteins that have resisted decomposition in the digestive processes; and, second, the proteins that have been synthesized in the cells of bacteria. Over one-half of the nitrogen may be present as synthesized protein, and this form is readily broken down when added to soils, so that the nitrogen is readily available to plants. However, the solid excrement contains large quantities of lignin, and it is believed that this lignin combines with protein, forming a lignin-protein complex which renders the protein less available. In other words, a large share of the organic matter in the feces is humified; a compound is formed similar to the humus that is formed in soils. As much as 50 per cent of the organic matter in the solid excrement may be in the humified state, and the nitrogen contained therein is only slowly available to plants when added to soils.

The liquid fraction or urine contains those plant nutrients which have been digested and utilized in the animal body and are later excreted. All the plant nutrients in this fraction are soluble and either are directly available to plants or readily become so. The liquid portion of manure differs from the solid not only in regard to the availability of nutrients but also in its low content of phosphorus and in its high content of potassium and nitrogen. (The distribution of plant nutrients between liquid and solid portions of manure is shown in Fig. 49). In general, the more digestible the food consumed by an animal, the greater is the proportion of its plant nutrients that appears in the urine. Furthermore, as a rule, the higher the food is in nitrogen, the greater the digestibility of the nitrogen and the larger the amounts of nitrogen in the urine.

When voided, the nitrogen in urine exists largely as urea—hippuric and uric acids. These compounds are not volatile at ordinary temperatures, but manure contains organisms capable of rapidly breaking these compounds down with the formation of ammonia, which combines with water and carbon dioxide to make ammonium carbonate. This compound is unstable; even in solution it tends to decompose with the loss of ammonia, especially with higher temperatures [$(NH_4)_2CO_3 \rightarrow 2NH_3 + H_2O + CO_2$]. This compound may lose all its ammonia on drying. The unstable nature of the nitrogen in urine presents a major problem in the handling of manure.

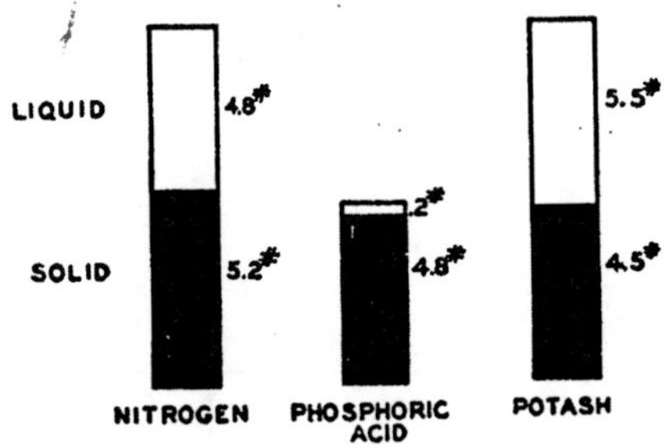

FIG. 49. Distribution of plant nutrients between liquid and solid portions of a ton of average farm manure.

Quantity and Composition of Excrements. Many factors influence the quantity of manure voided and its composition, such as (1) the kind and age of the animal, (2) the kind and the amount of feed consumed, (3) the condition of the animal, and (4) the milk produced or the work performed by the animal. Wide variations are often found in the manure of animals even of a given class. Animals of different ages and doing different kinds of work require different amounts and proportions of nutrients to maintain them. A young animal, for example, which is building muscle and bone needs considerable phosphorus, nitrogen, lime, and other elements, and the manure produced by such animals will contain much less of these elements. Since the composition of manure is so variable, data such as those presented in Table 26 can only be approximate.

The urine makes up only 20 per cent of the total weight of the excrement of horses but 40 per cent of that from hogs. These represent the two extremes. Since the urine makes up only 20 to 40 per cent of the total weight of manure in any animal, and yet contains approximately two-thirds of the total potash and one-half of the nitrogen, it is evident that pound for pound the urine is more concentrated and hence more valuable than the solid portion.

On a percentage basis the nitrogen in the urine, in most animals, is considerably more concentrated than in the feces. The phosphorus,

TABLE 26

QUANTITY AND COMPOSITION OF FRESH MANURE EXCRETED BY VARIOUS KINDS OF FARM ANIMALS *

Animal	Excrement	Pounds per Ton	Percentage Water	Pounds Nitrogen	Pounds P_2O_5	Pounds K_2O	Tons Excreted † per Year
	Liquid	400	..	5.4	Trace	6.0	....
Horse	Solid	1,600	..	8.8	4.8	6.4	....
Total		2,000	78	14.2	4.8	12.4	9.0
	Liquid	600	..	4.8	Trace	8.1	....
Cow	Solid	1,400	..	4.9	2.8	1.4	....
Total		2,000	86	9.7	2.8	9.5	13.5
	Liquid	800	..	4.0	0.8	3.6	....
Swine	Solid	1,200	..	3.6	6.0	4.8	....
Total		2,000	87	7.6	6.8	8.4	15.3
	Liquid	660	..	9.9	0.3	13.8	....
Sheep	Solid	1,340	..	10.7	6.7	6.0	....
Total		2,000	68	20.6	7.0	19.8	6.3
Poultry							
Total		2,000	55	20.0	16.0	8.0	4.3

* Compiled from "Fertilizers and Crop Production," Van Slyke, 1932, pp. 218 and 220. Orange Judd Publishing Co.

† Clear manure without bedding; tons excreted by 1,000 pounds of live weight of various animals.

although existing in relatively small quantities, is found largely in the solid. Potash, however, is more concentrated in the liquid than in the solid.

It is interesting to note that water is one of the most variable constituents in manure. Poultry and sheep manure, on a tonnage basis, contain much greater quantities of plant food than any of the other manures; yet these two types of manure contain much less water. The different manures, on a wet basis, listed in Table 26, contain from 22 pounds of plant food per ton in cow manure to 47.4 pounds in sheep manure. If the composition of the total manure excreted by each of the different animals per 1,000 pounds live weight is calculated on a dry-weight basis, the manures would not vary so greatly in their plant-food content. Horse and sheep manures are considerably drier than cow or hog manures, which partially explains why the former heat more quickly than the latter in storage and the reason they are preferred in the preparation of hotbeds.

Fertilizing Constituents Recovered from Feed. The fertilizing constituents in feed consumed by animals are not lost from the animal body in any measurable quantity except through excretion in the urine and feces. Therefore, all the fertilizing constituents that are not built into animal tissue or secreted in milk are excreted in manure. The recovery of nutrients may be practically complete if the animal is not gaining or losing weight, producing milk, or supplying nutrients for devel-

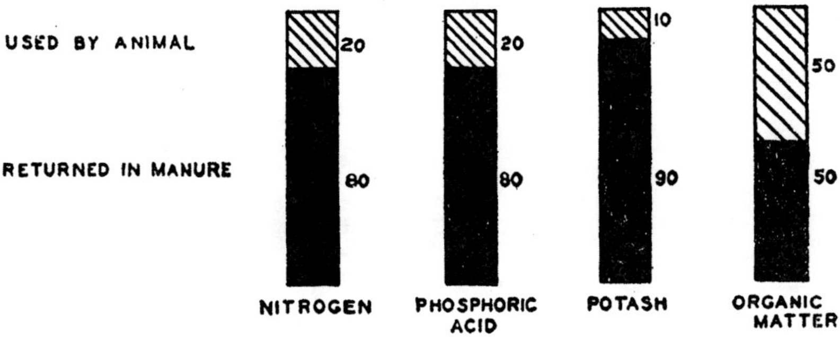

FIG. 50. Average proportion of plant nutrients and organic matter in feed consumed by animals which is excreted in manure.

oping young. Furthermore, animals that are not secreting milk and are losing weight may excrete more nitrogen than is contained in their feed as a result of animal-tissue breakdown. It has been indicated that young, growing animals and milking cows remove larger quantities of nutrients from feeds than do other classes of livestock. In producing 1,000 pounds of milk about 6.0 pounds of nitrogen, 1.7 pounds of phosphoric acid, and 1.7 pounds of potash are required; and about 2.5 pounds of nitrogen, 1.5 pounds of phosphoric acid, and 0.2 pounds of potash are required to produce a 100-pound increase in the weight of a calf.

On an average, farm animals return in the manure from the feed consumed 75 to 80 per cent of the nitrogen, 80 per cent of the phosphoric acid, 85 to 90 per cent of the potash, and 40 to 50 per cent of the organic matter (Fig. 50). Variations in the recovery of nutrients from feeds are due to different conditions, such as the kind and concentration of feed, age of animal, kind of animal, milk production, and work performed by the animal. Consequently, values may vary considerably from those given above.

THE DECOMPOSITION OF MANURE

Many farm conditions do not permit the daily hauling of manure direct to field; consequently it must be stored and, when handled in this way, marked changes occur in the manure. Owing to the heterogeneous nature of both the organic and inorganic constituents, the numerous kinds of microorganisms inhabiting manure, and the varying environmental conditions under which it is stored, a wide variety of chemical changes takes place. Some of the more important questions dealing with the decomposition of manure in storage may be listed as follows:

Questions:

1. What changes are produced in the nitrogen compounds?
2. What changes are produced in the carbohydrates?
3. Do the mineral constituents undergo pronounced changes?
4. Does humus formation accompany the decomposition of manure?
5. What is the value of fresh compared to well-rotted manure?

Changes Produced in the Nitrogen Compounds. The solid fraction of fresh manure as it comes from the stable consists of decayed or partially decayed plant materials with perhaps small quantities of broken-down animal tissue and bacterial cells. Usually this is mixed with litter and the entire mass is moistened with urine; it offers an excellent opportunity for many microorganisms to function. The rate and nature of changes affected by these organisms are determined to a large extent by the degree of aeration in the manure heap.

When manure is first produced, it is exposed to the air, rapid aerobic decomposition occurs with the liberation of considerable heat and carbon dioxide gas, and numerous changes occur in the nitrogen constituents. As the manure becomes more compacted, and if kept moist, oxygen is excluded and the decomposition becomes anaerobic. With the change from aerobic to anaerobic conditions, the rate of decomposition becomes slower, the temperature is lowered, and the nature of the decomposition products is materially different.

The urine fraction is relatively high in nitrogen, which exists chiefly in the form of urea. In decomposing, urea combines with water to form ammonium carbonate, which in turn combines with carbonic acid to form ammonium bicarbonate. These carbonate compounds are unstable and tend to lose gaseous ammonia and carbon dioxide into the atmosphere. The odor of ammonia gas in horse stables indicates the volatile nature of these ammonium compounds. This tendency is more pronounced with increases in temperature, alkalinity, and drying. The decomposition of urea is materially reduced if the reaction is kept below *p*H 7.0. Under favorable conditions the change of urea to ammonium carbonate takes place very rapidly and may be almost complete within three or four days.

Organic nitrogen compounds other than urea occur chiefly in the undigested food protein and synthesized microbial protein of the solid excrement. They are both more or less resistant to biological decomposition although some changes occur under both aerobic and anaerobic conditions. As decomposition proceeds, some of the nitrogen is released as ammonia and some is synthesized into microbial protein, with a net loss therefore in total organic nitrogen. The foul-smelling compounds, skatole, indole, hydrogen sulphide, and amines are protein-decomposition products.

Nitrification in manure as a general rule is probably not an important process. A good supply of oxygen is essential for active nitrification; yet well-aerated manure heaps often develop high temperatures fatal to the nitrifying organisms. Therefore nitrates are most likely to be produced in the outer parts of the heap where it is loose, reasonably dry, and relatively cool.

Nitrification presents the opportunity for denitrification, the formation by microorganisms of nitrogen gas from nitrites and nitrates. It has been suggested that when nitrates are formed in the outer portions of the manure heap, they are transferred by leaching to the interior of the heap, where they are reduced to nitrogen gas under anaerobic conditions.

Changes Produced in the Carbohydrates. The solid fraction of manure contains considerable quantities of cellulose, hemicellulose, lignin, and possibly lignin-protein complexes that escaped digestion and therefore represent the more-resistant fractions of biological decomposition. Litters such as straw and corn stover, commonly used as bedding, contain relatively large amounts of celluloses and hemicelluloses and smaller amounts of sugars, starches, and proteins, all of which are rather easily decomposable.

The rate of decomposition and the nature of the products formed from carbohydrates depend largely upon the degree of aeration. Decomposition is most rapid with a fairly high temperature, a good supply of air, ample moisture, and when the manure heap is loose and exposed to the open air. Under such conditions the decomposition of carbohydrates begins immediately unless delayed somewhat by a highly alkaline condition brought about by the decomposition of urea to ammonium carbonate. The principal change in the carbohydrates is induced by biological oxidation, the chief end products being carbon dioxide and water.

At first, when the more readily decomposable carbohydrates are being attacked, much energy is released in the form of heat, which may raise the temperature of the manure heap considerably. Since carbon dioxide, a volatile gas, is one of the chief carbohydrate-decomposition products, it is evident that as decomposition occurs there is a decrease in total dry matter. This decrease is rapid at first but slows up as the supply of readily decomposable carbohydrate is exhausted.

Anaerobic decomposition of carbohydrates is much slower than aerobic decomposition with the formation of different kinds of intermediate and end products. It is known as an acid type of fermentation, with the formation of carbon dioxide, methane, and hydrogen as end products and of various organic acids like acetic and butyric as intermediate products. These acids tend to combine with ammonia, lower the *p*H of the manure, and thus reduce ammonia losses. Anaerobic decomposition is accompanied by only slight increases in temperature and small losses of dry matter.

Changes in the Mineral Constituents. Phosphorus is excreted largely in the feces, and it exists in both organic and inorganic combinations. During the biological decomposition of manure, organic phosphorus compounds are broken down with the formation of inorganic phosphates, which are usually considered more available. Decomposition affects the phosphorus compounds in manure mainly in a favorable way, and there is rarely, if ever, any danger of loss as the result of the decomposition processes.

A large part of the potassium of manure is voided in the urine and of the total present in mixed manure, most of it is soluble in water or at least in an easily available form. The potassium in the liquid portion usually is converted into potassium carbonate and may be active in neutralizing acids formed by the decomposition of carbohydrates. Insoluble potassium, calcium, and magnesium compounds, although occurring in only small quantities in manure, are changed during decomposition into more readily available forms. These compounds undergo no loss as the result of decomposition changes.

Humus Formation in Manure. During decomposition, the fibrous portions of manure are disintegrated into a dark-brown or black mass of fine, soft humus material, with the eventual disappearance of all traces of the original structures. This change is accomplished by a loss of carbon and hydrogen in the form of gaseous compounds. One-half of the original dry matter of manure may be lost in this manner. This explains the shrinkage in bulk of manure during decomposition.

The black colloidal organic substances in soils, commonly referred to as humus, are probably compounds essentially of lignin and protein, the latter being synthesized chiefly by microbial action. As has been mentioned (p. 262) one-half of the organic matter of feces may exist in a form similar to soil humus. Since considerable quantities of lignin are contained in both feces and bedding and since microbial protein may be synthesized in the rotting process, an increase in humus content can be expected to occur in manure.

Although there may be an increase in the percentage of humus material on decomposition, there is always a loss in the total dry matter, with the net result that the absolute amount of humus usually shows a loss. Thus humus production is offset by decomposition losses of the original humus in fresh manure, with the result that the total humus content may not be increased but may even be decreased.

Value of Fresh vs. Rotted Manure. In speaking of a rotted manure, it is usually inferred that the original structure of the materials has disappeared more or less completely. In this connection, if we assume that the fresh manure is a normal mixture of urine and feces and that conditions have been controlled, fresh and rotted manure differ in composition as follows:

1. Rotted manure is richer in plant-food constituents. This concentration of plant nutrients is due to shrinkage in dry weight, which would automatically raise the level of plant food. One ton of fresh manure may lose one-half its weight in the rotting process. The loss occurs largely in the organic nitrogen-free constituents which contain no plant food.

2. The nitrogen in fresh manure is more largely soluble. The decrease in soluble nitrogen is brought about by microorganisms in the synthesis of their body tissues during active decomposition of the organic constituents of the manure. There is considerable utilization of urinary nitrogen in the formation of complex proteins during the decomposition of manure. Unless rotting conditions are carefully controlled, there may be considerable loss in total nitrogen through volatilization.

3. The solubility of phosphorus is greater in decomposed manure. If no leaching occurs, there is no change in the total quantities of phosphorus or potassium.

That the increase in concentration of plant nutrients in rotted manure is obtained at the expense of large losses of organic matter and considerable losses of nitrogen, especially available nitrogen, has not been fully appreciated. In general it can be said that benefits derived by the rotting process may be more than offset by losses. If manure can be applied to the field daily, there is little justification for allowing it to rot except for special uses.

LOSSES OCCURRING WITH MANURE

Common methods of careless handling about the barn and delayed application of manure result in the dissipation of a large share of the value of manure before it reaches the field. The degree to which manure decreases in value is determined largely by the kind of manure and by the methods used in collecting, storing, and applying it. Thus several processes contribute to losses of the fertilizing constituents and organic matter in manure. Pertinent facts about losses from manure are brought out in the answers to several questions.

Questions:

1. Do significant losses of the liquid portion of manure occur?
2. What losses occur by leaching?
3. Are volatilization losses important?
4. How can losses, due to scattering about barns and lots, be reduced or eliminated?

Losses of the Liquid Manure. The loss of urine from manure is serious from the standpoint of plant-nutrient content. This loss occurs largely from failure to use ample bedding, leakage through stable floors, seepage into earth floors, or drainage from manure heaps.

On the basis of the total plant food in manure, about 50 per cent of the value is in the urine. It can be readily understood that large quantities of plant nutrients are lost and that the portions so lost, as already pointed out, are the most readily available forms of plant nutrients in manure. It is probably safe to assume that more than one-half the liquid is lost on many farms. In terms of dollars and cents, this represents a very significant loss to the farmer.

Leaching Losses. Frequently manure is thrown from stables into an adjoining lot where it is unprotected from rain; in some instances it is thrown directly under the eaves of the barn. Leaching losses are greater when the manure is thrown in small, loose, and open piles. Very little excuse can be offered for storing manure in unprotected piles in the barnyard.

In the course of six months, manure which is exposed to rain may lose more than half its fertilizing value, depending upon the quantity of rain and the nature of the manure pile. The materials leached out are the most readily soluble and therefore the most quickly available plant nutrients. Leaching losses are confined not alone to the urine fraction but also to the nitrogen, phosphorus, and potassium compounds of the solid portion. Under alkaline conditions considerable quantities of soluble organic matter may leach out also; this is indicated by the dark color of leachings from manure piles.

Losses by Volatilization. Losses incurred by volatilization fall principally on nitrogen and organic matter. Large quantities of ammonia are produced in manure from urea and other nitrogenous compounds, and in the earlier stages of manure decomposition the ammonia is combined largely with carbonic acid as ammonium carbonate and bicarbonate. These ammonium compounds are rather unstable, and gaseous ammonia may be readily liberated. The tendency to lose ammonia nitrogen increases with the increase in concentration of ammonium carbonate and with the increase in temperature.

At ordinary temperatures little or no loss of ammonia from manure occurs at *p*H 7.0 or below. High temperatures, produced by aerobic decomposition in a loose manure pile, are conducive to very rapid loss of ammonia.

Freezing also tends to increase the loss of ammonia by increasing the concentration of the solution through the crystallization of the water. This loss may be considerable when manure is spread and becomes frozen.

Air movement greatly affects the loss of ammonia. Wind movements hasten evaporation of water, which decreases the capacity of the water to hold ammonia. Thus manure which is permitted to dry out may lose appreciable quantities of ammonia. This fact emphasizes the importance of ammonia loss due to air circulation in loosely piled manure heaps and in overfermented manure that is forked frequently. Losses may also be considerable if manure is spread and permitted to dry before plowing under.

It has been pointed out (p. 267) that, when manure decomposes, it suffers important losses in organic matter and that these losses occur mainly in the carbohydrate constituents. One of the important end products of carbohydrate decomposition is carbon dioxide, which is lost largely from manure by volatilization. Shrinkage that accompanies the partial decomposition of manure is evidence of organic matter loss.

Little or no loss of phosphoric acid or potash from manure occurs by volatilization.

Loss Due to Scattering. Another source of loss is that caused by the manure's being tramped into the mud of lots or dropped in roads or waste places where it cannot be picked up and returned to the land. Loss suffered by manure piled in lots may occur through the scattering of the material by chickens, hogs, and other livestock.

Some of this loss is unavoidable but certainly a part of it can be prevented. Such losses may be prevented or reduced by the direct hauling of the manure to the field or by protecting it from such mechanical scattering. Piling manure in straight-sided, compact, and flat-topped piles may also reduce such losses.

METHODS OF HANDLING MANURE

The primary object in the care and handling of farm manure is to prevent loss of plant nutrients as far as possible. Even under the most favorable conditions, it is practically impossible to prevent some loss of nitrogen and also some loss of organic matter. However, there is no difficulty in conserving all the phosphorus and potassium compounds because they are not volatile.

Although some loss of organic matter and nitrogen is inevitable under practical methods, such losses may be kept at a minimum with good management practices. Some of the more obvious ways of reducing losses of plant-food constituents in manure are indicated by the questions which follow. For the most part, these have been suggested by the discussion in the preceding section.

Questions:

1. Does the use of litter decrease losses from manure?
2. Is it wise to haul manure directly from the stable to the field?
3. What are the proper methods for storing manure?
4. Are preservatives effective in decreasing plant nutrient losses from manure?

The Use of Litter. Bedding or litter is used primarily to furnish clean and comfortable places for animals to lie down. In relation to the value of manure, bedding is used principally for the following purposes: (1) to prevent loss of urine by drainage; (2) to make manure easier to handle; (3) to adsorb and hold plant nutrients; and (4) to increase organic matter and plant-nutrient content.

The principal value of litter as a preserving agent lies in its ability to absorb the urine which in general carries more than one-half the fertilizing value of manure. The efficiency of various litters can be stated in terms of the amount of liquid which a given weight of material will absorb. Ordinary straw can take up two or three times its weight of water. Cut straw can take up about five times its weight of water, whereas good peat moss can absorb as much as ten times its weight of water.

The ability of litters to adsorb or fix both ammonia and potash protects them against loss by leaching, and the ammonia so fixed is less easily volatilized. This effect is very pronounced with such material as peat but little fixation occurs with wood shavings, sawdust, corn stover, and straw.

Litter contributes some fertilizing constituents but usually in relatively small amounts, and the nutrients frequently are of low availability. Litter may also affect the process of biological decomposition of manures. The tendency for manure to heat, which increases ammonia loss by volatilization, and the change of inorganic to organic nitrogen are determined to a large extent by the readily decomposable material contained in the litter. Excesses of highly carbonaceous litters are to be avoided in order to prevent a decreased availability of nitrogen.

The minimum desirable amount of bedding to use is that quantity required to absorb all the liquid. This quantity will depend largely upon the kind of litter used and the animal to be bedded. In general, straw, equal to 25 per cent of the weight of excrement, will be sufficient to absorb the free liquid. The requirement of unchopped straw per head of livestock per day is for cattle, 9 pounds; horses, 10 to 12 pounds; sheep, 1 pound; and hogs, 1½ pounds. It is highly important that ample bedding be used; skimping in the use of bedding is false economy.

Hauling Manure Directly to the Field. Many farmers haul manure daily to the field, which is usually a good practice. Manure spread on the land and worked into the soil is perhaps in its safest place. The soil has the capacity rapidly to fix large quantities of plant nutrients carried in the manure. Very little loss as a rule occurs when manure is hauled directly to the field, except where it is spread on hillsides or on frozen ground, where it may be washed down the slope. If the liquid fraction has been thoroughly absorbed by bedding materials, there will be little loss by runoff. At any rate, it is better to have the liquid go into the soil for plant use than to let it soak in the soil of the barnyard or wash away in the drainage.

If manure is spread daily and is not worked into the soil, considerable loss of volatile ammonia may occur, owing to drying winds or freezing weather, or both, unless sufficient rain has fallen to wash the ammonia into the soil. To decrease these losses (due to drying or freezing), the land should be disked after spreading if possible.

Daily hauling of manure generally is regarded as the ideal method for preventing losses of manure although, as indicated above, there is considerable evidence to show that appreciable losses occur where spread manure remains unincorporated with the soil. However, since it is often impractical to apply manure daily during certain seasons of the year, it is necessary to resort to some storage on most farms.

If field conditions make it impossible to spread manure at certain times of the year, it can be hauled directly to the field and placed in flat-topped, straight-sided piles of 8 to 10 loads per pile. This prevents scattering and washing and involves extra handling but usually is profitable.

The Storage of Manure. In storing manure all practical precautions should be taken to keep losses at a minimum, and consideration must be given to costs involved. Handling manure not only increases the expense but also decreases the value of manure by exposing it to the air, thereby increasing decomposition losses.

Good storage of manure makes provision for keeping the manure heap (1) thoroughly compact, (2) sufficiently moist but not too wet, (3) under cover or shelter, and (4) undisturbed during storage.

Permitting the accumulation of manure under the feet of animals in stalls or in covered lots or pens is an efficient and practical means of storage. If plenty of bedding is used, the liquid manure will be absorbed and the tramping of the animals keeps the manure compact and fermentation losses are kept at a minimum. This type of storage also permits considerable flexibility in time of hauling. Some of the liquid manure may be lost when earth floors are used, and sometimes it may be economical to provide concrete floors to prevent this loss.

At least a temporary storage of manure is necessary on most farms. A common and wasteful method of storage is to place the manure in open piles in the barnlot or under the eaves of the barn. The principal menace of this type of storage is leaching. Storing manure under cover, in a closed shed, is the most practical method of preventing these losses. Such a shed should have a water-tight floor, four continuous walls, and a roof. The manure should be kept compact during storage.

On many farms it is necessary to resort to outdoor open storage. The manure should be placed in compact piles with straight sides and flat top and in such a position that it will receive neither roof nor surface water. While this method usually results in considerable loss of plant food, it is reduced if the manure is piled properly.

Use of Preservatives. Chemical preservatives are added to manure to decrease nitrogen losses. Their action may be due either to their prevention of the biological decomposition of urea and other nitrogenous compounds or to the conversion of volatile nitrogen compounds into non-volatile salts. To be most effective, these preservatives must be brought into contact with the liquid manure as soon as voided because nitrogen losses may begin immediately.

Strong acids, such as phosphoric, sulphuric, and hydrochloric, are effective preservatives. These chemicals make the manure acid, prevent urea decomposition, and change the ammonia compounds to non-volatile salts. Although these acids are effective preservatives, their cost and the difficulties encountered in handling them limit their practical use. It would appear, however, that sulphuric and phosphoric acids, particularly the latter, offer considerable possibilities. Phosphoric acid not only conserves nitrogen but also increases the plant-nutrient content of manure by supplying phosphorus, the element which is deficient in manure. The reversion of the phosphorus in phosphoric acid to an unavailable form does not accompany the fixation of ammonia.

Calcium salts of strong acids [$CaSO_4$, $CaCl_2$ and $Ca(NO_3)_2$] are of some value as preservatives. The reaction of $CaSO_4$ with $(NH_4)_2CO_3$ is

$$CaSO_4 + (NH_4)_2CO_3 \rightleftharpoons CaCO_3 + (NH_4)_2SO_4$$

Their effect is that of changing the ammonia to a stable salt; and so long as the manure is moist, volatilization of ammonia will not occur, but if the manure becomes dry, the chemical reaction is reversed and loss of ammonia may occur.

Superphosphate is used rather extensively as a preservative of manure. Its effect is similar to that of calcium sulphate (gypsum) although its action is more complex on account of the content of monocalcium phosphate in superphosphate. Ordinary grades of superphosphate (16 to 20 per cent) contain 40 to 50 per cent of gypsum, and they are more effective in preventing loss of ammonia, especially when manure becomes dry, than either gypsum or monocalcium phosphate used separately. The reaction of superphosphate with ammonium carbonate may be indicated as follows:

$$CaH_4(PO_4)_2 + 2CaSO_4 + 2(NH_4)_2CO_3 \rightarrow Ca_3(PO_4)_2 + 2(NH_4)_2SO_4 + 2H_2O + 2CO_2$$

In this reaction tricalcium phosphate is formed which does not react with ammonium sulphate on drying so that loss of ammonia is prevented.

It is usually recommended that 1 to 2 pounds of 20 per cent superphosphate be used per day (24 hours) per horse or cow, and amounts for other animals in proportion to their weight. With horses the superphosphate is perhaps best applied just before cleaning the stables; if the manure is allowed to accumulate in barns or lots, usually it is preferable to apply the phosphate just before each bedding. Dusting superphosphate in the gutter of the dairy stable after it has been cleaned is very satisfactory; it acts as a good absorbent of the liquid.

Peat not only has a high liquid absorption capacity but also has a preservative action on manure. The peat may function as a preservative by absorbing ammonia nitrogen, by keeping the manure acid (in low-lime peats), or by virtue of the fact that it causes little heating when peat litter manures are stored. Where bedding materials must be purchased, it would be well to give peat worthy consideration.

FIELD MANAGEMENT OF MANURE

Good management of manure does not stop at the barn or feeding pens; it extends to the field as well. The method of using manure in the field, with reference to crop, kind of soil, use of commercial fertilizers, and time, method, and rate of application, has much to do with getting the most benefit out of each ton. Questions most commonly raised in discussions of methods of handling manure are proposed here.

Questions:

1. When should manure be applied, by what methods, and at what rate to give best results?
2. Should manure be supplemented with commercial fertilizers?
3. What crops respond best to farm manure?

Application of Manure. Prompt spreading of manure is generally considered best, although when in good storage it is likely to lose less value than if spread on the field without being plowed under or worked into the soil immediately. Losses of applied manure may occur in three ways: (1) volatilization of ammonia nitrogen as a result of drying or freezing; (2) surface runoff water carrying soluble portions of all three nutrients; or (3) leaching of nutrients.

Time of Application. Much experimental work with commercial fertilizers indicates that their effectiveness is decreased if they are applied a considerable time in advance of seeding the crop. This effect is attributed to leaching losses and to the fixation in less-soluble forms of the plant nutrients by the soil. Fresh and properly stored manures contain large amounts of soluble nutrients, and these same principles apply. Manure applied to corn land in the spring is likely to give greater returns than a like amount of manure applied in the fall. This effect will probably apply only to the crop of the first year and not on the later crops. If manure has already lost its readily soluble nutrients, the time of application would be of less concern.

Little loss of nutrients from manure may be expected when applied on the level, medium to heavy textured soils. These soils are safe for fall and winter spreading of manure although they will fix considerable quantities of the nutrients in less available forms. It is not advisable to place manure on sandy soils or on hilly fields much ahead of plowing time because of leaching and erosion losses. However, it is to be remembered that manure applied on sloping fields decreases erosion losses which counterbalances to some extent the loss of nutrients from the manure.

Method of Application. In applying manure broadcast, it is important to get an even distribution over the field. As a rule it is advisable to plow manure under or to disk it in immediately after spreading. The depth to which it should be incorporated depends upon the climate, the type of soil, and the kind of manure. In general, manure should be incorporated so that sufficient air will be present for normal decomposition and deep enough to prevent losses from drying. Deeper incorporations would be more desirable on light, porous soils than on heavy or wet soils. The most important thing, in regard to the method and depth of incorporation, is the provision for completely covering the manure.

Although not substantiated by reliable evidence, it is commonly recommended that manure be thoroughly mixed with the surface soil; disking before plowing is one method of attaining this end. There is some evidence to indicate that a higher efficiency can be obtained from localized placement (row or hill application) than from the broadcast application. Lack of necessary equipment for making such applications limits the use of this method at present, however.

A surface covering of manure, as a top dressing on small grains in which a legume seeding is made, has been found an effective method in many areas. The top dressing improves the moisture conditions and the tilth in the top soil layer more than sufficiently to compensate for the losses that may occur from the manure in using this method. Although this is an effective method of improving stands of legume seedings in small grains, larger yields of small grains are often obtained from applying the manure before plowing than from top-dressing the crop.

Rate of Application. The rate at which manure should be spread will usually depend upon the amount produced on the farm. As much of the cultivated land should be covered as possible. It is usually better to cover an entire field with a light application than to give only a part of the field a heavy coating. In other words, the crop response from 50 tons over 10 acres proves larger than from 50 tons on 5 acres. It is wise to gauge the rate of application so as to extend the acreage covered for each crop. If possible, every available acre should be covered each year.

Frequent small applications are more efficient than a single large application; returns per ton of manure are greater. This implies that two applications per rotation are better than one.

Supplement Manure with Commercial Fertilizers. The content of phosphorus in manure is low as compared with nitrogen and potassium; for many soils and crops it is not a properly balanced fertilizer. For this reason it is usually advisable to supplement manure with a phosphatic fertilizer. The phosphorus may either be added with the manure as it is spread or applied to the succeeding crops at planting time. The practice of supplementing manure with phosphate is known as manure reinforcement. When, however, the addition of nitrogen is the chief objective in the application of the manure, as when used in orchards or as a top dressing on pastures, reinforcement may be unnecessary.

The conservation and wise use of farm manure alone will obviously not result in the maintenance of the plant nutrients in the soil. Since only a portion of the plant nutrients removed by crops is returned to the soil in manure, even when all the crops are fed on the farm, commercial fertilizers and lime, as needed, must accompany the use of manure if the proper level and balance of plant nutrients in the soil are to be maintained.

Crops Which Respond Best to Farm Manures. Although all crops respond to manuring, the intertilled crops and grasses show most stimulation, and it is logical in the use and application of manure to favor those crops which are most responsive. Manure is essentially a nitrogen and potash fertilizer and, since corn, cotton, tobacco, potatoes, and other commonly grown intertilled crops are heavy users of nitrogen and potash, it is obvious that these crops should be particularly benefited. When these crops are grown, they should receive at least a part of the manure that is produced on the farm.

Because of the tendency for oats to lodge when too much nitrogen is available, manure generally should not be used directly for this crop. Oats are seeded usually after some cultivated crop and feed quite satisfactorily on the residual manure which had been applied for the cultivated crop.

Manure can generally be used to an advantage as a top dressing on fall wheat. It not only stimulates and protects the wheat but also greatly benefits the legume seeding that may be made in the wheat.

Manure is of value in orchards, especially on sandy soils where nitrogen is of much concern.

Where permanent pasture is a large enterprise, a liberal share of the manure can generally be used to an advantage on the grass. Yearly dressings of finely divided manure, particularly if reinforced with phosphorus, have given excellent response.

Heavy applications of manure have proved desirable for truck, garden, and greenhouse crops. It is an excellent fertilizer material for intensive cropping operations.

FERTILIZING PROPERTIES OF MANURE

Until the introduction of commercial fertilizers and the general utilization of leguminous crops, manure had always been the main source of supply of plant nutrients. In many instances this is still true. The beneficial effects produced by manure on plant growth are believed to be due for the most part to its content of nitrogen, phosphorus, and potassium. The growth-promoting influences of manure may in part be due to the presence of "minor" elements, organic matter, microorganisms, or to the presence of growth-regulating substances of a vitamin-like nature. As compared to ordinary grades of commercial fertilizer, manure is low in nutrient content. Stated in fertilizer terms, it has on the average the analysis 0.5–0.25–0.5, containing 1¼ units of plant nutrients as compared to about 20 units in ordinary mixed fertilizers. Although the nutrient content of manure is low, the quantity of nutrients added per acre in a normal rate of application of manure is usually fifty to one hundred times the quantity applied as commercial fertilizer. This difference, however, is offset to some extent by the lower availability of some of the manure constituents. Several questions follow which may be of assistance in studying the fertilizer value of manure.

Questions:

1. How does the availability of the plant nutrients in manure compare with that of the nutrients in commercial fertilizers?
2. What is the fertilizing value of manure compared to that of commercial fertilizers?
3. Are the growth-promoting effects of manure confined to its nitrogen, phosphorus, and potassium content?
4. Does manure have significant residual effects in soil?

The Availability of Fertilizing Constituents of Manure. The nitrogen of the liquid fraction of manure is about equal in availability to that of mineral fertilizers like ammonium sulphate and sodium nitrate. The nitrogen of the solid fraction, however, is of very low availability. Experimental results reported show that less than 10 per cent of the nitrogen in the solid portion is recovered in the first crop, whereas over 70 per cent of the nitrogen from the urine, when applied with the solid, may be recovered in the first crop. As would be expected, the residual effect, measured by the recovery of nitrogen, is greater for the solid than the liquid. The availability of nitrogen in ordinary farm manures may be expected to average about 25 or 30 per cent of that of mineral nitrogen fertilizers, depending upon the kind of manure and the changes taking place in handling and storage.

The availability of the phosphoric acid and the potash of manure is essentially equal to that of mineral fertilizers. A number of investigators have reported a higher recovery by plants of phosphoric acid with manure than with chemical fertilizers.

The Fertilizing Value of Manure Compared to That of Commercial Fertilizers. The value of a ton of farm manure in terms of dollars and cents has been the subject of much discussion. On account of the variable nature of manure, even though the amounts of plant nutrients present are known, it is difficult to place a value on the manure which shall express its worth to the farmer in the growing of crops.

Commercial values assigned manures usually are based solely upon the quantities of plant nutrients contained and do not take into consideration the availability of the nutrients or the value of the organic matter present. Since the composition of stable manure is so variable, any estimate of value nearer than perhaps 50 per cent should be regarded largely as guesswork.

Crop increases per pound of nutrients applied frequently are somewhat higher with chemical fertilizers than with manure, perhaps owing to the unbalanced plant-nutrient ratio of manure for most crops and the lower availability of the nitrogen in manure. Obviously, the nature of the crop and soil conditions will affect any such comparison. The idea that applications of manures tend to increase yields more than commercial fertilizers in unfavorable seasons does not appear to be substantiated by experimental evidence. There have been no significant differences in the yield of corn and wheat on comparably manured and chemically fertilized plats in long-time experiments at the Ohio station.

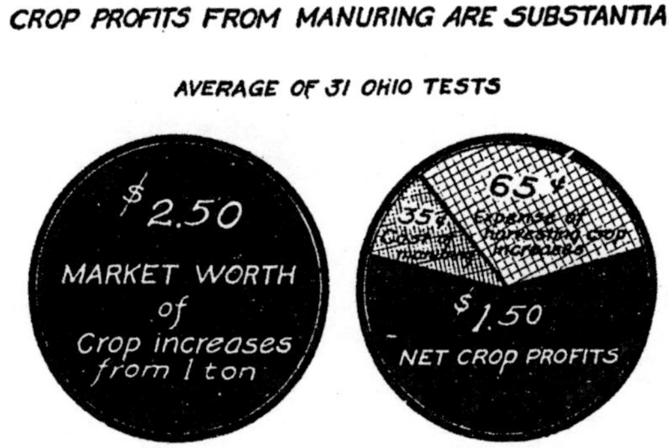

FIG. 51. The value of crop increase from the application of one ton of average manure. Note that net value of the increase is one dollar less than the gross value of the increase because of the cost of spreading the manure and of harvesting the additional crop. When carefully stored manure was used, the value of the crop increase was about one dollar greater. [From Ohio Ext. Ser., Bul. 131, by J. A. Slipher.]

Growth-Promoting Effects of Manure Other Than Those Attributed to Its Content of Nitrogen, Phosphorus, and Potassium. Although crop increases per pound of nutrients applied are no larger and frequently are less with manure than with chemical fertilizers, it cannot be concluded that manure has no value other than its direct fertilizing effect. It would appear that the relatively low availability of a considerable portion of the nitrogen of manure should decrease the value of manure compared with chemicals, below that usually found. This may be offset partially by certain effects other than those produced by nitrogen, phosphorus, and potassium.

The benefits from *heavy* applications of manure upon the physical properties of the soil frequently have been shown and generally are recognized by farmers. These effects, however, are not realized to any measurable degree under most farm conditions, especially on normally well aerated soils, because of the relatively small amounts of manure applied, compared to the weight of the plow layer of soil.

It is believed the addition of farm manures favorably influences the microbiological flora of the soil. Large numbers of microorganisms are introduced by manure, which may hasten desirable biological changes in the soil. Certain organisms, thus added, convert organic nitrogen into ammonia; others change ammonia into nitrate nitrogen; and still others decompose carbohydrate materials. Manures also supply food materials for various desirable organisms already present in the soil. An increased production of carbon dioxide occurs in manured soils, and this extra carbon dioxide may be of value in increasing crop growth.

The organic matter is an important factor in giving added value to manure in the soil aside from its direct effects on the physical and biological properties. When fresh manure is applied, the organic matter is gradually decomposed in the soil and its decomposition products tend to increase the availability of the mineral plant-food constituents in the soil. It has been demonstrated that the soluble humates in manure increase the solubility and mobility of mineral phosphates, and it has been suggested that the water-soluble and colloidal constituents of manure have a definite value aside from the accompanying nutrient salts.

There is also the probability that manure carries small quantities of certain growth-promoting substances, similar in nature to hormones and vitamins in animals. Thiamine chloride (vitamin B_1), creatinine, and other growth-regulating substances are known to be present in manure in low concentrations. Whether these substances contribute anything of practical value to manure cannot be said now.

In addition to nitrogen, phosphorus, and potassium, manure contains calcium, magnesium, sulphur, iron, manganese, and other elements that no doubt have some fertilizing value. It is thus evident that manure may have value which cannot be attributed to its content of nitrogen, phosphorus, and potassium.

Residual Effects of Manure. An application of manure usually shows a favorable influence on crop yields for several years. Owing to the slower availability of its nitrogen and to its contribution to the soil's organic supply, the beneficial effects of manure are distributed over a longer time than those of chemical fertilizers and, furthermore, with repeated applications the cumulative effects tend to be larger. Striking results in showing the long-continued effects have been obtained (Fig. 52) by making heavy applications of manure for several years in succession and then discontinuing the application.

The lasting effect is due in part to the slow availability of certain plant-food constituents in manure, in part to the fact that a portion of the organic matter may remain for several years and aid in increasing the soluble plant nutrients in the soil, and in part to greater quantities of roots and stubble of the larger crops grown.

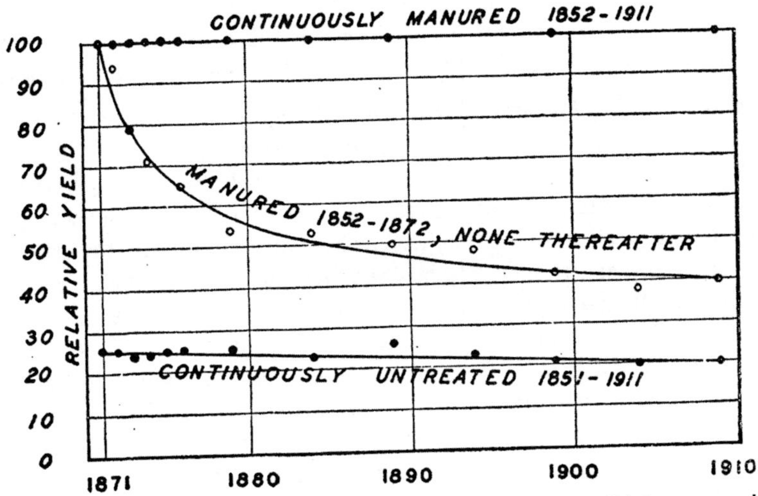

FIG. 52. Residual effects of heavy applications of manure (14 tons annually) on the yield of barley grown continuously on the soil of the Rothamsted Experiment Station in England. [From Ohio Exp. Sta. Bul. 605.]

EFFECTS OF MANURE UPON THE SOIL

Many chemical, physical, and biological properties of the soil are influenced by applications of manure. Perhaps the most important immediate effects are chemical in that the quantities of available plant nutrients are increased. Immediate physical effects are less marked although increases in aeration and the infiltration of water on heavy soils are observed. Increased crop growth, resulting from manure applications, tends to promote granulation and aids in retaining a good physical condition by protecting the soil from beating rains. The addition of manure also causes a rapid increase in the biological activity of the soil. The more important effects, however, are the result of long-repeated heavy applications of manure. The increased stable-humus supply of the soil serves as a storehouse for nitrogen and as an agent promoting desirable physical and biological properties.

The following discussion of this topic should provide answers to the questions listed.

Questions:

1. Does applying manure affect the supply of organic matter and nitrogen?
2. Do applications of manure influence the mineral composition of soils?
3. Are the physical properties of the soil influenced greatly by the normal rates of applying manure?
4. How does the application of manure aid in decreasing soil erosion?
5. What is the role of farm manure in the maintenance of soil fertility?

Effect on Soil Organic Matter and Nitrogen. The value of organic matter in soils is generally recognized, and dark-colored soils are usually considered to be superior to lighter-colored ones not so well supplied with organic matter. This fact, together with the observation that under most cropping conditions there is a gradual and continuous decrease in organic content, has resulted in placing much emphasis upon the humus-supplying power of manure. Available data indicate that this effect of manure has perhaps been exaggerated. The quantities of manure available in practice are normally too small to do more than retard the decline in soil organic matter. Therefore, it is more nearly correct to think of the normal applications of manure as a means of aiding in the maintenance of the soil organic matter supply rather than as a means of increasing it.

Changes in organic matter content of soils, attributable to manure, are affected in two ways: the actual residue of the manure applied, and the contribution by the roots and stubbles of the increased crops produced by the manure. The value of manure for conserving soil humus and nitrogen depends to a large extent upon the crops used in the rotation. If used on cultivated crops such as potatoes, corn, or tobacco, the gains from increases in roots and stubble residue may be very small, but with such crops as the sod-forming legumes its maximum indirect effect is realized.

In regard to the effect of manure on soil nitrogen, it is to be kept in mind that only 75 or 80 per cent of the nitrogen of the food of animals can be found in the fresh manure, and ordinarily a considerable portion is lost in handling. Although it is appreciated that crops raised on a farm and fed to animals of the same farm do not make so exhaustive a drain on soil fertility as when the crops are removed from the farm, yet under these conditions there can be no increase in soil nitrogen unless a large share of the acreage is devoted to leguminous crops.

Even under a livestock system of farming there can be no effective increase in the soil nitrogen supply, unless much feed is purchased to be fed on the farm, a considerable acreage of leguminous crops is grown or else commercial nitrogen is purchased.

The Mineral Composition of the Soil as Affected by the Application of Manure. The extent of the changes produced in the mineral composition of the soil depends upon the kind and amount of manure applied and on the accompanying losses of mineral elements from the soil. There can be no increase in the total mineral content of soils with a livestock system of farming where all the feed is grown and fed on the farm. In order that the total quantities of mineral elements be increased, it is necessary that they be purchased either in the fertilizer bag or in the form of feed to be fed on the farm. Obviously, the mineral content of a particular field may be increased by concentrating the applications of manure to that field, but that is merely taking fertility from one field and putting it on another.

Data from the Ohio station, covering 32 years of cropping, show that manure increased considerably the amounts of active calcium, magnesium, potassium, sodium, and manganese in the soil, compared to the quantities contained in the unmanured soil. In most cases, however, even the heavier application (16 tons per 5-year rotation) of manure did not maintain the supply of these elements at the original level. The lime requirement was not much affected by manure, but all plots were considerably more acid than at the beginning. The exchange capacity of the soil was increased.

Effect of Manure on the Physical Properties of the Soil. Some of the more important benefits from manure are brought about indirectly. The humus which manure supplies improves the tilth or physical condition of the soil, the water-holding capacity, the permeability of the soil to water, aeration, and temperature relations. The physical properties of soils, which are either too heavy or too light, can especially be improved by an increase in the humus content, which can be effected to some extent by manuring.

With the usual limited quantities of manure available on general farms, marked changes in the physical properties of the soil are not realized. However, the physical effects of incorporating coarse organic matter with the soil are important even with moderate applications. Furthermore, manure, when used as a top dressing, protects the soil from beating rains, decreases evaporation losses of water, and appreciably improves the tilth of heavy soils.

Effect of Manure on Soil Erosion. Incorporating manure with soil may be effective in reducing soil erosion by increasing the permeability of the soil to water, thus decreasing runoff losses, and by increasing the density of the vegetative cover, which in turn decreases rate of surface runoff and increases water penetration. A large number of experiments has been reported, showing that manure is effective in reducing both water and soil losses.

Farm Manure and the Maintenance of Soil Fertility. As the average fertility of our soils has decreased, the saving and efficient use of manure have become a matter of vital importance in our system of agriculture. For many years manuring has been known to be perhaps the most logical and practicable method of aiding in the maintenance of soil productivity. Most cultivated soils give marked increases in yield where manure is applied, and this has brought about an increasing interest in the handling and care of this valuable product. Manure is no longer considered something to dispose of, but it is generally recognized as a most valuable by-product of the farm.

The greatest returns from manure are realized only when it is used in conjunction with other good soil-management practices. Proper methods employed in the handling, care, and use of farm manure are not sufficient in themselves to maintain soil fertility. Manure cannot be substituted for good cropping systems, lime, fertilizers, good seed, or proper drainage and cultural practices. Maximum returns from manure are assured only where it is well cared for, efficiently handled, and wisely used in combination with the other good soil-management practices.

CHAPTER XI

NUTRIENT REQUIREMENT OF PLANTS

The growth and development of plants are determined by numerous factors of soil and climate and by those factors inherent in the plants themselves. Some of these factors are under the control of man, but many are not. Man has little control over air, light, and temperature, for example, but can influence the supply of plant nutrients in the soil. He may increase the supply of available nutrients by modifying soil conditions through good management or by making additions in the form of fertilizers. Anyone dealing directly with the growth of plants is particularly concerned with their nutrient requirements, and in this connection the following topics will be discussed.

Objectives:

A. Elements used by plants.

B. Effects of nitrogen, phosphorus, and potassium on plants and the quantities removed by crops.

C. Determining soil-nutrient deficiencies.

ELEMENTS USED BY PLANTS

If a soil is to produce crops successfully, it must have, among other things, an adequate supply of all the necessary nutrients which plants take from the soil. Not only must required nutrient elements be present in forms that plants can use but also there should be at least a rough balance between them in accord with the amounts needed by plants. If any of these elements is lacking or if it is present in improper proportions, normal plant growth will not occur. Elements required by plants are designated as *essential* or *indispensable*, and in giving consideration to them the following questions arise.

Questions:

1. From what sources do plants obtain their nutrients?
2. Which elements are essential?
3. Which of the essential elements receive major consideration?
4. Of what importance are the minor elements?

Sources of the Plant Nutrients. Plants get their nutrient elements from three sources, that is, air, water, and soil. Carbon and some of the oxygen are obtained from the air; whereas hydrogen, some oxygen, and possibly some carbon are taken from the soil solution. Legumes inoculated with effective nodule bacteria obtain some of their nitrogen from the air. Besides those mentioned the nutrient elements must be taken from the soil by crops.

The Essential Plant-Nutrient Elements. The ten chemical elements required in rather large quantities for growth of *all* crops are carbon, hydrogen, oxygen, phosphorus, potassium, nitrogen, sulphur, calcium, iron, and magnesium. Plant growth will be limited by a deficient available supply of any one of these elements regardless of the abundance of the remaining ones. The experimental work until 1900 indicated that only these ten elements were required by plants, and they have been considered until rather recently the only ones necessary for the normal growth and development of plants. But with the aid of improved technique, investigators have added to the original list several elements, including copper, boron, manganese, and zinc, which are undoubtedly essential although needed in small quantities. In addition to the elements listed which are needed by all plants, certain others like sodium, chlorine, and possibly iodine are needed by some but not necessarily by all plants. The supply of these elements in soils in forms available for plant use becomes, therefore, a very important consideration.

Although more than thirty different elements have been detected in the analysis of plants, the mere presence of one of them in a plant does not prove that it aids in the development of that plant or that it is indispensable for plant growth. The necessity for any particular element can be determined only by carefully conducted experiments.

Major Plant-Nutrient Elements. Plants are made up largely of carbon, hydrogen, and oxygen, but some of the elements which occur only in relatively small amounts are just as essential to their growth and development as those which compose the greater part of the plant tissue. Obviously, it is incorrect to speak of one *essential element* as being more important than any other *essential element* without specifying the viewpoint under consideration, but from the agricultural point of view or from the farmer's standpoint, the elements nitrogen, phosphorus, and potassium are of major concern. (The importance of calcium is discussed on pp. 119 and 120.) Years of experimental work and practical experience have proved that the available supplies of these elements in soils are more likely to be insufficient for maximum plant growth than are the supplies of the other essential elements. In fact, there are very few soils which are not deficient in one or more of these elements. They assume agricultural importance by virtue of the fact that (1) they may be rapidly removed or lost from soils, (2) they may exist largely in unavailable forms in the soil, and (3) the only way to increase the phosphorus and potassium content of a soil is to buy them in some form or other. Consequently, the fertilizer companies have made these three elements (N, P, and K) the basis of the composition of commercial fertilizers which may supply any one, two, or all three of them.

Minor Elements. It has already been mentioned that plants must have several elements other than nitrogen, phosphorus, and potassium for proper growth. In speaking of fertilizers, then, additional elements are referred to frequently as minor elements either because they are needed by plants in small quantities or because fewer soils are deficient in them, although in some soils a deficiency of one or more elements materially limits crop production. However, undue publicity has undoubtedly exaggerated the general need for certain of these elements in fertilizer materials, and often the impression has been made in the minds of many people that soils in general are deficient in these elements.

Minor or secondary elements, such as magnesium, manganese, sulphur, calcium, zinc, copper, boron, iron, and perhaps others, are essential, but except in certain specific cases either they are supplied abundantly by the materials commonly used in making mixed fertilizers or they are present in sufficient quantities in the soil. The addition of these elements to all fertilizer mixtures without regard to the crop or soil is unwarranted and sometimes may prove actually to be injurious.

Under conditions where a minor element deficiency occurs, two important questions come to our minds: (1) Why are such deficiencies appearing now rather than at some earlier time; and (2) how do these elements act in promoting growth? There are perhaps several reasons why more minor element deficiencies are coming to attention now than at any previous time. The continued use of land for cultivated crops, the use of higher-analysis fertilizers (more pure salts), and the diminishing use of organic nitrogen fertilizers (plant or animal origin) have done much to hasten the occurrence of minor element deficiencies. It is also true that deficiencies occurred in some instances at earlier times, but the symptoms were not recognized.

The function of all these minor elements is not well understood, but it is believed that they affect plant growth in one or perhaps more of the following ways: (1) used directly as constituents of plant tissue; (2) act as catalysts or stimulants; (3) effect oxidation-reduction processes in the plant; (4) may aid in regulating the acid content of the plant; (5) may affect the plant osmotically; (6) may affect the entrance into the plant of other elements; or (7) may aid plant growth by providing a more favorable environment for the plant roots.

Calcium supply is frequently a limiting factor in acid soils. It is the important constituent of lime and is usually not classed as a fertilizer, but where the content of soluble calcium in soils is so low as to supply plants inadequately, then it should be considered a fertilizer. Calcium performs many functions in the soil and plant which are of great fundamental significance (p. 118). It is generally present in large quantities in mixed fertilizers of the ordinary grades.

Magnesium is most likely to be deficient on sandy soils and particularly in seasons of heavy rainfall. Magnesium deficiencies have been observed frequently in the soils of the south Atlantic coastal plains and to some extent on the lighter soils in other parts of the country. This element will be found in variable quantities in most fertilizers because of the frequent use of dolomitic limestone in their manufacture. Magnesium should be included in some form available to plants in nearly all fertilizers used on the soils along the Atlantic coast.

Manganese usually is present in sufficient quantities in most soils and, it need not be added, in fertilizer mixtures except in soils where a definite deficiency has been noted. The amount of manganese required by plants is small, and a deficiency in this element is most likely to be found in alkaline soils, especially those which originally were acid and have since been heavily limed. Manganese usually is needed for certain crops on alkaline muck soils.

Copper, boron, and *zinc* are sometimes found in insufficient quantities in soils, especially in the southeastern and in some of the eastern states. The yields of several crops on different soil types have been increased considerably by the addition of one or more of these three elements. A number of crops grown on certain muck soils, particularly those with an acid reaction, are greatly benefited by applications of copper sulphate. For soils in which deficiencies occur, these elements should be added in fertilizers in amounts found necessary by field tests.

Sulphur is used in large amounts by most plants but is usually found in considerable quantities in the soil. In areas close to industrial centers sufficient sulphur to supply crops is brought down by rain and snow from the atmosphere. The ordinary grades of superphosphate contain 40 to 50 per cent of calcium sulphate, and several other fertilizers and fertilizer ingredients also contain sulphur. They are all important sources of sulphur in the fertilizer-consuming areas of the United States. This element frequently is used as a corrective agent in overcoming an unfavorable soil condition known as alkali. As a result of the oxidation of applied sulphur to sulphuric acid, the *p*H of the soil is lowered.

Iron has been recognized as an essential element for a long time, and soils usually contain sufficient quantities of iron for normal plant growth. Its availability varies widely with the degree of soil aeration, being higher under anaerobic conditions. Occasionally soils are found which are deficient in available iron and, on soils containing considerable quantities of lime, the movement and activity of the iron within plants are reduced in some manner by the presence of excess calcium.

EFFECTS OF NITROGEN, PHOSPHORUS, AND POTASSIUM ON PLANTS AND THE QUANTITIES REMOVED BY CROPS

A knowledge of the various ways in which the nutrient elements are of value in affecting the growth of plants is of practical interest and importance. Each nutrient performs definite duties within the plant and no one nutrient can be completely substituted for another. Although each element performs certain specific functions, they must all work together to produce the best results. It is to be remembered, therefore, that the effect of any particular nutrient on plant growth is governed by the supply of the other essential elements, and hence the effects of any one cannot be interpreted on the basis of the activity of that element alone. In this section we are concerned with the three elements (nitrogen, phosphorus, and potassium), and answers to the following questions will be sought.

Questions:

1. What are the effects of nitrogen on plant growth?
2. In what respects do the influences of phosphorus differ from those of nitrogen?
3. Are noticeable effects on plants produced by potassium?
4. Do crops remove appreciable quantities of nitrogen, phosphorus, and potassium from soils?

The Effects of Nitrogen. Perhaps no element has received so much attention as has nitrogen in studies relative to plant nutrition. It is found in greater quantities in young, growing parts of plants than in the older tissues and is especially abundant in the leaves and seeds. Nitrogen is a constituent of every living cell, and hence its contribution to plant and animal life is evident. Its importance in crop production is emphasized by the knowledge that this element generally occurs only in small quantities in soils in available forms, is used by crops in large quantities, is easily lost from soils by leaching and erosion, and of the three elements (nitrogen, phosphorus, and potassium) is the most expensive to buy in the form of commercial fertilizers.

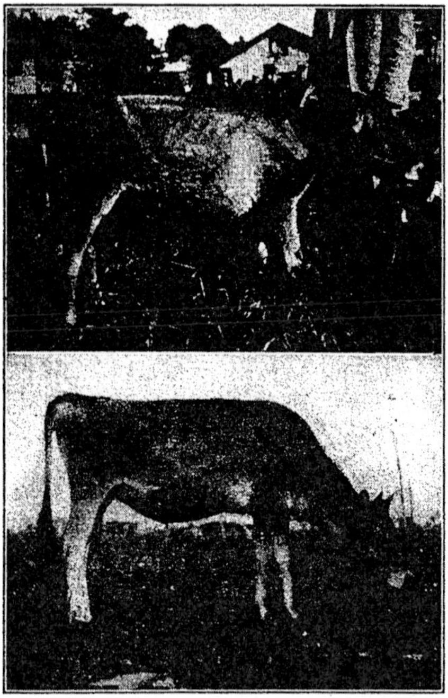

FIG. 53. In different soil areas animals may suffer from a deficiency of different nutrients in the locally grown feed. "A calf showing an advanced stage of salt sick or nutritional anemia. Note emaciation, lack of condition as indicated by the hair, appearance of the eye, and that the animal shows evidence of diarrhoea" The animal "recovered when given access to the iron-copper supplement." [Courtesy of Fla. Exp. Sta.]

Effect on Growth and Color. A deficiency of nitrogen is evinced by the plants' turning yellow and by a slow and stunted growth. An abundance of nitrogen promotes rapid growth with a greater development of dark-green leaves and stems. Although one of the most important functions of nitrogen is the encouragement of above-ground vegetative growth, this growth cannot take place except in the presence of adequate quantities of available phosphorus, potassium, and other essential elements.

Effect on Maturity. An ample supply of available nitrogen during the early life of the plant may stimulate growth and result in earlier maturity. However, the presence of an excess of nitrogen throughout the growing season of the plant frequently prolongs the growth period, and the plant may fail to mature properly. This effect is especially significant for certain crops in regions having a short growing season or in those areas where an early-fall freeze may do great damage to fruit trees whose season's growth period has been prolonged.

FIG. 54. Two views of the same cow. Above, when fed a ration deficient in cobalt. Below, after a cobalt supplement was included in the ration. [Courtesy of Dairy Department, Michigan State College.]

Effect on Quality and Disease Resistance. A large supply of available nitrogen encourages the production of soft, succulent tissue which is susceptible to mechanical injury and the attack of disease. Either effect may decrease the quality of the crop. However, the development of softness in the tissues may be desirable or undesirable, depending on the kind of crop. For vegetables used for their leaves, pronounced succulence, tenderness, and crispness are desired. Other vegetables and some fruits may have their keeping and shipping qualities impaired when they are grown with an excess of available nitrogen. An excess of nitrogen may encourage lodging in grains, which frequently decreases the quality, but a normal amount of nitrogen usually increases plumpness in grains.

The Effects of Phosphorus. The total supply of phosphorus in soils is relatively small and the available supply usually falls short of crop requirements. Phosphorus is used more generally on mineral soils than is any other fertilizing element, and it is usually the first element that becomes deficient after a soil is brought under cultivation. The importance of phosphorus in plant and animal nutrition is perhaps well recognized by every one. It is present in seeds in larger amounts than in any other part of plants although it is found extensively in the young growing parts. Like nitrogen it is a constituent of every living cell.

Sufficient quantities of available phosphorus are necessary for normal transformations of carbohydrates in plants—the changing of starches to sugars, for example. Phosphorus is necessary also for the assimilation of fats in plants, and apparently it increases the efficiency of the chloroplastic mechanisms. Phosphorus enters into the composition of nuclear compounds in cells. The presence of phosphorus aids the plant in taking up potassium and tends to counteract the effects of excess nitrogen. An excess of phosphorus in proportion to the supplies of other required nutrients may decrease yields, especially on the lighter soils. This is believed to be caused by a hastening of the maturation processes and by the consequent reduction in vegetative growth. The exact functions of phosphorus in plant nutrition in all their relationships are not well understood. However, it is known that phosphorus produces several important effects on crop growth which are of practical interest. The effects of phosphorus on plants are less striking than those of nitrogen and are often difficult to recognize by casual observation.

Effect on Root Development. The most obvious effect of phosphorus is usually on the root system of plants. Phosphorus-starved plants tend to have a stunted root system which decreases their feeding zone, and thus the plants are less able to withstand adverse conditions. The stimulation of root development, due to the addition of available phosphorus, is especially valuable on the heavier soils where root development is naturally restricted.

Effect on Growth and Color. Phosphorus hastens the ripening processes of plants. In the presence of sufficient available phosphorus, seed formation begins sooner and crops may mature several days earlier than where phosphorus is deficient. Phosphorus is essential for seed formation and its effect on the maturity of crops may be explained on this basis. In certain climatic regions the importance of hastening maturity is obvious. Since phosphorus is essential for cell division, plants starving for this element are stunted in growth.

Effect on Quality and Composition. Phosphorus-starved plants may mature late and be injured by frost, thus lowering the quality. Hay and other forage crops grown on phosphorus-deficient soils contain reduced amounts of this element and so are of inferior feeding value. The presence of this element seems to favor good pollination, which affects quality, in corn. It strengthens the straw and stems and thus decreases lodging, decreases the ratio of straw to grain in cereals, and increases total yield. In general, phosphorus improves the quality of plants and plant products by stimulating the production of a more vigorous plant growth, making it more disease resistant.

The Effects of Potassium. The total content of potassium in soils is usually high, but because of the small amounts available to plants it is of major concern to farmers. Potassium is usually needed more often and in larger quantities on the light soils (sandy soils) and mucks than on the heavy soils. The total potassium in sandy soils frequently is low and that of muck often reaches a very low level. Potassium, expressed as K_2O (potash), makes up about 40 per cent of the ash of most plants, and it is not localized in any part of the plant to the extent that phosphorus is, although in some crops it may tend to accumulate in the leaf and stem rather than in the grain.

With the passing of time the need for potassium in our cultivated soils is becoming more acute. Potash deficiency symptoms in plants are appearing in widespread areas. This element plays an important part in many of the vital physiological processes in the plant although the exact nature of the mechanism by which potassium functions is not definitely known. It is essential in all cell metabolic processes and apparently has a specific role in influencing the uptake of certain other mineral elements, in regulating the rate of respiration, in affecting the rate of transpiration, and perhaps also in influencing the action of enzymes.

Potassium encourages the development of the root system of plants. An excess of this element may delay maturity although in general it has a balancing effect on both nitrogen and phosphorus, with respect to the maturation processes. The intake and retention of water seem to be regulated to some extent by the quantity of potassium present, thus affecting the resistance of the plant to injury from drought and frost.

Effect on Synthetic Activities. One of the most important functions usually attributed to the element potassium is its effect on the plant synthesis of carbohydrates and proteins. Potassium is essential for the production of starch, sugar, and other carbohydrates and in the translocation of starch and other materials within the plant. Some investigators are of the opinion that potassium aids in the reduction of nitrates in the plant preparatory to protein synthesis. It is also believed essential in the development of chlorophyll and in the synthesis of oils or fats and albuminoids. Potassium is not found in permanent organic combinations with these compounds, but it is believed essential for the production of them all.

Effect on Vigor and Disease Resistance. Potassium appears to improve the general tone and vigor of the plant, which in turn permits the plant to be more disease resistant. As a rule, crops which do not receive sufficient potassium are more susceptible to disease. This effect is especially noticeable where the plant receives excess nitrogen.

Effect on Quality and Composition. Potassium increases plumpness in grains, producing greater test weight per bushel, and it makes the stalk and straw of plants more rigid, thus preventing lodging to a certain extent. Ears of corn, produced on potassium-deficient soils, frequently are chaffy, tapering at the tips, and the kernels are loose on the cob because they are not well filled with starch.

Where a deficiency exists, the addition of potassium affects the quality of tobacco, improves the quality in potatoes, and increases the sugar content of sugar beets. Many other illustrations might be given, indicating the effect of potassium on quality and composition of crops.

Quantities of Nutrients in Crops. Most of the plant nutrients removed from a virgin soil by the native forest or prairie vegetation are ultimately returned to the soil. Such soils in their natural, undisturbed condition are covered at least in part with a layer of organic matter in its various stages of decay and, as decomposition proceeds, the nutrients in the organic matter are released and are again available for plant use. A decrease in soil fertility under virgin conditions is thus not likely to occur, owing to the return in time of almost all the vegetative growth.

When these soils are brought under cultivation, the condition is entirely changed because frequently most of the crop produced is removed from the soil permanently. Under a livestock system of farming where the crop is fed on the farm, a considerable portion of the nutrients removed can be returned to the soil in the form of

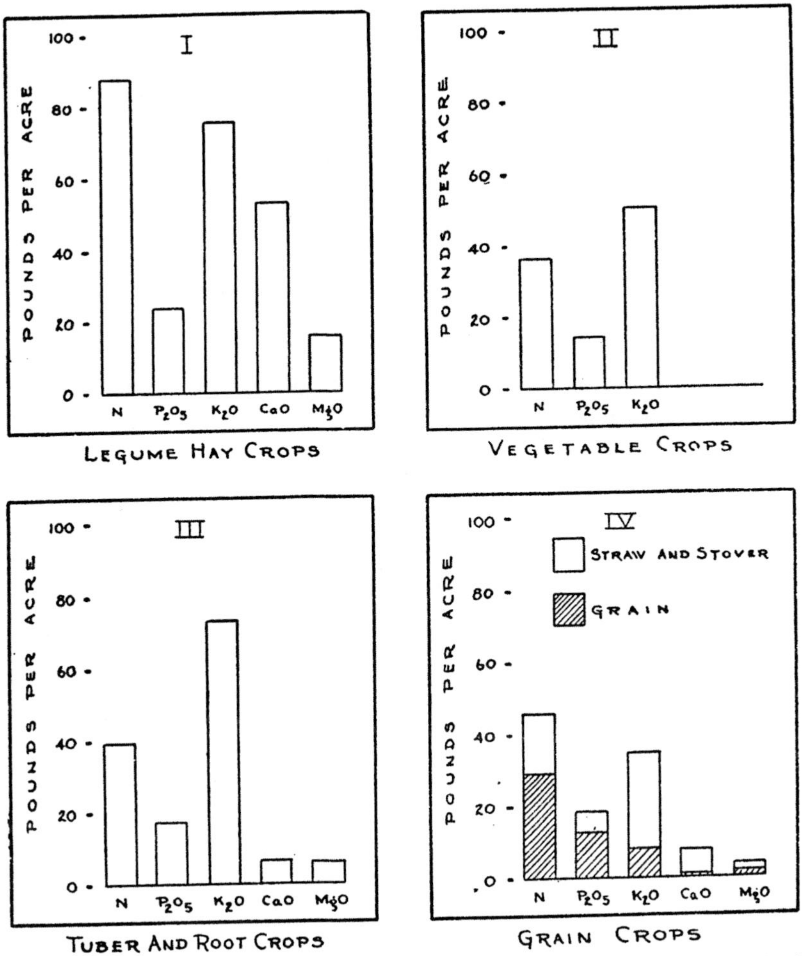

FIG. 55. The relative and also total quantities of plant nutrients contained in average yields of different crop groups vary greatly.

manure. But it is to be remembered that when any product is sold off the farm, whether it is livestock, wool, milk, beans, wheat, cotton, tobacco, corn, sugar beets, or potatoes, it represents a loss to the farm of the plant nutrients it contains.

The nutrient content of several farm products is given in Table 27. In considering these quantities of nutrients, it is important to keep in

TABLE 27

QUANTITIES OF NUTRIENT ELEMENTS CONTAINED IN FARM PRODUCTS *

Crop	Amount	Pounds				
		Nitrogen (N)	Phosphorus (P_2O_5)	Potash (K_2O)	Calcium (CaO)	Magnesium (MgO)
Apple pomace, fresh (21.1% d.m.)	2,000 lb.	4.2	6.4	2.0	0.4	....
Apples	300 bu.	6.0	3.0	15.0	1.2	1.7
Barley, grain	35 bu.	29.4	12.6	8.4	0.7	2.0
straw	1,600 lb.	13.4	4.5	24.6	3.7	1.1
Total		42.8	17.1	33.0	4.4	3.1
Barley, malt	2,000 lb.	40.6	8.4	7.4	1.2	....
Beans, seed	25 bu.	60.0	18.0	19.5	3.0	2.6
straw	2,000 lb.	28.0	6.0	38.0	....	....
Total		88.0	24.0	57.5	....	....
Beets, common, roots only	25,000 lb.	62.5	25.0	125.0	5.0	5.0
mangles, roots only	40,000 lb.	60.0	40.0	140.0	....	....
sugar, tops removed	10 tons	36.6	13.4	56.2	....	....
Blackberries	4,000 qt.	11.0	3.0	12.0	....	....
Bluegrass, Kentucky	2 tons	53.0	21.6	84.0	16.7	....
Buckwheat, grain	20 bu.	15.0	6.0	3.0	....	....
straw	5,000 lb.	62.5	7.5	57.5	34.5	6.0
Total		77.5	13.5	60.5	....	....
Cabbage heads	20,000 lb.	60.0	20.0	80.0	18.0	4.0
Carrots	10,000 lb.	23.0	13.0	53.0	6.0	2.0
Cauliflower heads	15,000 lb.	42.0	15.0	50.0	6.0	3.0
Celery tops	10,000 lb.	25.0	20.0	75.0	....	....
Cherries	8,000 lb.	16.0	4.0	20.0	....	....
Corn, grain	50 bu.	46.4	18.2	11.0	0.4	3.0
stover	3,000 lb.	30.0	9.0	42.0	14.2	2.5
cobs	500 lb.	2.0	0.4	2.2	0.1	0.1
Total		78.4	27.6	55.2	14.7	5.6
Cotton, lint	500 lb.	1.0	0.7	3.0	0.7	....
seed	1,000 lb.	38.5	14.8	12.0	2.8	....
stalks and leaves	2,000 lb.	28.5	8.0	19.2	16.0	....
Total		68.0	23.5	34.2	19.5	....
Cream (30%)	1,000 lb.	3.9	1.1	1.2	0.9	0.1
Cucumbers, fruit	100 bu.	5.5	3.3	11.0	....	....
Distillers' rye grains, dried (94.0% d.m.)	2,000 lb.	58.0	8.6	0.8	....	....
Eggs	100 lb.	2.3	0.4	0.15	....	....
Fat calf	200 lb.	4.9	3.0	0.4	2.4	0.1
lamb	80 lb.	1.6	1.0	0.1	0.7	....
pig	200 lb.	3.5	1.3	0.3	0.9	....
steer	1,000 lb.	23.3	15.5	1.8	12.8	0.4
Flax, grain	15 bu.	30.5	12.5	8.2	2.8	....
straw	1,800 lb.	20.5	3.4	18.8	13.2	....
Total		51.0	15.9	27.0	16.0	....
Grape pomace, fresh	2,000 lb.	19.0	2.8	12.6	....	....
Grapes	6,000 lb.	9.0	6.0	18.0	0.4	1.2
Hay, alfalfa	3 tons	147.0	30.0	126.0	83.5	21.3
cowpea	2 tons	100.0	22.0	70.0	50.0	....
lespedeza	2 tons	78.0	41.0	81.6	55.6	....
millet	3 tons	75.0	27.0	90.0	23.0	....
red clover	2 tons	84.0	20.0	80.0	45.7	10.8
soy bean	2 tons	92.0	28.0	44.0	49.2	15.5

TABLE 27 (*Continued*)

QUANTITIES OF NUTRIENT ELEMENTS CONTAINED IN FARM PRODUCTS*

Crop	Amount	Pounds				
		Nitrogen (N)	Phosphorus (P_2O_5)	Potash (K_2O)	Calcium (CaO)	Magnesium (MgO)
Hay, sweet clover	2 tons	80.0	22.0	74.0	31.4	7.4
timothy	2 tons	50.0	22.0	40.0	7.1	4.1
Leaves, scarlet oak	2,000 lb.	31.4	5.8	38.6	43.0	...
red and black oak	2,000 lb.	32.6	5.0	39.8	46.2	...
white oak	2,000 lb.	31.8	3.6	29.2	53.4	...
beech	2,000 lb.	35.8	3.0	26.4	41.6	...
sugar maple	2,000 lb.	28.0	2.8	41.6	92.6	...
red maple	2,000 lb.	11.2	8.8	12.2	32.0	...
Malt sprouts, dried (92.2% d.m.)	2,000 lb.	84.4	14.2	30.4	4.8	...
Milk	1,000 lb.	6.0	1.7	1.7	1.1	0.1
Muskmelons, fruit	5 tons	22.0	8.0	40.0	....	...
Needles, Jack pine	2,000 lb.	11.6	1.8	3.8	17.2	...
Norway pine	2,000 lb.	13.4	3.0	5.8	27.0	...
white pine	2,000 lb.	12.8	3.6	4.0	30.4	...
Oats, grain	50 bu.	32.0	13.0	9.6	1.6	1.9
straw	2,500 lb.	16.0	5.0	31.2	7.5	3.5
Total		48.0	18.0	40.8	9.1	5.4
Onions, bulbs	300 bu.	39.3	15.4	37.6	18.8	3.4
Parsnips	6 tons	26.4	24.0	78.0	9.6	4.8
Peaches	400 bu.	22.0	11.0	45.5	....	...
Pears	300 bu.	7.5	3.0	15.0	2.9	1.4
Peas, canning, grain	20 bu.	44.0	10.3	12.0	2.8	...
straw	3,000 lb.	30.0	5.7	31.8	59.0	...
Total		74.0	16.0	43.8	61.8	...
Peppermint	2,000 lb.	7.0	4.0	13.0	....	...
Plums	200 bu.	15.3	5.5	21.3	....	...
Potatoes	150 bu.	31.5	13.5	45.0	1.8	2.7
Raspberries	4,000 qt.	10.5	4.5	12.0	....	...
Rutabagas	10 tons	40.0	24.0	100.0	12.0	4.0
Rye, grain	20 bu.	19.1	9.8	6.7	0.5	1.3
straw	2,000 lb.	10.0	6.0	17.0	4.4	1.4
Total		29.1	15.8	23.7	4.9	2.7
Soy beans, grain	15 bu.	79.5	27.0	30.0	9.9	12.6
straw	3,000 lb.	19.2	3.9	18.6	....	...
Total		98.7	30.9	48.6	....	...
Strawberries, fruit only	6,000 qt.	13.5	10.3	27.0	1.4	...
Sweet corn, ears	4,000 lb.	18.0	8.0	12.0	....	...
Tobacco, leaves	1,500 lb.	41.0	6.8	72.0	57.0	...
stalks	1,250 lb.	26.0	5.7	39.6	9.7	...
Total		67.0	12.5	111.6	66.7	...
Turnips, common	10 tons	50.0	20.0	90.0	10.0	2.0
Wheat, grain	25 bu.	30.0	12.8	6.0	0.6	2.0
straw	2,500 lb.	12.5	3.8	15.0	5.2	1.5
Total		42.5	16.6	21.0	5.8	3.5
Wool, unwashed	1,000 lb.	54.0	0.7	56.2	1.3	0.2

* The data in this table are drawn or computed from data from many sources, as "Fertilizers and Crop Production," Van Slyke; "Feeds and Feeding," Morrison; and other books and bulletins.

Note: d.m. means dry matter.

mind that they are taken from the more readily available supply in the soil. Furthermore, the quantities removed by a single crop may seem rather small in some instances; but when the quantities contained in all the crops of a rotation are totaled or when the amounts removed by crops for several years are considered, the necessity of supplying plant nutrients in the form of fertilizer and manure, to maintain soil productivity, is apparent.

The quantities of elements given in Table 27 do not represent the total quantities that crops require during growth, but rather the quantities contained in the harvested material. Roots and other portions of the plant which may not be harvested require considerable quantities of nutrients. Furthermore, with the advancing maturity of the plant certain nutrient elements, such as potassium or calcium, may pass out of the plant roots into the soil.

Although a knowledge of the amount of plant food removed from a soil by the crops of any particular rotation will not tell definitely the fertilizer needs of the soil, this information, taken in conjunction with a knowledge of the plant-food condition in the soil, will give some indication of the amount of fertilizer and of the analyses to use. Obviously, the selling of plant food from the farm cannot go on indefinitely if soil productivity is to be maintained.

DETERMINING SOIL-NUTRIENT DEFICIENCIES

Since the growth of crops is influenced by rainfall, drainage, temperature, rotation, soil reaction, previous soil treatment, and bacterial activity, in addition to the supply of nutrients, it is a very difficult matter accurately to determine the soil conditions that require the application of fertilizers. It is of interest and of much practical concern in determining the kind and amount of fertilizer to use in any soil management system to have a knowledge of the total quantities and of the proportion of nutrient elements removed from an acre of soil by good yields of crops. This factor must be considered in dealing with the problem of soil-productivity maintenance. In any case, the practical problem in regard to the conditions that require fertilizer application is to know before hand or at planting time the kind and quantity of fertilizer to use to give the most satisfactory yields. In working toward that goal, due consideration must be given to the crops and to the soil; they are of prime concern. Even after taking these things into consideration, the results obtained from fertilizer applications are frequently affected and to a large extent determined by climatic conditions.

Not one of the several laboratory methods that have been developed, based on chemical or biological principles, is able to answer the practical question of just what fertilizer to apply to a given soil to get the maximum yield of a certain crop under field conditions. This should not be expected, for the supply of nutrients is only one of the many factors that affect plant growth. The fact remains, however, that certain chemical tests aid in giving us a better understanding of the possible supply of nutrients and permit the determination of the presence of certain harmful substances. Some questions which come up in connection with the determination of nutrient deficiencies in soils are suggested.

Questions:

1. Of what value is a total chemical analysis of a soil?
2. Do the rapid chemical tests aid in determining soil deficiencies?
3. What are the limitations in determining these deficiencies by analyzing the ash of plants?
4. What are the possibilities and limitations of testing the sap of plants and plant tissue?
5. Can plant symptoms be used as a guide in diagnosing soil-nutrient needs?
6. What objections may be raised to the use of pot tests?
7. Why are field tests regarded as the most reliable method?

Total Chemical Analysis of the Soil. The plant-nutrient content of a soil is usually divided into two arbitrary groups: the potential supply and the available supply. The potential supply is determined by a total chemical analysis of the soil and informs us very little regarding the fertilizer needs of the soil. It has already been emphasized that the growth of plants is determined by the nature of the chemical changes taking place in the soil and plant, and not by the total or potential plant-nutrient content of the soil. A total chemical analysis of a soil tells us little regarding the rate at which plant nutrients will become available during the growing season, although if good growing conditions prevail, a soil analyzing high in plant nutrients would generally be expected to produce more available nutrients during the growing season than a soil low in total plant nutrients. However, a total chemical analysis of soils is of value to the soil scientist and hence serves a useful purpose although it is of little service to the farmer.

Rapid Tests. Various methods have been developed to measure the so-called available[1] plant-nutrient supply in the soil and attempts have been made to interpret the test results in terms of the fertilizer requirements of the soil. To diagnose soil conditions by means of chemical methods is an attempt to place the matter on a basis which may apply to all soils. Although these tests are of value in determining the chemical condition of nutrients in a soil at any particular time, none of them can be relied upon entirely as a basis for making fertilizer recommendations. It is impossible by any known chemical or biological soil test to imitate the action of higher plants in obtaining nutrients from the soil and so to determine the available plant-nutrient supply.

Many systems of rapid chemical tests have recently been developed for determining nutrient deficiencies, nutrient levels, and the presence or absence of harmful substances in soils, the last being particularly important from the standpoint of greenhouse soils. The main difference in these various methods lies in the strength and nature of the extracting reagent used. Such tests should be looked upon as guides in determining fertilizer needs but should not be relied upon entirely. The proper interpretation of results obtained by using these rapid methods is in many cases rather difficult and should be attempted only by experienced persons.

Testing the Ash of Plants. It is generally understood among investigators that chemical analyses of plant ash cannot be depended upon to indicate the need for fertilizers. The proportion of various elements that may be found in plants depends upon numerous and varied factors, such as (1) general climatic conditions, (2) species of plant, (3) nature of root system, (4) stage of maturity of plant, (5) methods of cultivation, and (6) nature of soil. In spite of the fact that analyses of plant ash cannot be used in general as a basis for fertilizer recommendations, they may be used as an aid in explaining unproductive areas in certain localities. The relation of phosphate content of feeds to the phosphate content of soils, on the one hand,

[1] Nutrients that are in a condition suitable for immediate use by plants, or those in a state that can be readily changed under good soil conditions to suitable forms for plant use, are commonly called *available*. Those compounds or materials which are not in form for immediate use and are not readily changed to usable forms are called *unavailable*. The line of division between availability and unavailability is quite arbitrary, to say the least, and it is impossible always to distinguish between the two since so many factors must be taken into consideration.

and the health and nutrition of animals on the other, may be given as an example.

Sap and Tissue Tests. Methods involving the testing of the expressed sap of plants and of plant tissue have been developed with the idea of aiding in making fertilizer recommendations. Obviously, several limitations have presented themselves in these attempts. These tests are of value in that they indicate the content of elements in soluble mineral forms in the plant, and before an element can be of direct benefit it must be taken up by the plant. In this respect they are superior to soil tests which do not tell the amounts of nutrients a plant will absorb. However, tests of the expressed sap of plants do not permit interpretation at seeding time, which is important from the standpoint of determining the immediate fertilizer requirements. These test results may indicate certain limiting elements which may be used as a guide in making fertilizer recommendations the following year on the same crop and perhaps be of value in indicating the kind of fertilizer to apply as a top dressing or side dressing the current season.

In regard to tests on the sap and tissue of the plant, the matter of determining the part of the plant to test and the stage of maturity of the plant when samples should be taken must be considered. Regardless of the various limitations of these rapid plant tests and also of the rapid soil tests, when the two are used simultaneously they offer much information regarding the chemical condition of the soil and are most valuable guides in making fertilizer recommendations.

Deficiency Symptoms. When plants are starving for any particular nutrient (nutrients are discussed elsewhere), characteristic symptoms usually appear on these plants. If crops are not vigorous and healthy, it is important to know and understand the cause. If the unhealthy appearance is due to disease, then it may be possible to save the crop by spraying, or if it is a nutrient deficiency, fertilizers may be applied as a top dressing or a side dressing in time to save the crop. These deficiency symptoms appear only when the supply of a particular element is so low that the plant can no longer function normally, and in these cases it will usually be profitable to apply fertilizer long before the symptoms indicating acute starvation actually appear.

Nitrogen. The need for more nitrogen is indicated by a light-green to yellow appearance of the leaves. As a rule, the older bottom leaves start to turn light green, then get yellow at the tip. The entire leaf may turn yellow even though the tissues are alive and turgid. In the corn plant the yellowing extends up the midrib of the leaf, with the outer edges remaining green the longest. A nitrogen-starved cucumber may have a small or pointed blossom end; a deficiency of nitrogen may cause the kernels of cereals to become shriveled and light in weight. In fruit trees the early shedding of leaves, death of lateral buds, poor set of fruit, and development of unusually colored fruit are indications of a lack of nitrogen. Extreme nitrogen starvation is most likely to occur in sandy soils or waterlogged soils, although heavy soils with low humus content usually need additional nitrogen.

Phosphorus. If phosphorus is deficient, cell division in plants is retarded and growth is stunted. These stunted plants usually have a rather dark green color and some plants may form a reddish or purplish coloration on the leaves and stems. Bronze-purple-colored leaves sometimes are observed at the top of new shoots of phosphorus-starved apple trees. In the absence of sufficient phosphorus, general maturity of the crop and seed formation are usually delayed. With corn, poor pollination frequently is associated with phosphorus starvation. Perhaps the most characteristic symptom of phosphorus deficiency, considering plants in general, is the stunted growth.

Potassium. A deficiency of potassium usually shows up as a "leaf scorch" in most plants. Corn indicates a need for potassium by a yellowing of the tips and margins of the lower leaves. This coloration does not move up the midrib as with a nitrogen deficiency but gradually spreads upward and inward from the leaf tip and edges. This leaf scorch is frequently mistaken for "burning" or "firing," owing to a deficiency of moisture during dry weather. When insufficiently supplied with potassium, alfalfa frequently develops a series of pale-yellow or white spots near the margin of the older leaves. These spots gradually enlarge until the entire margin of the leaf has become discolored. The leaf edges finally will dry up and curl under. Potato plants indicate a potassium deficiency by a marginal scorch of the lower leaves, and frequently the areas between the veins of potato leaves bulge out, giving a wrinkled appearance. A cucumber, starved for potassium, grows with a small stem end.

Boron. Many physiological diseases of plants, such as the internal cork of apples, yellows of alfalfa, top rot of tobacco, cracked stem of celery, and heart rot and girdle of beets, are associated with a deficiency of boron. In sugar beets, boron deficiency appears as a stunting and curling or twisting of the petioles, associated with a crinkling of the heart leaves. They have unusually dark green and thicker leaves which wilt more rapidly under drought conditions. The older leaves frequently become chlorotic, and rotting of the beets, starting in the crown, may occur. Girdle or canker of table beets occurs as a cracking of the outer skin of the beets near the soil surface, followed by a breakdown of the root tissue.

Copper. There is little evidence to indicate that copper is lacking in soils except those of a high organic content. Growth abnormalities of many plants that are produced on peat and muck soils have been corrected by the application of copper compounds; and in general the more acid the muck, the greater the number of crops which show a response to it. Copper additions produce better color of such crops as onions, spinach, lettuce, and carrots, increases sugar content of beets and carrots, and improves the flavor of most crops. The addition of copper to acid mucks prevents the premature dying of onions.

The upper leaves of tobacco plants, grown without copper, are unable to maintain their vigor, and wilt badly. The leaves are permanently wilted and do not gain turgidity even in the presence of sufficient moisture. It has also been observed that the amount of seed set is reduced and the seed stock is unable to stand erect if copper is limiting.

Magnesium. Magnesium is a constituent of chlorophyll, and a deficiency is shown in crops by a characteristic discoloration of leaves; in some instances a premature defoliation of the plant occurs. The chlorosis of tobacco, known as "sand drown," is due to a magnesium deficiency. Cotton plants, suffering from a lack of this element, produce purplish-red leaves with green veins. Leaves of sorghum and corn become striped; the veins remain green but the areas between the veins become purple in sorghum and yellow in corn. The lower leaves of the plant are affected first. In legumes, the deficiency is shown by chlorotic leaves.

Manganese. In the absence of sufficient manganese, tomato, bean, oat, and tobacco plants are dwarfed. Associated with this dwarfing is a chlorosis of the upper leaves of the plant, and the leaves become spotted. The tissues of the spots die and frequently drop out. The "gray speck" of oats has been attributed to a shortage of this element in some soils.

It has frequently been observed that the leaves of onions, growing in alkaline muck soils, become dwarfed and curled during growth and that the bulbs remain immature at harvest time. On similar soils celery becomes yellow; spinach, lettuce, and potatoes are chlorotic and frequently unmarketable.

Zinc. Zinc deficiencies frequently have been encountered in Florida. Pecan rosette, the yellows of walnut trees, the mottle leaf of citrus, the little leaf of the stone fruits and grapes, white bud of corn, and the bronzing of the leaves of tung trees are all ascribed to zinc deficiencies. In tobacco plants, a zinc deficiency is characterized by a spotting of the lower leaves, and in extreme cases almost total collapse of the leaf tissue may occur.

It should be pointed out that these deficiency symptoms are not always easily diagnosed. Some of them might be mistaken for discoloration or abnormal characteristics produced by diseases or they may be due to a deficiency of some other element or factor of plant growth. But on careful diagnosis some of these symptoms may indicate rather definitely the need for certain elements.

Pot Tests. Pot tests with many modifications have been used by various investigators. Essentially these tests consist in filling a number of pots with soil material and in adding various fertilizer salts. The need for fertilizer is indicated by the growth of the crop, by amount of dry matter produced, or by analysis of the plant ash. These tests are usually conducted under greenhouse or laboratory conditions and permit the control of moisture supply, temperature, and other factors that cannot be controlled in conducting field tests. Nevertheless, the results have to be interpreted in terms of field conditions if they are to be of value to farmers. This constitutes the chief objection to the method. Regardless of this criticism, pot tests generally are regarded as ranking second to field experiments for determining the fertilizer needs of a soil.

Field Experiments. The most reliable method of determining a soil's need for fertilizer is properly conducted field tests. The reliability of this method lies in the fact that the tests are made under field conditions and on soils in their natural positions. Obviously the plan of any field experiment is determined by its purpose, but such tests usually are made on a series of plots of equal size by treating the plots in different ways with various kinds and amounts of fertilizer. In the older experiments every third plot is used as a check plot and receives no fertilizer. For most reliable results each treatment should be repeated several times on plots located in different parts of the experimental area. Increases in yield of the treated over the untreated plots indicate the need for fertilizer. To carry on long-time field experiments properly requires an enormous amount of time and a considerable expenditure of money. Results obtained from field experiments are invaluable, and such experiments have become the standard method of agricultural experiment stations.

One of the best places to carry on these field experiments is the individual farm. A farmer, for example, can well afford to try fertilizers in an experimental way and to determine for himself if they pay on his particular soils and in his own soil-management system. Each farm is a particular problem in itself.

CHAPTER XII

FERTILIZERS AND FERTILIZER MATERIALS

Fertilizers, in a broad sense, include all materials that are added to soils to increase the growth or yield of crops. However, the meaning of the term fertilizer frequently is restricted to apply only to artificially prepared materials containing plant nutrients. Although fertilizers may affect the soil and plant growth in a number of different ways, they are used primarily to increase the supply of available plant nutrients in the soil and are used also to *balance* the plant-nutrient ratio.

It is customary to speak of fertilizers and fertilizer mixtures as containing nitrogen, phosphoric acid (P_2O_5), and potash (K_2O) instead of nitrogen (N), phosphorus (P), and potassium (K). The early chemists calculated the results of their fertilizer analyses in terms of NH_3, P_2O_5, and K_2O, even though these compounds as such were not actually present in the materials. This method, based on the custom of the earlier practice, is used today to express the chemical analyses of fertilizers, with the exception of the values for nitrogen which have recently been changed and are now expressed in all the states as nitrogen rather than ammonia.

Fertilizers do not consist of the elements nitrogen, phosphorus, and potassium as such, but they are combined with other elements to form either organic or inorganic compounds. Fertilizer materials are classed as nitrogenous, phosphatic, or potassic, depending on whether their principal constituent is nitrogen, phosphorus, or potassium, although some materials can be placed in more than one of these classes. Thus the following subjects necessarily present themselves.

Objectives:

A. Fertilizer materials supplying nitrogen.
B. Phosphatic fertilizer materials.
C. Potassium fertilizers.
D. Mixed fertilizers.

FERTILIZING MATERIALS SUPPLYING NITROGEN

The atmosphere is the original source of all nitrogen. In the free state, nitrogen is a very inert element. Chemical methods, however, have been perfected whereby this element can readily be combined with others into substances which may be used in the manufacture of fertilizers and for other industrial purposes, and now the possibilities for the production of synthetic[1] nitrogen fertilizer materials are unlimited.

Natural processes have resulted in the accumulation of large quantities of combined nitrogen in soil, coal, nitrate beds, plants, and animals. Relative costs, no doubt, will determine which of the various sources of nitrogen will be drawn on for fertilizer purposes in the future.

Materials used to supply nitrogen are called *nitrogen carriers,* and they may be classified in various ways. A convenient classification is one based on the origin or on the nature of the material. There are thus recognized three general groups of nitrogen carriers: (1) the non-synthetic organic materials, (2) the non-synthetic inorganic materials, and (3) the synthetic nitrogenous materials. Several questions may be kept in mind in studying nitrogenous fertilizers.

Questions:

1. What are the sources and characteristics of the non-synthetic organic nitrogen carriers?
2. What are the more important non-synthetic inorganic nitrogen carriers?
3. Which of the synthetic nitrogenous fertilizers are most commonly used?
4. Are the various forms of nitrogen equally efficient in increasing crop production?
5. How much of the nitrogen applied as fertilizers is recovered by the crop?

Non-Synthetic Organic Nitrogen Carriers. Included in this group are (1) those by-products of animal origin coming from the meat and fish packing and rendering industries, (2) materials of plant origin derived for the most part from vegetable oil industries, and (3) other materials originating from both plants and animals. Several of the more common non-synthetic organic nitrogen carriers with their approximate content of nitrogen, phosphoric acid (P_2O_5) and potash (K_2O) are given in Table 28. In the early days of the fertilizer industry the carriers of nitrogen were principally the wastes and by-products of other industries, but since the higher grades of these

[1] Refers to manufactured materials utilizing nitrogen from the air.

TABLE 28

THE PRINCIPAL NON-SYNTHETIC ORGANIC FERTILIZER MATERIALS

Material	Percentage Nitrogen	Percentage P_2O_5	Percentage K_2O
Dried blood	8.0 to 14.0	0.3 to 1.5	0.5 to 0.8
Animal tankage	5.0 to 10.0	3.0 to 13.0	small amounts
Garbage tankage	2.0 to 4.0	1.0 to 3.0	0.5 to 1.5
Process tankages	6.5 to 10.0	variable	small amounts
Fish scrap, dried	6.5 to 10.0	4.0 to 8.0	small amounts
Sewage sludge, ordinary	1.6 to 3.3	1.0	small amounts
Sewage sludge, activated	4.1 to 7.5	2.5 to 4.0	0.75
Cottonseed meal	6.0 to 9.0	2.0 to 3.0	1.0 to 2.0
Bone meals	0.7 to 5.3	17.0 to 30.0	...
Castor pomace	4.0 to 7.0	1.0 to 1.5	1.0 to 1.5
Cocoa shell meal	2.5	1.0	2.5
Tobacco stems	1.3 to 1.6	0.9	4.0 to 9.0
Sheep or cow manure, dried and pulverized	1.0 to 2.0	1.0 to 2.0	2.0 to 3.0
Poultry manure, dried and pulverized	5.0 to 6.0	2.0 to 3.0	1.0 to 2.0

by-product materials are now used in the manufacture of commercial feeds, cost largely prohibits their use as fertilizers. Organic nitrogen as a rule is more expensive than inorganic nitrogen. In 1913 approximately 52 per cent of the nitrogen in fertilizers was supplied by organic materials, but at present probably less than 10 per cent of the fertilizer nitrogen comes from these sources.

The non-synthetic organic materials have the power to take up considerable water and still retain good drilling qualities. They are useful in preventing hardening or lumping when used in mixed fertilizers and by virtue of this property they are often referred to as driers or conditioners. The nitrogen in these products is not water soluble although it becomes available to crops rather easily under good growing conditions, through biological processes which convert it into ammonia and nitrates.

Non-Synthetic Inorganic Materials. The non-synthetic inorganic materials may be divided into two groups: (1) those obtained from natural salt deposits, such as the Chilean deposits of $NaNO_3$, and (2) those secured as a by-product [$(NH_4)_2SO_4$], for example, which is obtained in the process of coking coal. The United States is the largest producer of by-product nitrogen of any country in the world. There are over 150 by-product nitrogen plants (including both coke and gas plants) operating in the United States. This is one of the cheapest and most important sources of nitrogen for fertilizers.

Synthetic Nitrogen Materials. This group includes those fertilizers containing nitrogen taken from the air by any of the chemical nitrogen-fixing processes. The products separate themselves into two groups: (1) organic and (2) inorganic. Urea $[CO(NH_2)_2]$ and calcium cyanamide ($CaCN_2$) belong to the former group, and materials such as $NaNO_3$, $(NH_4)_2SO_4$, and $Ca(NO_3)_2$ are placed in the latter group. The nitrogen content of several synthetic nitrogen materials is presented in Table 29.

Relative Merits of the Various Forms of Nitrogen. Frequent attemps have been made to determine the relative efficiency of nitrogen fertilizers by applying equal quantities of nitrogen per acre, in the various materials, for a given crop. The yield obtained from the use of $NaNO_3$ has commonly been used as a standard for comparison. For example, if $NaNO_3$ produced 20 bushels of wheat per acre and some other fertilizer produced only 18 bushels per acre, then the latter fertilizer would be considered 90 per cent as efficient as the former. Since so many factors, such as temperature and moisture conditions, soil reaction, leaching, kind of crop, and time and method of application affect the action of any nitrogen fertilizer, relative fertilizing values so obtained may be misleading.

Nitrogen, in the nitrate form, is readily soluble in water and is rather quickly utilized by most crops. Nitrates are easily leached from the soil by rains because of their high solubility and because they are not fixed or held in the soil to any appreciable extent. Although the ammonia form of nitrogen is soluble in water, it is not leached out as readily as nitrates because rather large quantities of ammonia can be adsorbed and held by the soil. A number of different crops can use ammoniacal nitrogen, but most of it is converted to nitrates, owing to the process of nitrification, before plants take it up. In the non-synthetic organic nitrogenous fertilizers the nitrogen exists in the form of complex organic compounds such as proteins, which for the most part are insoluble in water. These substances as such cannot be used by plants but must first be broken down to more simple compounds. This means slower availability of the nitrogen but the furnishing of a more or less continuous supply of available nitrogen throughout the growing season, which is often an important factor in successful plant growth. However, as pointed out elsewhere, the higher cost of

TABLE 29

THE PRINCIPAL COMMERCIAL FERTILIZER MATERIALS SUPPLYING NITROGEN OF SYNTHETIC ORIGIN *

SYNTHETIC INORGANIC NITROGEN CARRIERS

Nitrogen carrier	Percentage nitrogen	Remarks
Sulphate of ammonia	20.5	Also obtained as a by-product from coke ovens
Nitrate of soda	15.5 to 16.5	Is also a naturally occurring salt found in Chile
Calcium nitrate	13.0 to 15.5	Also called nitrate of lime
Cal-Nitro	16.0 to 20.5	Ammonium nitrate mixed with $CaCO_3$
Calurea	34.0	A compound of calcium nitrate and urea
Ammonium sulphate—Nitrate	26.0	A double salt of $(NH_4)_2SO_4$ and NH_4NO_3
Ammonium nitrate	35.0	Has undesirable physical properties
Ammonium chloride	26.2	Similar in fertilizer properties to $(NH_4)_2SO_4$
Anhydrous liquid ammonia	82.2	Used for ammoniating purposes
Ammonia liquor	20.5 to 23.5	Formed when ammonia is absorbed in water
Urea ammonia liquor A	45.5	Solution of urea in crude ammonia liquor
Urea ammonia liquor B	45.3	Same as above except more urea and less ammonia
Crude nitrogen solution	44.4	Sodium nitrate and ammonia dissolved in water
Nitrogen solution II	37.5	Ammonium nitrate and ammonia dissolved in water

SYNTHETIC ORGANIC NITROGEN CARRIERS

Nitrogen carrier	Percentage nitrogen	Remarks
Calcium cyanamide	21.0 to 24.0	Sold under the trade name "Cyanamid"
Urea	46.0	This (also Cyanamid N) is classed as non-proteid organic nitrogen

* Other nitrogen carriers are listed with the mixed fertilizers in Table 32.

nitrogen in these organic materials has greatly reduced the demand for them. The nitrogen in the synthetic organic materials (urea and calcium cyanamide) is largely soluble in water, and under good soil conditions it is rapidly changed to ammonia and nitrates.

Ammonium sulphate and sodium nitrate are without question the most widely used nitrogenous fertilizer materials at the present time, and this well indicates the value placed on these materials. In this group of highly valuable nitrogen materials should also be included many of the newer synthetic materials more recently proposed for fertilizer use.

The Recovery of Applied Nitrogen. The question frequently is asked, how much of the nitrogen applied as fertilizers is recovered by the crop? That is a question which cannot be answered specifically because the amount recovered will vary with the kind of fertilizer, soil type, climatic conditions, and plant. But in general about 50 per cent of the nitrogen added as $(NH_4)_2SO_4$ or $NaNO_3$ will be recovered by the crop. The remainder of the nitrogen is lost from the soil by leaching and erosion, and a small amount may be changed to organic forms by the synthesizing action of microorganisms and retained by the soil. The application of ammonia and nitrate forms of nitrogen in the usual amounts will not result in any great accumulation of soil nitrogen directly but through their effect in increasing the amount of organic residues the nitrogen of the soil may be increased. Nitrogen of chemical fertilizers cannot be held by soils for any great length of time except as it is fixed in complex organic compounds.

PHOSPHATIC FERTILIZER MATERIALS

The amount of phosphorus removed from the soils of the United States by erosion, leaching, crops, and grazing greatly exceeds the amount added to the soil in fertilizers, manures, and crop residues. There are no natural processes by which the phosphorus content of our soils can be increased appreciably once it is depleted. In order to increase the phosphorus content of soils, we must buy phosphorus either in the form of fertilizers or in feed. The phosphorus problem, then, differs from the nitrogen problem in that we cannot obtain this element from the air but must eventually buy it if we continue to sell phosphorus from our farms. If purchased in feed for livestock, a portion of the phosphorus contained therein is added to the soil. In general, however, addition of phosphorus must be made largely in the form of fertilizers which are derived chiefly from phosphate rock. In considering the subject of phosphatic fertilizer materials, the following questions arise.

Questions:

1. What are the sources of the phosphatic materials and how are the various carriers prepared?
2. Is the phosphorus of the different carriers equally available to plants?
3. What is meant by the reversion and fixation of phosphates in soils?
4. How can the solubility of phosphorus which is fixed in soils be increased?

Carriers of Phosphorus and Their Sources. The principal sources of phosphorus are the natural deposits of phosphorus-bearing rocks,

TABLE 30

THE PRINCIPAL PHOSPHATIC FERTILIZER MATERIALS *

Material	Percentage Available P_2O_5	Remarks
Ground phosphate rock	25 to 35 †	Effectiveness depends largely on degree of fineness
Superphosphate, ordinary	16 to 20	Made by treating ground phosphate rock with H_2SO_4
Double, treble, or triple superphosphate	40 to 50	Made by treating ground phosphate rock with liquid H_3PO_4
Basic slag	5 to 20	By-product obtained in the manufacture of steel
Bone meals	17 to 30 †	Includes raw as well as steamed bone meals
Precipitated bone	- 40	By-product from manufacture of glue stock from bones
Bone black	32 to 35	Bones heated in closed retorts
Bone ash	32 to 38	Bones burned with free access of air
Calcined phosphate	34	Prepared by heating a mixture of finely ground phosphate rock with alkali salts
Calcium metaphosphate	63	Prepared by treating phosphate rock with hot, gaseous phosphorus (P)
Colloidal phosphate	18 to 23 †	A finely divided, relatively low-grade rock phosphate or phosphatic clay

* See Tables 28 and 32 for other phosphorus carriers.

† Total phosphoric acid rather than available.

iron ores, and animal bones. The phosphatic fertilizers made from these materials may be classified as (1) the natural phosphates, including materials such as ground rock phosphate and bone meal; (2) the treated natural phosphates, such as bone black, bone ash, superphosphates, calcined phosphates, and calcium metaphosphates; (3) the by-product phosphates, such as basic slag (bone meal could also be placed in this group); and (4) the so-called chemical phosphates, such as ammoniated and nitrated superphosphates, potassium phosphate, potassium metaphosphate, Ammophos, Leunaphos, Nitro-

FIG. 56. Rock phosphate is mined from surface deposits by hydraulic pressure. It is then washed and crushed in preparation for acidulation with sulphuric acid. [Courtesy of Agricultural Bureau of American Agr. Chemical Co.]

phoska, and similar compounds. A list of some of the more common phosphate fertilizers together with their approximate content of phosphorus is given in Table 30.

Rock Phosphate. The United States possesses the largest known deposits of phosphate rock in the world. These deposits are believed to represent 40 per cent of the total known reserves. Extensive deposits of phosphate rock occur in some of the western states (Idaho, Montana, Utah, and Wyoming), and somewhat less extensive deposits occur in the southeastern part of the United States (Florida, Tennessee, South Carolina, Kentucky, and Arkansas). Although the southeastern deposits are somewhat smaller in extent compared to those of the West, the former are much more favorably located with respect to the heavy-consuming areas of fertilizers.

The principal constituent of American phosphate rock is fluorapatite ($9CaO \cdot 3P_2O_5 \cdot CaF_2$). This material sometimes is called phosphate of lime and for convenience is frequently expressed as tricalcium phosphate [$Ca_3(PO_4)_2$]. Raw rock phosphate, on account of its insolubility, is not used extensively as a fertilizer. To be effective it is very important that it be finely ground and thoroughly mixed with the soil. Its availability is greatly increased by the presence of decaying organic matter, which exerts a dissolving effect on the phosphate owing to the continual production of CO_2. Manure, or green manure, therefore, aids in making the phosphorus available, and when used under favorable conditions rock phosphate yields reasonably good returns. Rock phosphate should not be used on soils low in organic matter or in those sections of the country where a long and expensive freight haul is necessary.

Bone Meals. Various kinds of bone meal are on the market as fertilizers, the principal ones being raw bone meal, steamed bone meal, and precipitated bone meal. Raw bone meal results from the fine grinding of bones without any other treatment. Steamed bone meal is produced by grinding animal bones that have been steamed under pressure. The phosphorus is more quickly available in the latter product because of the finer grinding and the removal of the fat in the steaming process. Precipitated bone is a by-product obtained in the manufacture of glue from bones. Bone meals are expensive forms of phosphoric acid in comparison with superphosphates.

Superphosphate. Superphosphate is by far the most important source of available phosphoric acid. More than 90 per cent of all the phosphoric acid used in fertilizers in this country is supplied by superphosphate. It is the most important single fertilizer material, both from the standpoint of total tonnage and also from that of quantity of plant food supplied. By treating finely ground raw rock phosphate with approximately an equal weight of sulphuric acid, a superphosphate containing 18 to 19 per cent of phosphoric acid (P_2O_5) is obtained. This is reduced to the 16 per cent grade by adding filler or is made up to 20 per cent by the addition of treble superphosphate containing 40 to 50 per cent phosphoric acid. (It is possible to produce a 20 per cent product directly by selecting a high-grade raw phosphate rock.) The resulting product of treating raw rock phosphate with sulphuric acid is largely a mixture of monocalcium phosphate [$CaH_4(PO_4)_2$] and calcium sulphate ($CaSO_4$) in about equal proportions. The acid treatment of raw rock phosphate thus changes the relatively insoluble tricalcium phosphate into an available form, monocalcium phosphate. The ordinary grades of superphosphate contain, in addition to phosphoric acid, significant quantities of soluble calcium, sulphur, and other impurities.

Treble Superphosphate. Treble superphosphate (40 to 50 per cent available P_2O_5) looks very much like the ordinary grades of superphosphate, but differs from them in that it contains very little gypsum ($CaSO_4$). In making these higher-grade superphosphates, liquid phosphoric acid (H_3PO_4) is used instead of sulphuric acid. In general, the high-grade superphosphates are just as effective as the ordinary grades and are especially well suited for use in the production of concentrated fertilizer mixtures.

Calcined Phosphate. The term calcined phosphate refers to the product formed by heating ground phosphate rock to high temperatures in the presence of water vapor, or it is made by heating to a high temperature a mixture of ground phosphate rock and an alkali. This heat treatment increases the availability of the phosphorus, and the calcined phosphates are considered good sources of phosphorus for crop plants.

Calcium Metaphosphate. Calcium metaphosphate is a new phosphate fertilizer containing the equivalent of about 62 per cent available phosphoric acid. It is made by treating phosphate rock with hot, gaseous phosphorus (P). The gaseous phosphorus is obtained by treating phosphate rock with silica (sand) and coke in an electric furnace. The hot gas is passed directly in contact with phosphate rock at a high temperature. The reaction is vigorous. The mass melts and the resulting slag is withdrawn from the furnace, allowed to cool, and when finely ground is ready for fertilizer use. It appears to be a very good source of available phosphoric acid for plants.

Basic Slag. Basic slag, sometimes referred to as Thomas phosphate, ranks next to superphosphate as the world's leading phosphatic fertilizer. It is produced as a by-product of the iron industry. Phosphorus is contained in certain iron ores, and steel made from them is brittle if the phosphorus is not largely removed. In producing the slag, a blast of air is blown through the molten iron, containing phosphorus in a converter which is lined with lime. The phosphorus is oxidized and the oxide unites with the lime. The resulting mass, being lighter than iron, rises to the surface, is drawn off, cooled, and finely ground so that most of it will pass through a 100-mesh screen. The phosphorus in slag is soluble in citric acid and is considered available to crops.

Ammoniated Superphosphate. This material is made by treating superphosphate with anhydrous or aqueous ammonia. Other materials such as solutions of urea and ammonia, nitrate of soda and ammonia, and ammonium nitrate and ammonia have been used for ammoniating superphosphate. The ammoniation of superphosphate is recognized as an excellent process by virtue of the fact that the added ammonia may neutralize any excess acidity in the superphosphate. The process makes use of a cheap source of nitrogen, and the nitrogen content of the fertilizer is increased without appreciably increasing the bulk of material.

Phosphate of Potash. Phosphate of potash, made by treating monocalcium phosphate $[CaH_4(PO_4)_2]$ with potassium hydroxide (KOH), is not on the fertilizer market in large quantities, but this material offers possibilities of becoming an important source of phosphate and potash.

Other Chemical Phosphates. In some of the other chemical phosphates such as Ammophos, Leunaphos, and Nitrophoska, the ammonia is combined with phosphoric acid (H_3PO_4) instead of with sulphuric acid. These are all high-grade mixed fertilizers, carrying 40 to over 60 units of plant food.

Relative Availability of the Phosphate Carriers. The phosphorus compounds contained in phosphatic fertilizers have been arbitrarily placed into three groups, based on their ease of solubility, that is (1) the water soluble, (2) the citrate soluble and, (3) the insoluble. The compounds falling into the first two groups make up the available phosphoric acid content of any particular fertilizer, and the insoluble compounds are considered unavailable. Monocalcium phosphate (the main phosphorus constituent of superphosphate) and the phosphate compounds of potassium and ammonium are soluble in water and are considered readily available. The chemical phosphates are included in this group. Dicalcium phosphate and the phosphorus in calcined phosphate, calcium metaphosphate, and precipitated bone are soluble in neutral ammonium citrate, and the phosphorus in basic slag is soluble in citric acid. Therefore the phosphorus in these materials is classed as available. On the other hand, the phosphorus in materials such as bone meal and rock phosphate is neither water soluble nor citrate soluble and is thus classified as unavailable, although when these two materials are applied to soils the phosphorus in the former becomes available much more rapidly than the phosphorus in the latter.

The availability of applied phosphorus to crops, however, obviously cannot be determined by the arbitrary solubility tests referred to above. When phosphorus fertilizers are added to soils, the amount that will be or become available for crop use will vary with the type of soil (*p*H, lime content, organic content), the kind of plant, seasonal conditions, and also with the kind and amount of fertilizer applied.

When phosphorus fertilizer is applied to soils, the phosphorus soon combines with various soil constituents to form new compounds, some of which are only slightly available to plants. Therefore, only a portion of the phosphorus added is available for crop use, in some cases not more than 20 per cent being recovered by the first crop. The processes by which this element is combined with soil constituents are known as phosphate fixation, and this phenomenon is of considerable practical importance because it affects the efficiency of phosphorus fertilizers on different soil types.

Reversion and Fixation of Phosphates in Soils. The changing of phosphorus to a less soluble form is frequently referred to as reversion of phosphate. The changing of monocalcium phosphate to dicalcium phosphate or dicalcium phosphate to tricalcium phosphate or the changing of either dicalcium or monocalcium phosphate to aluminum or iron phosphate are examples of phosphate reversion. The process of reversion may be considered the same as the fixation of phosphate in soils. Fixation is rapid and the nature of the compounds formed is obviously determined by the nature of the soil. The phosphorus may be changed to compounds of calcium and magnesium; it may combine with organic matter, be assimilated by microorganisms, unite with iron and aluminum compounds; or it may be adsorbed by the clay complex in a form such that a part of the phosphorus is exchangeable with other anions.

Soils vary greatly in their capacity to fix phosphate. The organic content of the soil, the *p*H, the amount of soluble calcium and magnesium, the amount and chemical nature of the clay colloids are all important factors affecting the phosphate-fixing power of soils and the nature of the compounds formed. The availability of fixed phosphate depends upon the nature of the compounds formed. For example, when fixed as compounds of calcium and magnesium, phosphorus is generally considered to be more available than when fixed as compounds of iron and aluminum. The amount applied and the elapsed time since its application are important factors affecting the extent of phosphorus fixation.

The fixation of phosphate in soils is beneficial in that, when fixed, it does not leach out of the soil. But the process is detrimental in that the phosphate may not be readily available to crops and, furthermore, fixation prevents the penetration of the fertilizer into the soil. As a result, phosphates applied as top dressings become fixed in the soil surface, and the downward movement into the root zone is very slow. This slow downward movement is an important problem from the standpoint of fertilizing permanent pastures and orchards, for the effects of the phosphate are minimized.

Experiments have indicated that many plants are able to make a more efficient use of phosphorus fertilizer if it is placed in bands to the side of the seed instead of being broadcast. Yields of crops are usually much better from row or hill application than from the use of the same amount of fertilizer applied broadcast because the phosphorus is kept in a more available condition as it comes in contact with a much smaller amount of soil. The finer a fertilizer is ground, the more rapid and more complete is the fixation of phosphate. Thus the use of granular or pellet forms of fertilizer should result in less fixation and an increased efficiency of the fertilizer.

Increasing the Solubility of Fixed Phosphorus. There are various kinds of organic and inorganic acids produced in soils which are considered important agents in making fixed phosphorus more available. The more important of these acids are H_2CO_3, H_2SO_4, and HNO_3. Using tricalcium phosphate and carbonic acid, the following equation illustrates how these acids aid in phosphate solubility:

$$Ca_3(PO_4)_2 + 2H_2CO_3 \rightleftharpoons Ca_2H_2\ (PO_4)_2 + Ca(HCO_3)_2$$

Phosphate compounds of iron and aluminum frequently are encountered in acid soils, and the application of lime to these soils causes a gradual change of the phosphates to calcium phosphates.

Thus it can be said that the addition of any organic material to the soil that increases the production of carbonic, nitric, or sulphuric acid will favor the availability of phosphates. Furthermore, the practice of liming acid soils will aid in bringing about the solubility of phosphate compounds, but overliming may temporarily decrease the availability of applied phosphates by causing the soluble phosphates to revert to the tricalcium-phosphate form.

POTASSIUM FERTILIZERS

The supply of potassium in soils is frequently twenty or more times as great as the total supply of either nitrogen or phosphorus. Muck and peat soils offer an exception to this statement, and the supply of potassium in light sandy soils is much less than in heavy soils. In the majority of soils, however, the potash problem is not one of total supply, but it is one of availability, and often the quantity of available potassium is so small that it is necessary to supplement it with potash in the form of commercial fertilizers in order to obtain maximum yields.

The soils in the newer agricultural areas of the United States are as a rule not in need of potash, but with the continued use of these soils for the production of crops a deficiency of potash will appear sooner or later. The same situation exists with potassium as with phosphorus in that there are no natural processes whereby this element is returned in appreciable quantities to the soil. In other words, the farmer must ultimately buy potash because the native supply in the soil cannot furnish crops with sufficient available potassium indefinitely. Until recently the farmers of the United States were largely dependent on Europe for their supply of potash, but now all that is consumed as fertilizer in this country can be produced here. In relation to the potash situation, the following questions may be raised.

Questions:

1. What are our important domestic sources of potash?
2. Where are the important foreign sources of potash located?
3. What are the principal carriers of potassium?
4. Are the different carriers of potassium equally efficient?
5. Does fixation of potassium occur in soils?

Domestic Sources of Potash. At the beginning of the first World War the potash supply from Europe to the United States was cut off and the price advanced over 1,000 per cent, reaching as high as $500 per ton. This situation did much in stimulating the United States to make every effort possible to provide for an adequate supply of domestic potash. As a result of considerable exploration and research, important sources and deposits of potash have been discovered in the United States and as far as this fertilizer element is concerned, this country is self-sufficient.

Kelp. This is a seaweed found along the Pacific coast which contains 11 to 12 per cent of K_2O. Extraction of the potash is not commercially profitable at the present time.

Surface Deposits of Rocks and Minerals. Surface deposits of potash-bearing minerals and rocks are widely distributed over the United States. Some of the more common of these minerals and rocks are (1) alunite (5 to 10 per cent K_2O), found in abundance in the states of Utah, Colorado, and Nevada; (2) brown mica (about 10 per cent K_2O), found in the leucite-bearing rocks of Wyoming; (3) leucite (16 to 21 per cent K_2O), found in Wyoming; (4) green sand (6 to 8 per cent K_2O) found in New Jersey; and (5) potash shales (8 to 9 per cent K_2O), found in Georgia. The potash in all these minerals is insoluble and can be made available only by rather expensive chemical processes. These materials will not be used extensively until more economical methods are developed for the recovery of the potash in available forms.

Saline Lakes. These lakes constitute valuable sources of potash in the United States which can be recovered by rather simple chemical processes. Shallow lakes of potassium-bearing brine occur in Ne-

FIG. 57. The porous crystalline mass which fills Searles Lake in California. The brine from this crystal bed is pumped out, and from it are recovered various grades of potash and borax and also lithium chloride. [Courtesy of American Potash Institute.]

braska, Utah, Nevada, and Southern California. Large quantities of this element are being extracted each year from the brine of Searles Lake in California. This constitutes one of our most important sources of fertilizer potash.

Industrial Wastes. Considerable quantities of potash can be recovered from certain industrial wastes. Sugar beets and sugar cane contain much potassium and, in the manufacturing of sugar, molasses, and alcohol, a large quantity can be recovered from them. Furnace and flue dusts offer a possible source of potassium. Iron ore and coke, for example, contain different amounts of potash and many tons of it could be recovered annually from the blast furnaces of the country.

Underground Deposits. In recent years underground deposits of crude potassium salts of commercial importance have been found in the United States, particularly in the states of Texas and New Mexico. The mines located in the vicinity of Carlsbad, New Mexico, supply a considerable proportion of our domestic potash. In this region there are 33 square miles of proved area of commercial potash salts. Seventy counties of Texas are underlain at different depths by beds of potassium-bearing salts that vary greatly in thickness and purity. The latter deposits are not being exploited at the present time.

FIG. 58. Crushing and loading crude potash salts in one of the mines at Carlsbad, New Mexico. [Courtesy of American Potash Institute.]

Wood Ashes and Tobacco By-Products. These products are also potential sources of potash, but, as with other miscellaneous materials, their potash content is too low to justify their extended use. Where these materials can be used locally or where transportation expense is not excessive, they are of considerable value. At one time (early part of the 17th Century) wood ashes were exported from the United States for fertilizing purposes.

Foreign Sources of Potash. The largest known deposits of materials containing potassium occur in Germany and France. Less-extensive deposits are found in Poland, Spain, and Italy. Germany and France have controlled the world's potash market in the past because of their tremendous deposits and because they were the first to develop their potash resources extensively. Furthermore, the deposits are favorably located, being near the large fertilizer-consuming areas of Europe and in a position to take advantage, in peace time, of cheap water-transportation rates to the United States.

The potassium compounds, largely chlorides, in these deep deposits (approximately 1,200 to 3,500 feet) are more or less impure, being mixed with chlorides and sulphates of calcium, sodium, and magnesium. After the salts are removed from the mine, they are ground, dissolved in water and allowed to crystallize for concentration and purification. Some of the materials used for fertilizers are the crude products, as taken from the mine and then treated to fine grinding, and to some extent the crude salts are mixed with the refined products.

Carriers of Potassium. The principal potash carriers, as used commercially, are potassium chloride (commonly called muriate of potash), potassium sulphate (often referred to as sulphate of potash), potassium nitrate, and manure salts. Manure salt is a semi-refined product, composed chiefly of potassium chloride, magnesium sulphate, and sodium chloride. In some potash-bearing materials, as in wood ashes and molasses waste, the potassium may exist in the carbonate form, but only very small quantities of these substances are used for fertilizing purposes. The principal potassium fertilizer materials and their approximate potash contents are given in Table 31. Potassium chloride is used much more widely than any other forms because of its lower cost.

Relative Efficiency of the Different Forms of Potash. All the potash fertilizer salts are soluble in water and are considered to be readily available. In general it can be said that there is very little difference in their effects on crop production except in rather special cases. Discrimination is sometimes made against carriers having a high percentage of chlorine for special crops like potatoes and tobacco. The sulphate, or carbonate if available, is usually preferred for tobacco, especially where large amounts are to be added, because a crop of superior burning quality is produced. The muriate is just as efficient as the sulphate for potato production, according to most of the experimental evidence.

Potash Fixation in Soils. When a potash salt is added to soil, a part of the potassium is adsorbed or fixed by the soil in more or less unavailable compounds. The amount so fixed is usually directly proportional to the amount of colloidal matter in the soil, being greatest in the heavy soils and least in the light or sandy soils. In general, potassium is not as readily fixed as phosphates but is more readily

TABLE 31

THE PRINCIPAL POTASSIC FERTILIZER MATERIALS *

Material	Percentage K_2O	Remarks
Potassium chloride	48.0 to 62.0	(Muriate of potash) The most commonly used potassium fertilizer salt
Potassium sulphate	47.0 to 50.0	(Sulphate of potash) Usually sells for higher price than KCl
Potassium carbonate	15.0 to 50.0	May be obtained from wood ashes, sugar residue, cottonseed hulls, or other plant materials
Manure salts	19.0 to 32.0	Crude potash salts, mostly in the form of KCl
Kainite	14.0 to 22.0	A semi-refined product of potash manufacture, corresponding to the 20 per cent manure salts
Sulphate of potash magnesia	25.0 to 27.0	A double sulphate of potash and magnesia
Cement dust	6.0 to 9.0	A by-product from cement manufacture
Tobacco stems	4.0 to 9.0	Includes tobacco stalks and tobacco wastes
Hardwood ashes	1.5 to 8.0	The potash exists principally as the carbonate
Cotton-hull ashes	10.0 to 45.0	The supply of this material is very limited

* See Tables 28 and 32 for other potash carriers.

fixed than most nitrogen salts. The potassium ion, on being adsorbed by the colloids, displaces some other ion such as calcium, magnesium, or sodium. Fixed potassium moves slowly, if at all, in the soil. In light soils the movement may be appreciable, but in heavy soils the movement usually is restricted to the upper few inches of soil. A soil's ability to adsorb and hold potash is of great importance because it serves to decrease leaching and provides for a more continuous supply of available potash. Plants are able to utilize the more easily exchangeable adsorbed potassium.

MIXED FERTILIZERS

Soils vary greatly in their ability to supply crops with available nutrients, and the mineral requirements of different crops are also quite variable. In order to supply nutrient deficiencies in soils and to meet the various requirements of different crops, fertilizers containing two or more essential elements are prepared in many different grades. They are known as mixed fertilizers and are made by mixing two or more of the separate fertilizer carriers. Thus some mixed fertilizers contain only two of the three primary plant-nutrient elements and are called incomplete fertilizers. The most commonly used incomplete mixtures contain phosphorus and potassium but no nitrogen. Mixed fertilizers containing all three of the primary fertilizing elements are called complete mixtures. About two-thirds of the commercial fertilizers sold in the United States are in the form of mixed goods, and most of these mixtures are complete fertilizers. Less than 5 per cent of all mixed fertilizers used are incomplete. In dealing with this subject, the following questions are discussed.

Questions:

1. What advantages are there in using mixed fertilizers rather than using the different materials separately?
2. How are mixed fertilizers prepared?
3. What change has occurred in the strength of fertilizer mixtures used in the United States during the past twenty or thirty years?
4. Are low-analysis fertilizers more efficient than either high-analysis or concentrated materials?

Advantages in Using Mixed Fertilizers. The use of mixed fertilizers has several advantages over applying the different materials separately. In the first place, the cost of applying a mixture is less than that of applying the various carriers individually; second, mixed fertilizers, as a rule, have better drilling qualities than the individual carriers; and third, the use of mixed fertilizers requires less care on the part of the farmer to insure a properly balanced fertilizer for his particular conditions than when the individual materials are applied separately. Mixed fertilizers can be secured from local dealers at such times and in such quantities as the farmer may desire, and it is usually possible to obtain the grade of fertilizer best suited to the particular crops in the locality. Another possible advantage of using mixed fertilizers is the ease with which the acidity of certain fertilizer materials may be controlled. The acidity may be more conveniently and effectively controlled by adding lime to the mixture than by applying it separately in the field.

Mixed fertilizers are made in various grades in order to adjust or bring about a proper balance of the primary fertilizer elements in the soil. If the soil requires only one plant-nutrient element, obviously the advantages listed above do not hold.

Preparation of Mixed Fertilizers. The preparation of most mixed fertilizers is a relatively simple operation. It consists essentially in mixing suitable fertilizer materials in the correct proportion to give the desired grade or analysis. There are many mixing plants in the United States, and they vary greatly in size and in the nature of the equipment used.

The processes involved in the preparation of mixed fertilizers are assembling, pulverizing, acidulating, and mixing. Many of the raw materials require grinding, and often the raw materials, particularly the carriers of phosphorus, must be treated with sulphuric acid in order to increase the solubility of the plant nutrients prior to mixing. After acidulation the mass is permitted to stand in a large pile for a considerable time to "cure" before being ground.

There are four types of commercial fertilizer-mixing plants: first, the complete plant that makes its own sulphuric acid, acidulates its phosphate, and then makes the final fertilizer mix; second, the plant that buys its supply of sulphuric acid, but grinds its own phosphate rock and acidulates it; third, the plant that makes only its nitrogen carriers and buys its other materials; fourth, the dry-mixing plant which buys all its materials ready for mixing.

Many of the standard grades of fertilizers contain a total of only 1,400 to 1,800 pounds of the different carriers per ton. The other 200 to 600 pounds is composed of some material known as filler. Sand was formerly the most commonly used material for this purpose, but today dolomitic limestone is extensively used. Muck and various other organic materials sometimes are used as filler, and in addition they take up moisture and tend to prevent the mixture from "setting up" or becoming hard on standing.

Fertilizers in storage tend to assume an unfavorable mechanical condition largely as a result of "setting up," which is essentially a cementation as in plaster of Paris, or it may be due to the surface tension effects of moisture which forms films around the particles of fertilizer.

Superphosphates or other materials which have an excess of sulphuric acid, those which were improperly cured, are greatly improved in their drilling qualities by the use of dolomitic limestone as a filler which neutralizes any excess acidity.

The drilling qualities of certain mixed fertilizers made from fertilizer salts which take up moisture readily can be appreciably increased by adding an organic filler such as muck.

Strength of Fertilizer Mixtures. Fertilizer materials now contain a much higher proportion of the plant-nutrient elements than formerly. When these higher grade materials are used in mixed fertilizers, it obviously becomes necessary to increase or raise the analysis of the mixture or to dilute it with filler if the analysis is to remain the same. The use of higher-grade fertilizers rather than the use of mixtures diluted with filler affords several economic advantages. (See p. 346.)

In certain sections of the country there has been a marked increase during the last twenty or thirty years in the strength of the mixed fertilizers used, but in other sections the largest demand is still for the older or low-grade mixtures. Since more-concentrated raw materials are now being used in mixed fertilizers than formerly, there has been an increase in the use of filler in the lower grade mixed fertilizers. The net result in the United States during the past twenty years has been about a 20 per cent increase in the total plant-nutrient content of the average mixed fertilizer and approximately a 65 per cent increase in the use of filler.

Various suggestions have been made for grouping or classifying mixed fertilizers according to their plant-nutrient content, but as yet no standard method has been adopted for this grouping. Differences of opinion exist as to the number of groups that should be made, the limits of each group, and also as to the names that should be applied to each group. The most commonly used classification divides mixed fertilizers into three arbitrary groups, known as low-analysis, high-analysis, and concentrated fertilizers. The first group includes all fertilizer mixtures containing less than 20 per cent of plant nutrients; the second group, those containing 20 to 30 per cent; and the third group, those with 30 per cent or more. It is not to be inferred that the fertilizers belonging to any one of these groups are radically different from those of any other group. This grouping of mixed fertilizers has reference only to the plant-nutrient content and should not be interpreted as indicating in any way the general properties of the particular fertilizers.

Efficiency of Different Grades of Mixed Fertilizers. The efficiency of a fertilizer is determined by the uniformity with which it can be distributed and by its quality or its chemical composition. The drillability of a fertilizer is determined by its physical properties, the two most important of which are the relative sizes of particles and their moisture content or hygroscopic properties. A fertilizer is said to be of good quality if it does not cause injury to plants when applied in the usual manner, does not leave any harmful residual effect in the soil, and contains the nutrient elements in proper balance. High-grade mixed fertilizers are now on the market which meet these requirements, and they are being made in granular form to insure good drillability.

TABLE 32

THE ANALYSIS OF SOME MIXED FERTILIZERS

Material	Percentage Nitrogen	Percentage Available P_2O_5	Percentage K_2O	Remarks
Potassium nitrate	14	0	45	Occurs only as very small natural deposits
Potassium ammonium nitrate	16	0	27	Essentially a mixture of KCl and NH_4NO_3
Nitrate of soda potash	14 to 15	0	10 to 13	A constituent of Chilean salt deposits
Potassium phosphate	0	32 to 53	30 to 50	Made by combining KOH and $CaH_4(PO_4)_2$
Potassium metaphosphate	0	60	40	Prepared from pure KCl and H_3PO_4
Monoammonium phosphate	11	60	0	Sold largely as a constituent of Ammophos
Diammonium phosphate	21	53	0	Has a tendency to lose ammonia due to its alkalinity. Best used with other materials
Ammoniated superphosphate	3	16	0	Analysis will vary, depending upon the amount of ammonia added
Nitrated superphosphate	5 to 6	16 to 17	0	Superphosphate treated with nitrogen solution II (See Table 29)
Ammophos A	11	45 to 48	0	Largely a mixture of ammonium phosphate and ammonium sulphate
Ammophos B	16.5	20	0	
Ammophoska *	12	24	12	A mixture of Ammophos and K_2SO_4
Nitrophoska *	15	30	15	There are several different grades
Leunaphos	20	20	0	Essentially a mixture of diammonium phosphate and ammonium sulphate
Leunaphoska	10	10	13	A mixture of Leunaphos and a potash salt

* The trade name for a series of highly concentrated complete fertilizers.

Thus it can be said that a decided increase can be made in the plant-nutrient content of many of the popular low-analysis mixtures without bringing about any marked changes in their chemical or physical properties. In general, a pound of plant nutrients in high-analysis fertilizer mixtures is just as efficient as a pound of plant nutrients in low-analysis fertilizers.

CHAPTER XIII

FERTILIZER PRACTICES

The practice of adding materials to soils to improve their productive capacity dates back several hundred years B.C. The use of animal manures, marl, chalk, and wood ashes was practiced by the Chinese, Greeks, and Romans more than one thousand years ago. The Egyptians have fertilized their soils with settlings from the overflow of the Nile River since the beginning of agriculture in that area. When the white man came to America he found the Indians putting a fish in each hill of corn. George Washington, in the absence of commercial fertilizers, ran experiments with manure, marl, gypsum, common salt, and green manures.

Although crops have been fertilized in this country for many years, the use of commercial fertilizers dates from about 1830, when Chilean nitrate was first imported. The importation of guanos from Peru, beginning in approximately 1840, did much to acquaint farmers with the general value of fertilizers and to stimulate the rapid development of the fertilizer industry. The production of mixed chemical fertilizers was started in Baltimore in 1850, and the establishment of other plants along the Atlantic coast soon followed. The use of by-products from meat-packing plants, the residues from plant-oil industries, and scrap from fish canneries and salting plants filled an important place in the early years of the fertilizer industry. Since the development of methods for the fixation of atmospheric nitrogen, the fertilizer industry has changed from a salvage outlet for the waste products of other industries to one of the largest chemical industries.

Farmers in this country were slow at first to use fertilizers because of the large areas of fertile virgin land available for crop production. As the fertility of the older farm soils became depleted, the early farmers moved westward to new territories containing soils of high producing power. But as time went by and the soils in these newer regions became more depleted, the use of fertilizer increased rapidly. At the present time the value of commercial fertilizers in farm practice is generally recognized.

Fertilizers should be regarded as additions of plant nutrients for immediate results more than as soil-building materials, although their proper use in adequate quantities may contribute directly or indirectly to an increased supply of nutrients in the soil and an increased humus content. The productive capacity of nearly all soils under cultivation in humid regions can be improved by the use of well-chosen fertilizers, and their use should receive consideration in any good system of soil management. In regard to fertilizer practices the topics listed below are particularly significant and are discussed on the following pages.

Objectives:

A. Effects of fertilizers on certain soil properties.
B. Effects of fertilizers on crops.
C. Laws controlling fertilizer sales.
D. Home-mixing fertilizers.
E. The purchase and use of fertilizers.

EFFECTS OF FERTILIZERS ON SOILS

Soils are complex chemical, physical, and biological systems and the continued use of fertilizers, especially in large quantities, may produce marked changes in these systems. The beneficial effects of certain fertilizers or the harmful effects of others are no doubt due in part to the influence of the fertilizers on the chemical, physical or biological properties of the soil. In other words, fertilizers may exert important actions on soils aside from merely increasing the supply of available nutrients. Some of the more important of them are discussed with the idea of gaining a better understanding of the use and function of fertilizers. Furthermore, it is necessary to have a knowledge of some of these effects for the proper interpretation of the results obtained from the application of fertilizers. The questions listed should be answered during the study of this subject.

Questions:

1. What important chemical changes in soils may be produced by the continued use of fertilizers?
2. In what ways may fertilizers affect the physical properties of soils?
3. Are microorganisms in any way influenced by the application of fertilizers?

Effects on Chemical Properties. During the last few years the effect of fertilizers on soil acidity has received a great deal of attention. There has been much agitation concerning the manufacture of fertilizers which are neutral, that is, fertilizers which leave neither an acid nor an alkaline residue in the soil. Many fertilizer companies believe the acid-producing effects should be corrected in the fertilizer itself by using ground dolomitic limestone as a filler and by using non-acid-forming materials to supply nitrogen. The amounts of lime so added to soils are too small to be effective in neutralizing an already acid soil, but the lime prevents any possible small increase in soil acidity from the action of the fertilizer. It is much more desirable, therefore, that fertilizer companies use dolomitic limestone as a filler instead of some inert material as sand.

A review of the investigations relative to the effect of fertilizers on soil acidity leads to the following conclusions:

1. The common potassium fertilizers, such as the muriate and sulphate of potash, have no permanent effect on soil acidity.

2. Superphosphates in general will have no permanent effect on soil reaction. Basic slag, bone meal, and rock phosphate have a tendency to correct or neutralize soil acidity.

3. Fertilizers containing nitrogen in the form of ammonia or in other forms subject to nitrification (being changed to nitrate) will produce acidity unless sufficient liming material is present in the fertilizer to neutralize the acid formed. Experimental fields that have received applications of sulphate of ammonia fairly regularly for many years without being treated with lime have become, in some cases, too sour to grow clover. This effect is more pronounced on the lighter soils such as sands and sandy loams. Fertilizers containing sulphate of ammonia should not be discriminated against, however, since the increased quantity of lime needed to keep the soil in condition for growing legumes in a normal system of soil management is of little practical significance.

4. Nitrogenous fertilizers in which the nitrogen is in the nitrate form and is combined with bases such as sodium or calcium will result, upon being utilized by plants, in decreased soil acidity. Some of the fertilizers in this group are nitrate of soda, calcium nitrate, cal-nitro, and calurea. Calcium cyanamide should be placed in this group although the nitrogen is not in the nitrate form, but the fertilizer carries a rather high content of lime. The acidity or basicity of several nitrogen fertilizers is given in Table 33.

5. In general, the systematic use of medium to large quantities of high-grade mixed fertilizers at suitable times in the rotation will not appreciably affect soil acidity.

The application of fertilizers may affect, either directly or indirectly, the availability of the soil nutrients. Acid-forming fertilizers in certain soils may decrease the available phosphorus and increase the available potassium. Basic fertilizers may result in an increase in both available phosphorus and available potassium. The addition of basic ions in fertilizers may release various cations from the exchange complex of the soil, making such ions more available and more subject to loss from leaching.

TABLE 33

EQUIVALENT ACIDITY OR BASICITY OF SEVERAL NITROGEN FERTILIZERS *

Fertilizer Material	Nitrogen Percentage	Net Equivalent Acidity or Basicity per Unit of Nitrogen, Pounds $CaCO_3$
Ammonium sulphate	21.1	107
Ammonium chloride	26.1	107
Ammophos	11.0	100
Urea	46.6	36
Sodium nitrate	16.4	36 †
Calcium cyanamide	22.0	57 †
Animal tankage	9.12	3
Garbage tankage	2.50	54 †
Cottonseed meal	6.76	29
Dried blood	13.00	35
Milorganite	7.00	34

* Determination of Equivalent Acidity and Basicity of Fertilizers, W. H. Pierre, "Ind. and Engg. Chem.," Vol. 5, 1933, p. 229.
† Basicity.

Another chemical effect produced in soils by fertilizers is that of increasing or conserving the reserve supply of nitrogen, phosphorus, and potassium. As has been pointed out (p. 331), fertilizers are primarily used to meet the current needs of crops by supplementing and balancing the soil-nutrient supply, but usually liberal applications of fertilizers will ultimately increase the reserve supply in the soil; this is particularly true of phosphorus. An accumulation of either nitrogen or potassium is usually not appreciable, unless the humus content of the soil is increased, because these elements suffer heavy losses by leaching. Any increase obtained in the total reserve supply of nitrogen, phosphorus, and potassium by the fertilizing programs followed by most farmers is insignificant.

Effect on Physical Properties. The physical properties of soils are determined largely by the amount and nature of the mineral colloids present and by the content of organic matter. The application of mineral fertilizers will not affect the quantity of soil colloids but may appreciably affect the chemical nature by regulating to some extent the exchangeable cations on the colloids. The kind and amount of exchangeable cations in turn exert a pronounced effect on the physical properties of the clay. For example, the continued use of large quantities of $NaNO_3$ may result in producing a soil with poor physical properties. The nitrate radical is adsorbed more by plants than is sodium. The excess sodium combines with carbonic acid (H_2CO_3) to form sodium carbonate (Na_2CO_3), which causes soil granules to disperse. In other words, sodium, on being adsorbed by the soil, causes the soil particles to deflocculate or disperse. This may ultimately produce a puddled soil. Acid-producing fertilizers also tend to disperse the soil by encouraging the losses of calcium and other basic ions. On the other hand, those fertilizers containing a good supply of calcium, magnesium, or potassium tend to promote granulation and thus make for a better physical condition of the soil. In this connection it is to be emphasized that no appreciable accumulation of sodium results from ordinary applications of sodium nitrate ($NaNO_3$).

Fertilizers may affect the physical properties of soils indirectly through their effect on the quantity of organic matter. In general, fertilizers increase the supply of organic matter by virtue of the fact that they produce more crop residue (roots, stubble, etc.) which is returned to the soil.

Effect on Biological Activities. Most soil organisms belong to the plant kingdom and it is not unreasonable to expect them to respond to applications of fertilizer. All microorganisms require nitrogen, phosphorus, and potassium for their growth and various metabolic functions, and they respond much the same way as do higher plants to applications of commercial fertilizers. Nitrogen-fixing organisms are able to utilize atmospheric nitrogen and consequently are not as a rule appreciably benefited by nitrogen fertilizers, but they do demand an adequate supply of phosphorus and potassium. After the application of fertilizers, the microorganisms first satisfy their needs, and crops feed on the nutrients left over.

Fertilizers may stimulate biological activity, thereby hastening the decomposition of organic matter, increasing ammonification, nitrification, and nitrogen fixation, and bringing about increases in the solution of soil minerals. Increased acidity as a result of fertilizer applications or of the destroying of the granular structure in soil may lead to a decreased biological activity. Certain parasitic soil organisms may be controlled by proper liming and fertilizer practice. It becomes quite evident that many of the observed beneficial effects of fertilizers on higher plants are brought about indirectly through their effect on the soil microorganisms. This feature of fertilizer practice is frequently overlooked.

EFFECTS OF FERTILIZERS ON CROPS

In addition to their effects on the growth and yield of crops, fertilizers may affect plants in other ways. For example, they may regulate the uptake of certain nutrients by plants, affect the toxicity of certain salts, modify the character of growth, or influence the quality of produce. Several questions that may be raised concerning the effects of fertilizer on plant growth are suggested.

Questions:

1. Do the ions of fertilizer salts affect the toxicity or the intake of other ions by plants?
2. In what ways may fertilizers affect the character of plant growth?
3. What can be said relative to the general influence of fertilizers on crop quality and composition?

Toxicity of Ions and Effects of Fertilizer on Uptake of Nutrients. A single fertilizer salt may be definitely toxic to plants, but in the presence of one or more other salts there is a tendency for each to counteract the toxicity of the other, thus providing a better nutritive balance. Calcium ions are very effective in decreasing the toxicity of other ions. A fertilizer, constituted to suit best the development of a plant, is called a balanced fertilizer, but because of so many variable factors it can be approximated only theoretically. The effect of certain ions on the uptake of certain other ions is a factor which may affect the efficiency of any particular fertilizer. For example, calcium or magnesium may favor the absorption of ammonia nitrogen or they may interfere with the absorption of nitrate nitrogen; high alkalinity may hinder absorption of manganese and iron by plants. It is impossible to forecast all the reactions that applied fertilizers undergo or the effects produced on other elements, but these various indirect effects must be reckoned with.

Effect of Fertilizers on Character of Plant Growth. One of the most important factors affecting the growth of plants is the weather (temperature and quantity and distribution of rainfall), and the response that any particular crop will make to the application of fertilizer is largely governed by the weather, particularly the moisture supply. In seasons when it is necessary to delay planting certain crops because of unfavorable weather conditions, the application of fertilizers may speed up the growth processes of the plant and thus offset

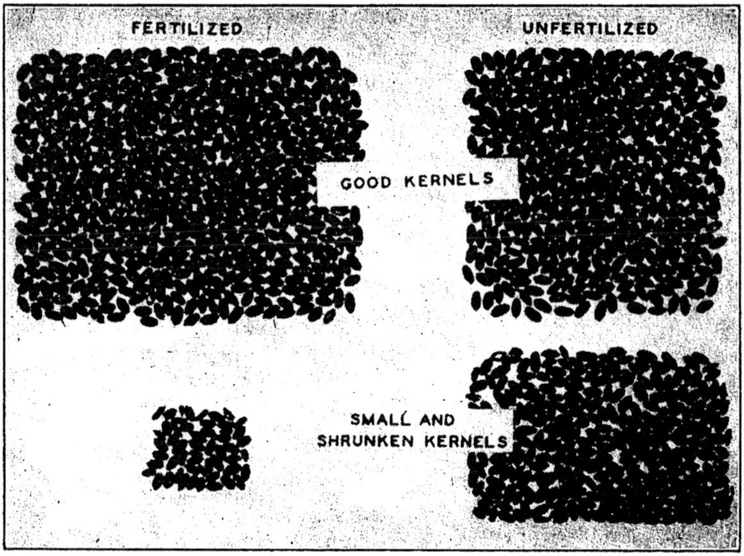

FIG. 59. Properly chosen fertilizer improves the quality of many crops. Left: 20 grams (549 kernels) of wheat grown on fertilized Brookston silt loam soil contained 53 shrunken grains. Right: 20 grams (617 kernels) of wheat from unfertilized soil contained 268 shrunken grains.

somewhat the unfavorable effects of the season. But, looking at the fertilizer-weather relationships from another viewpoint, it is frequently observed that fertilizers in general stimulate early crop growth and if dry weather prevails about midseason the fertilizer may result in decreased yields. This is brought about because the soil moisture is more rapidly exhausted by the fertilized crop through increased growth and greater leaf development.

The early growth of a crop should not be taken as a measure of the effect of a fertilizer on yield. At times fertilizers may stimulate early crop growth, but as the season advances this difference disappears, and at harvest no increase is found. In other instances, fertilizers may cause little effect on the rate of growth of certain crops, but at harvest a decided increase in yield is noted.

Effect of Fertilizers on Quality and Composition of Crops. Some of the more general effects of nitrogen, phosphorus, and potassium on the quality of crops have already been discussed (Chapter XI). The influence of fertilizers upon the composition of the mineral matter of plants is exceedingly complex. Their influence is affected by the variety of crop, climate, water supply and other environmental conditions. The influence which the presence of one element exerts upon the absorptive powers of the plant for other mineral elements of the soil or fertilizer also complicates this problem.

While it is possible to increase to a limited extent the content of mineral constituents of some crops by fertilization, one should not be misled by any exaggerated claims for mineralizing human and animal foods. For a more-detailed discussion of this subject, see p. 291.

LAWS CONTROLLING FERTILIZER SALES

In the early development of the fertilizer industry in the United States, many low-grade fertilizers were offered for sale in competition with those of higher grade. At that time no fertilizer laws existed for the protection of the farmer or the purchaser and the honest manufacturer from fraud. There were no effective methods available by which the farmer could be assured of the quality or composition of the different fertilizers, in advance of their use, which were offered for sale. This situation encouraged certain persons to make extravagant claims for materials having little or no fertilizing value. In order to protect the farmers against fraudulent goods and to protect the reputable manufacturer, nearly all the states have passed laws governing the sale of fertilizers; many states have provided for systems of inspection and analysis. The questions listed below may serve as guides in studying measures governing the sale of fertilizers.

Questions:

1. What is the general nature of the fertilizer laws?
   a. What is the order and form used in expressing the grade of a fertilizer in most states?
   b. What information should the guarantee include?
   c. What is meant by an open-formula guarantee?
2. How are fertilizers inspected and who makes the analysis?
3. Who is responsible for the enforcement of the fertilizer law?

General Nature of Fertilizer Laws. In general, the nature of the laws controlling fertilizer sales in the various states is similar. They all require periodical registration of brands or analyses offered for sale and accurate labeling of the bags or packages. Most of the states require that there be printed on each fertilizer bag, or on an attached tag, the following information:

1. Name, brand, or trademark.
2. Analysis (guarantee) or chemical composition.
3. Net weight of fertilizer.
4. Name and address of manufacturer.

As a rule, all fertilizers selling above a certain price require the payment of a license fee for each grade or analysis or payment of a tonnage tax so that the enforcement of the law is at least self-supporting.

The Order and Form Used in Expressing the Grade of a Fertilizer. For every fertilizer material there must be presented a guarantee as to its content of nutrient elements although the exact form of stating the guarantee is generally determined by the state in which the fertilizer is offered for sale. In all states the grade is expressed in the order $N–P_2O_5–K_2O$. The total nitrogen is expressed as elemental nitrogen (N); the phosphorus is expressed as available phosphoric acid (P_2O_5); and the potassium is given as water-soluble potash (K_2O). In some states the law specifies a certain minimum in the total plant-food content of mixed fertilizers; the required minimum ranges from 20 per cent in some states, to 12 per cent in others, although most of the states do not have minimum requirements.

Information Needed in the Guarantee. State laws vary considerably as to the information the guarantee must include. Some do not require all the information that may be desired, while others contain considerable information that is of little value and often is confusing to the farmer. For all practical purposes the analysis should include the following information:

1. Percentage of water-soluble nitrogen.
2. Percentage of water-insoluble nitrogen.
3. Percentage of available insoluble nitrogen.
4. Percentage of available phosphoric acid.
5. Percentage of water-soluble potash.

In some cases the percentages of minor elements may be desired.

The composition of a complete fertilizer, as represented by the analysis does not give all that is desired in regard to the source of materials. For example, it is possible for a manufacturer to mix high-analysis organic nitrogen compounds with low-grade water-insoluble organics and produce a mixture that may pass inspection in some states. This could be avoided by the use of an open-formula guarantee.

The Open-Formula Guarantee. The term formula refers to the quantity and grade of stock materials used in making a fertilizer mixture. When both the analysis and the formula are shown on the tag or bag, such a guarantee is called an open-formula guarantee.

The Inspection of Fertilizers. From time to time during the year the various fertilizers offered for sale are sampled, under official supervision, throughout the state. Fertilizers offered for sale may be sampled by inspectors any time during the year and at any point in the particular state. The samples are sent to the control laboratory and are analyzed to determine whether the goods are up to the guarantee. The results are checked against the guaranteed analysis and by this means the purchaser is protected from loss through the activities of unreliable companies. The inspection and analysis may be in the hands of the state department of agriculture, of the state agricultural experiment station, or of a state chemist. Once a year, in most states, the findings of the chemists are published in bulletin form, and copies of this bulletin may be obtained upon request by any person interested.

The Enforcement of the Fertilizer Laws. The enforcement of the fertilizer laws in each state is invested in a control official, as indicated above, who is usually associated with either the state department of agriculture or the state agricultural experiment station. When a fertilizer falls below the guarantee (allowing in most cases for a small variation from the guarantee), the manufacturer is subject to prosecution by the state courts. Penalties of varying degrees of severity are levied for failure to come up to the specified guarantee and for other violations of the fertilizer law. The analyses of most commercial fertilizers are generally equal to or greater than the guarantee. Although the farmer is protected by his state laws, the imposition of fines in controlling these laws is seldom necessary because the publicity derived from the annual publication of the findings of the control agency is usually a more effective means of checking fraudulent guarantees. Not only has publicity been effective in preventing fraud but it is also to the advantage of the fertilizer consumer and of the honest manufacturer to know that the guarantee has received an official sanction. Fertilizers purchased from outside the state by mail order are not subject to state inspection.

HOME-MIXING FERTILIZERS

Some farmers prefer to buy the raw materials and to mix their own fertilizers, and by so doing they are able to obtain any analysis they may desire. Fertilizers can be mixed satisfactorily on the farm if the farmer is willing to take the necessary time and precautions to do the job correctly. However, the preparation of a fertilizer which will give satisfaction from all standpoints requires considerable technical knowledge. Before undertaking home-mixing one should be familiar with the answers to the questions suggested.

Questions:

1. What factors should be considered in the home-mixing of fertilizers?
2. What steps are necessary in home-mixing?
3. How are fertilizer formulas calculated?
4. Are there any particular advantages in the farmer's mixing his own fertilizer?
5. What disadvantages are associated with home-mixing?

Factors To Consider in Home-Mixing. First of all, the farmer must decide on the kind and form of nutrients best suited to his soil conditions and crop requirements and then determine the separate materials needed. He must then investigate the cost of home-mixing (including cost of raw materials, hauling, and labor) compared to the cost of purchasing ready-mixed goods in order to determine if home-mixing is profitable. Most of the farmers home-mix because of the possible savings to be had, although occasionally it is practiced in order to secure a particular analysis which cannot be purchased locally.

In deciding on the analysis of fertilizers to be used, the farmer must consider soil conditions such as acidity, texture, organic content, drainage, and soluble-nutrient supply, as well as the cropping system used and the crop to be fertilized. An effort should be made to fit the fertilizer to the soil and crop—no single fertilizer analysis is best for all conditions.

Materials that go into mixed fertilizers cannot be mixed indiscriminately. Certain mixtures may cause the loss of nitrogen, the reversion of soluble phosphates, or they may produce a fertilizer which hardens on standing. It is usually a wise plan not to prepare home-mixed fertilizers until they are needed for immediate use unless great care has been used in selecting the materials to be mixed and unless satisfactory storage facilities are provided.

Attention should be given to uniformity in mixing the raw materials. Sometimes unsatisfactory results are obtained in the use of a fertilizer owing to a lack of thoroughness in mixing. This would result in an uneven distribution of the nutrients, and under such conditions best results cannot be expected. However, a uniform mixture can be made by the farmer, provided sufficient effort is put forth.

Necessary Steps Involved in Home-Mixing. The first important step in home-mixing fertilizers is to decide on the analysis. It is then necessary to determine the most desirable kinds of carriers to supply the required amounts of plant nutrients as demanded by the chosen analysis. The formula or recipe (quantity of each carrier required) for the mixture can now be calculated. It is at this stage in the home-mixing process that the farmer should determine if home-mixing will show a profit for that particular year. He will have to compare the cost of the raw materials plus his labor with the price of the ready-mixed goods in the community and then make his decision.

The next step is the actual purchase and mixing of the materials. The mixing process is very simple and requires no equipment other than a shovel, a screen, a tamping or grinding device, a homemade mixer or a tight floor, suitable for mixing, and a pair of scales. Weighing may be unnecessary since the pounds of fertilizer are shown on each bag, and the bags may be halved or quartered with sufficient accuracy. Any lumps in the fertilizers should be crushed and all material passed through a fine screen before mixing. This insures a more-thorough mixing and better drilling qualities.

If the mixture is to be stored for any length of time it may be necessary to add some sort of a drier or conditioner, such as dry saw-dust, muck, or soil, in order to prevent lumping. However, in mixing fertilizers on the farm for immediate use, it is usually unnecessary to add a filler, for a proportionately less quantity of fertilizer per acre can be used to supply the amount of plant food desired.

The Calculation of Fertilizer Formulas. In calculating formulas for homemade mixtures it is necessary to decide first what percentages of nitrogen, available phosphoric acid, and water-soluble potash are desired in the fertilizer mixture and then what materials are to be used in making the mixture.

For example, a farmer wishes to make one ton of a 2–12–6 fertilizer using the following ingredients:

Ammonium sulphate	20 per cent nitrogen
Superphosphate	20 per cent available phosphoric acid
Muriate of potash	50 per cent water-soluble potash

The problem is to find out how much of each of these materials is needed. This may be done by use of the following equation:

$$X = \frac{A \cdot B}{C}$$

in which X equals pounds of carrier required, A equals pounds of mixed fertilizer required, and, with nitrogen, B equals the percentage of nitrogen desired in the mixture and C equals the percentage of nitrogen in the carrier (ammonium sulphate). By substituting these values in the above equation, the result is easily determined.

$$X = \frac{2{,}000 \times 2}{20}$$

$X = 200$ pounds, the amount cf ammonium sulphate required

In a like manner the required amounts of superphosphate and muriate of potash may be determined.

$$X = \frac{2{,}000 \times 12}{20}$$

$X = 1{,}200$ pounds, the weight of superphosphate required

$$X = \frac{2{,}000 \times 6}{50}$$

$X = 240$ pounds, the weight of muriate of potash required

The total amount of materials used in this fertilizer mixture (200 + 1,200 + 240) equals 1,640 pounds. It is necessary to add 360 pounds of filler to make a ton of the required mixture.

The quantity of a carrier of a given composition that is required to furnish a given percentage of plant nutrient in a ton of mixed goods, where each nutrient represented in the analysis is obtained from a single carrier, is readily computed from Table 34.

Another type of problem that may need to be solved in home-mixing fertilizers is one in which a part of one or more of the nutrients is derived from two or more materials. For example, calculate the quantity of ingredients required for a ton of a 2–16–8, using the following materials. One-half of the nitrogen is to be derived from cottonseed meal and one-half from sodium nitrate.

TABLE 34

Table for Computing Pounds of Carrier Needed to Make 1 Ton of Fertilizer

Percentage of Plant Food Desired in Fertilizer	Percentage of Composition of Material (Carrier)													
	6	*8*	*10*	*12*	*14*	*16*	*18*	*20*	*22*	*24*	*48*	*50*	*60*	*62*
1	333	250	200	166	143	125	111	100	90	83	41	40	33	32
2	666	500	400	333	286	250	222	200	180	166	83	80	66	65
3	1,000	750	600	500	429	375	333	300	272	250	124	120	100	97
4	1,333	1,000	800	666	571	500	444	400	364	333	166	160	133	129
5	1,666	1,250	1,000	833	714	625	555	500	454	417	208	200	166	161
6	2,000	1,500	1,200	1,000	857	750	666	600	545	500	250	240	200	194
7		1,750	1,400	1,166	1,000	875	777	700	635	583	292	280	233	226
8		2,000	1,600	1,333	1,144	1,000	888	800	729	666	333	320	266	258
9			1,800	1,500	1,285	1,125	999	900	818	750	375	360	300	290
10			2,000	1,666	1,428	1,250	1,111	1,000	909	833	416	400	333	322
11				1,833	1,571	1,375	1,222	1,100	1,000	916	458	440	366	355
12				2,000	1,714	1,500	1,333	1,200	1,090	1,000	500	480	400	387
13					1,857	1,625	1,444	1,300	1,182	1,083	542	520	433	419
14					2,000	1,750	1,555	1,400	1,272	1,166	583	560	466	451
15						1,875	1,666	1,500	1,364	1,250	625	600	500	484
16						2,000	1,777	1,600	1,454	1,333	667	640	533	516

	Analysis
Cottonseed meal	6–3–2
Sodium nitrate	16–0–0
Superphosphate	0–40–0
Muriate of potash	0–0–60

$$X = \frac{2{,}000 \times 2 \times \frac{1}{2}}{6} = 333.3 \text{ pounds cottonseed meal}$$

$$X = \frac{2{,}000 \times 2 \times \frac{1}{2}}{16} = 125.0 \text{ pounds sodium nitrate}$$

$$X = \frac{2{,}000 \times 16 - (333.3 \times 3)}{40} = 775.0 \text{ pounds superphosphate}$$

$$X = \frac{2{,}000 \times 8 - (333.3 \times 2)}{60} = 255.5 \text{ pounds muriate of potash}$$

333.3 pounds cottonseed meal
125.0 pounds sodium nitrate
775.0 pounds superphosphate
255.5 pounds muriate of potash

1,488.8 pounds
511.2 pounds filler

2,000.0 pounds

Advantages in Home-Mixing. Home-mixing is done largely because of economy. There is usually some saving by this practice, which varies with the analysis to be mixed and the prevailing price of fertilizers, especially if the farmer's labor is not taken into consideration. Home mixtures frequently can be made with farm labor at a saving of 10 to 20 per cent.

Some farmers prefer to mix their own fertilizers so that they can obtain any particular analysis they want; moreover, they can choose their own ingredients. A particular or special analysis may be desired which cannot be purchased locally, and with this unusual condition it would be advantageous to home-mix.

Perhaps the greatest advantage in home-mixing fertilizer is that it impels the farmer to study fertilizers and their influence on his crops. In other words, home-mixing has a distinct educational value. The farmer must become familiar with the comparative availability and properties of various fertilizer materials; he will spend more time studying his soil and crop needs; and he should understand the residual effects of different fertilizer materials.

Disadvantages in Home-Mixing. The practice of home-mixing fertilizers has not become popular in this country largely because farmers do not like to assume the responsibility of buying and mixing the separate materials. Furthermore, it is much easier and more convenient to buy the ready-mixed goods. When it comes time to apply fertilizers, farmers are usually busy and hesitate to take the required time for mixing. If fertilizers are mixed in advance, then suitable facilities for storage must be provided.

In comparing the cost of home-mixed fertilizers with that of the factory-mixed goods, the time, trouble, and extra labor involved are frequently overlooked by the farmer. On most farms it is doubtful whether home-mixing should be attempted unless the work can be done when other farm work is not pressing. In most cases home-mixing is not economically worth while, especially where only a small amount of fertilizer is used, even though it is possible to cooperate with other farmers in purchasing the necessary ingredients (carriers) in larger quantities.

It is not always easy to obtain the necessary ingredients from local fertilizer dealers in certain areas because these raw materials are not always carried in stock.

It appears that the advantages of home-mixing are worthy of serious consideration only by those farmers who use several tons of fertilizer each year. Obviously, fertilizers should be bought by the method that will give the greatest value for each dollar expended.

THE PURCHASE AND USE OF FERTILIZERS

Good judgment is not exercised frequently by farmers in buying fertilizers. Too often the price per ton of a fertilizer is the factor governing the purchase rather than the cost per unit of plant nutrients. It is the duty of the purchaser to invest his money in the fertilizer from which he will receive the greatest return per dollar invested as far as the quantity and quality of the fertilizers are concerned. However, attention should be called to the fact that, unless the fertilizers are properly used, maximum benefits cannot be expected. It is the object, therefore, of this section to discuss some of the items that determine the price of fertilizers, how they should be purchased and used to obtain the greatest efficiency.

Questions:

1. What charges are included in the retail price of fertilizers?
2. Why should farmers purchase fertilizers on a cash basis?
3. What are some of the more important factors affecting efficiency in the use of fertilizers?

The Price of Fertilizers. The retail price of a fertilizer is the sum of four charges. First, the price of the plant nutrients contained; second, the transportation charges; third, the profits and selling charges of the dealers, manufacturers, and salesmen; and fourth, the cost of maintaining and operating the fertilizer factory and the mixing plant and other incidental items. The last three items may be grouped together and called "general expense," which constitutes a more or less fixed charge against each ton. This basic fixed cost is the average cost of manufacturing, bags, transportation, sales, and all other incidentals independent of the plant-nutrient content. The basic cost will vary from year to year and will also vary with the different sections of the country for any given year. The general-expense item amounts to a large percentage of the total price, particularly of the low-analysis fertilizers, and the basic price would be charged for any mixture handled in the usual way even if it consisted entirely of filler. A source of expense to the user of fertilizer is the purchase of low-analysis mixtures. It can be readily seen that the same amount of plant nutrients will be contained in either 1 ton of 4–16–4 or two tons of 2–8–2, but the consumer would have to pay the general expense charges on 2 tons if he purchased the 2–8–2, and on only 1 ton if he chose the 4–16–4. This, of course, would make the plant nutrients in the 2–8–2 mixture much more expensive even though it cost considerably less per ton. Therefore, it is more economical for a farmer to buy the higher grades of fertilizer, making sure that they contain plant nutrients in proper proportion to meet his needs.

One of the causes for the large overhead on each ton of fertilizer is the making of so large a number of analyses. Every time a mixture of different composition is desired, the mixing machinery must be allowed to run empty and readjustments and arrangements made, which results in the loss of much time and in an increase in the expense. There is no question that the number of analyses put on the market in most of the states could be greatly reduced and yet supply all the needs of farmers and other users of fertilizer.

Purchase Fertilizers on a Cash Basis. Farmers should purchase fertilizers on a cash basis because the cost of fertilizer credit usually is high. Fertilizer credit is different from credit on a piece of farm machinery. If payment is not made on the machinery, the dealer can repossess the implement, but with fertilizers it would be impossible. The question of the buyer's good intentions should not be open to discussion, but floods, drought, frost, sickness, or accidents may make payment utterly impossible. These factors combined make the cost

FIG. 60. An excellent device for placing fertilizer at one side of and below the seed. Note also the attachment for varying the depth of planting. [Courtesy of John Deere Co., Moline, Ill.]

of fertilizer credit unusually high. It would be profitable for one to borrow money from a bank or other agency at, say, 6 per cent for a short time in order to avoid payment of the high cost of fertilizer credit.

Most fertilizer companies recognize the advantage of prompt payment and encourage it by rather liberal cash discounts. If a farmer can use carload lots of fertilizer and fertilizer materials and can pay cash, he usually can effect considerable saving by making his purchase direct from the manufacturer, broker, or importer. Cooperative buying of fertilizers by farmers has been employed successfully in many sections, and in this way the small farmer can obtain his fertilizer at as low a price as the large farmer.

Efficiency in the Use of Fertilizers. Most soil types, particularly in the humid regions of the United States, will respond favorably to the use of well-chosen fertilizers. Fertilizers should not be expected to make up for every shortcoming of the soil and crop, such as poor seed, unadapted varieties of crops, unfavorable weather conditions, poor tillage practices, weeds, poor drainage, bad physical condition of the soil, low organic content, or insufficient lime. All these factors are important, affecting the efficiency of any fertilizer for any crop grown on any given soil. In other words, the proper use of fertilizer is only one, though a very important one, of the many phases of the scientific management of the soil.

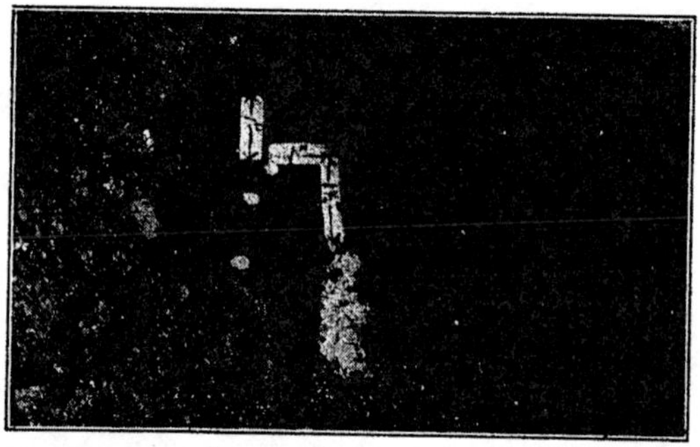

FIG. 61. For sugar beets and white pea beans the fertilizer should be placed about one inch to the side of and one and one-half inches deeper than the seed. Doubtless the same fertilizer placement will prove advantageous in the case of other row crops.

The most profitable return from fertilizer is nearly always obtained from those soils that are in the best physical condition for plant growth. The most profitable results from fertilizers cannot be expected when they are used on soils that are too heavy or too light, too compact or too loose, too dry or too wet.

Frequently fields having apparently the same natural characteristics respond differently to the application of a particular fertilizer for the same crop. This condition is often caused by differences in the previous treatment of the two soils. Variations in past cropping systems, tillage practices, and fertilizer and manure practices bring about differences in the productivity of fields which may be reflected in their response to present fertilization.

Another important factor affecting the efficiency of fertilizers is that of applying the fertilizer in the correct place in relation to the seed. This is probably as important as applying the right kind and amount of fertilizer. No single fertilizer-placement pattern has been found superior with all crops and under all conditions. Almost without exception, however, localized placement of the fertilizer in relation to the seed or plant has been most efficient. This usually means that the fertilizer is placed at the sides of the seed or plant rather than applied broadcast. Obviously with crops such as alfalfa, pastures, and orchards this method of fertilizer application cannot be followed.

FIG. 62. The effect of fertilizer placement on growth of pea seedlings. Left: 300 pounds 4–16–8 in contact with seed; Right: 300 pounds 4–16–8 in bands ½ inch away from seed and on the same level.

The primary object in the use of fertilizers is to produce a profit. Space does not permit a detailed discussion of all the factors affecting the efficiency of fertilizers when applied to soils; only a few of the more outstanding factors have been mentioned relative to the judicious use of fertilizers. For more specific information in regard to the best analysis and methods to use, the farmer should refer to fertilizer-recommendations bulletins, consult the county agricultural agent, or some other responsible person or agency. Most of the state agricultural experiment stations issue a fertilizer-recommendations bulletin, based on the results of field experiments, which offers the best information available. It may also be pointed out that there is close cooperation between the fertilizer industry and the experiment stations in encouraging the sale of recommended analyses. In several states, only the recommended grades are permitted to be sold.

CHAPTER XIV

SOIL FERTILITY MAINTENANCE AND PRODUCTIVITY RATING OF SOIL

Differences in the productive capacity of different kinds of soils, or land, are commonly recognized on the basis of experience in farming. The range in productive capacity of different soil types is very evident where the same crops are grown and the same general farming practices are followed; but even where practices are modified and crop adaptation is recognized, differences in yields, or net money income, still prevail on widely different soil types. Some system of expressing the difference in productivity of soil types or land areas is highly desirable. Furthermore, the importance of maintaining soil productivity, to farmers, to urban workers, and to the nation, is well recognized and is stressed in another section of this book. But the question may well be asked, how does a farmer know whether or not he is maintaining the productive capacity of his soil? It might be assumed that, if crop yields do not decrease, fertility is being maintained. This is a false assumption, however, for the use of better tillage methods, of more carefully selected seeds and seeds of improved crop varieties, and of more suitable rates, dates, and methods of planting may maintain or even increase yields for a period in spite of a lower soil fertility. The discussion of the problems of measuring fertility maintenance and of rating the productivity of land is the objective of this chapter.

Objectives:

A. Maintaining soil fertility.
B. Soil-productivity rating and land classification.

MAINTAINING SOIL FERTILITY

In recent years some system has been sought of determining if a given rotation or soil-management system is maintaining the productivity of the soil and, if not, how rapidly fertility is being lost. Workers at the experiment stations in Ohio and Missouri have taken the lead in the endeavor, and the systems proposed by them are discussed in connection with the following questions.

Questions:

1. What productivity indexes are given to different groups of crops?
2. On what basis are productivity indexes determined?
3. Why is so much emphasis placed on nitrogen in assigning index values to crops?
4. Of how much value is manure in maintaining production?
5. Do fertilizers contribute to increased productivity?
6. How detrimental is erosion?

Crop Indexes. The Ohio and Missouri systems give to each crop a productivity index based on the effect of growing that crop on the nitrogen or organic matter content of the soil. Crops grown in rows and intertilled are given the highest minus ratings, whereas leguminous crops which have a sodlike habit of growth are given the highest positive ratings. Grass sods are given +0.25 ratings because these plants harbor no nitrogen fixing bacteria on their roots, but they do reduce leaching and their culture entails no destruction of soil humus through tillage. When a legume is grown in association with a grass, a positive index is given in accord with the proportion of the legume present.

Basis for Crop Indexes. The productivity ratings assigned to different crops were arrived at through a study of data from the Ohio experiment station. These data show that when corn, oats, or wheat were grown year after year on the same land, the organic content of the soil was greatly decreased, especially for corn, and that the yields decreased correspondingly. When these crops were included in rotations with clover or alfalfa, the decrease in the nitrogen content of the soil was much less, as was also the decrease in crop yield; and, furthermore, the greater the number of years that clover or alfalfa occupied the soil in the rotation period, the more effective was the rotation in maintaining soil productivity. A rotation in Ohio, including 3 years of alfalfa and 1 each of corn and oats, showed no decrease in corn yield for a 20-year period and no loss of soil organic matter. On the other hand, a rotation consisting of 3 years of corn, 1 of clover, and 1 of wheat, resulted in a decrease in corn yield of 21.6 per cent in 20 years and a loss of 18 per cent of the soil nitrogen. Missouri found a direct relationship between the content of nitrogen in the plow soil of different soil types and the yield of corn. The results are presented in Table 35. The productivity indexes assigned to various crops by the Ohio experiment station are given in Table 36.

Nitrogen—the Key Element. The question may well be raised as to why soil nitrogen content is made the basis for the fertility ratings of crops in place of some other element required by plants like phos-

TABLE 35

The Relation of Total Nitrogen Content of Soils to Corn Yields *

Soil Type	Total Nitrogen (Pounds per Acre, 7 Inches Deep)	Average Yield of Corn (Bushels per Acre)	Pounds of Soil Nitrogen per Bushel of Corn
Marshall silt loam	3,630	38.6	94.04
Grundy silt loam	3,370	32.0	105.31
Eldon silt loam	3,160	31.2	101.28
Crawford silt loam	2,840	25.4	111.81
Cherokee silt loam	1,950	22.5	86.67
Gerald silt loam	1,890	19.0	99.47
Union silt loam	1,600	16.0	100.00
Average	2,634	26.4	99.92

* "Evaluating Annual Changes in Soil Productivity," A. W. Klemme and O. T. Coleman, Mo. Exp. Sta. Bul. 405, 1939, p. 4.

phorus. In this connection it must be remembered that nitrogen is accumulated in soils of humid regions almost entirely in the form of humus, which in turn is largely a product of plant-tissue decay together with the cells of microorganisms. In other words, to accumulate humus it is necessary for soil conditions to be favorable for a considerable if not luxuriant growth of plants including legumes. This growth in turn involves at least a moderate supply of all essential plant-food elements, a not-unfavorable soil reaction and drainage condition, and a reasonable amount of precipitation or irrigation water. On the whole, then, the nitrogen or humus content of a soil is a fairly accurate index of productivity.

Value of Manure. Manure contributes to soil productivity through both its humus content and the plant nutrients supplied which stimulate crop growth. The two contributions may be considered about equal in value in increasing the humus of the soil. The Missouri station determined the effect of a 6-ton annual application of manure for a period of 50 years on the nitrogen content of soil that was growing various sequences of crops. The total increase in nitrogen content of the soil was 41.1 per cent. As 300 tons of manure were applied, this makes an increase of 0.137 per cent of nitrogen for each ton of manure. A fertility index of 0.15 per cent is assigned for each ton of manure applied, as shown in Table 36.

TABLE 36

SOIL-PRODUCTIVITY INDEXES SUGGESTED BY THE OHIO EXPERIMENT STATION *

Crops on Rotated Land	Soil-Productivity Index (Changes Effected by One Year's Growth)
	Points or percentage
Corn, as grain or silage	−2.0
Potato, tobacco, and sugar beet	−2.0
Oats, wheat, barley, rye, and buckwheat	−1.0
Soy bean as hay or seed (straw not returned)	−0.5
Credit for crop residues: soy bean or wheat straw, cornstalks	+0.25
Alfalfa, for change effected by end of 1st hay year	+2.5
Alfalfa, for change effected during 2d hay year	+0.5
Alfalfa, for change effected during 3d hay year	0.0
Timothy and other grass sod	+0.25
Clover-timothy mixed, as hay or pasture	+1.25
Common clovers, as hay or pasture	+2.0
Sweet clover, crop plowed green in April or May	+2.5
	Points or percentage
For each ton of protected manure applied	+0.15
For each 200 pounds average commercial fertilizer applied	+0.15

Erosion Class †	Degree of Erosion	Modifying Factor for Erosion ‡		
		No special control methods	Farmed on contour	Strip cropped or terraced
1	little or none	0.00	0.00	0.00
2	slight	0.25	0.125	0.05
3	moderate	0.50	0.33	0.125
4	severe	1.00	0.80	0.30
5	very severe	2.00	Advisable to retire such crop land to permanent pasture or forests	

* "Our Heritage—The Soil," R. M. Salter, R. D. Lewis, and J. A. Slipher, Ohio Agr. Ext. Ser., Bul. 175, 1936, pp. 11, 14, and 15.

† Classes correspond to those recognized by the Soil Conservation Service in its erosion surveys.

‡ To apply this factor, multiply it by the sum of the negative crop-productivity values.

Fertilizer Application. A study of the effect of applications of fertilizer on the nitrogen content of soil has led the Ohio station to give a positive index value of 0.15 per cent for each 200 pounds of fertilizer containing approximately 20 per cent of plant nutrients.

Detrimental Action of Erosion. For many years the Missouri station has studied the loss of soil by erosion under different cropping systems on a field having a slope of 3.7 per cent. More recently many studies have been made by the Soil Conservation Service and by different experiment stations of erosion losses and of the effect of the loss of surface soil on productivity. A consideration of the available data has led to the selection of the factors contained in Table 36 for use in computing the detrimental effects of different degrees of erosion on soil productivity when the designated control measures are employed.

SOIL-PRODUCTIVITY RATING AND LAND CLASSIFICATION

Generally, in the past, differences in the productive capacity or economic value of soils have been expressed loosely and qualitatively by such terms as "good," "fair," and "poor" or by the use of the economist's terms "supermarginal," "marginal," and "submarginal." Frequently the sale price of land has been accepted as a measure of quality of land, notwithstanding that money value may be determined by other factors than crop productivity. However, there is a need for more-exact comparisons or for some quantitative scale of ratings to be used in farm appraisal work, assessment of land for tax purposes, and in making inventories of land resources, which in turn are basic for soil-conservation studies, rural zoning, and general land planning.

The desirability of having some scheme for the more-exact comparison of the value or quality of soil and land, especially if constructed on a mathematical basis and expressed by indexes, being accepted, questions arise about the criteria to be used and their validity. Some of these questions are stated below.

Questions:

1. Should soils be rated upon the basis of natural productivity alone, upon the basis of management, or both?
2. How is a standard of value selected for a mathematical rating of productivity of soils?
3. What are the criteria which can be used in the classification of land on an agricultural basis?

TABLE 37

DATA SHOWING METHOD OF CALCULATING ANNUAL SOIL-PRODUCTIVITY BALANCE AS APPLIED TO TWO DIFFERENT SCHEMES OF SOIL AND CROP MANAGEMENT *

SITUATION I Rotation: Corn, wheat, timothy Manure applied during the rotation equals 5 tons per rotated acre Fertilizer applied in rotation equals 200 pounds per rotated acre Class 2 erosion, no special control measures			SITUATION II Rotation: Corn, wheat, alfalfa, alfalfa Manure applied during the rotation equals 10 tons per rotated acre Fertilizer applied in rotation equals 400 pounds per rotated acre Class 2 erosion, strip cropping practiced		
Crop	Productivity value		Crop	Productivity value	
	Positive	Negative		Positive	Negative
Corn		−2.0	Corn		−2.0
Wheat		−1.0	Wheat		−1.0
Timothy	+0.25		Alfalfa (2 years)	+3.0	
Total	+0.25	−3.0	Total	+3.0	−3.0
Balance due to crop factor		−2.75	Balance due to crop factor	0.0	0.0
Reduction in balance for erosion (0.25 × −2.75)		−0.69	Reduction in balance for erosion (0.05 × −3.0)		−0.15
Gain in balance for manure (5 × 0.15)	+0.75		Gain in balance for manure (10 × 0.15)	+1.50	
Gain in balance for fertilizer (200 div. by 200) × 0.15	+0.15		Gain in balance for fertilizer (400 div. by 200) × 0.15	+0.30	
Total	+0.90	−3.44	Total	+1.80	−0.15
Soil Productivity Balance			Soil Productivity Balance		
(a) For rotation		−2.54	(a) For rotation	+1.65	
(b) Annual (−2.54 div. by 3)		−0.85	(b) Annual (+1.65 div. by 4)	+0.41	

* "Our Heritage—The Soil," R. M. Salter, R. D. Lewis, and J. A. Slipher, Ohio Agr. Ext. Ser. Bul. 175, 1936, p. 18.

Basis for Rating Soils. Soils can be rated solely upon the basis of natural productivity or inherent fertility, but only by an assumption that management factors are constant. If the basis of management is emphasized or carried to the extreme, a rating of soils is likely to become nothing more than a comparison of the efficiency of different management practices, or a comparison of the relative efficiency of individuals or classes of men. Since the effect of the natural character of soil cannot be entirely eliminated by management practices, and since the productive capacity of soils may be greatly increased by the use of amendments, such as commercial fertilizers and by irrigation, it follows that both the natural productivity and the productivity obtained by amendments should be taken into consideration in making an index rating.

Standard of Value. Obviously, some standard of value has to be established in order to compare soil types on a mathematical basis. For example, the figure 100 may be taken as a standard, and this may represent the average yield for a particular crop in a region where the crop is extensively grown and where average yields are high. For example, the standard for corn may be 50 bushels per acre; then, if the average yields for a particular soil type are estimated at 25 bushels, the productivity rating for that soil type is 50. Soils may also be given a "productivity grade." Productivity-grade numbers are obtained by a simple percentage weighting of the various crop indexes. Soils with a weighted average between 90 and 100 are rated 1; between 80 and 90, are rated 2; etc. This is a general agricultural rating and not one for a particular crop. The method, here simply outlined, is in substance that employed by the Division of Soil Survey of the United States Department of Agriculture.

The example of the application of the method is taken from a table of productivity ratings given in a Soil Survey report. ("Soil Survey of Ingham County, Michigan," United States Department of Agriculture, 1941). Two contrasting soil types have been selected from the complete table for the county.

Criteria for Rating Land. *Land* should be distinguished from *soil* or *soil type* since these words, as generally used, are not exactly the same in meaning. If an individual soil type happens to be very uniform in character and occurs in large separate bodies, it can be taken as a satisfactory unit for land classification; however, in many places soils of widely divergent chemical and physical character are closely associated in small bodies, and again configuration features of the

Soil	Crop-Productivity Index * for																												General Productivity Grade		Land Classification
	Corn (grain)		Corn (silage)		Wheat		Oats		Rye	Mixed timothy and clover		Red clover		Alfalfa		Beans		Sugar beets		Potatoes		Vegetables (leafy)		Vegetables		Small fruits	Apples	Permanent pasture			
	A	B	A	B	A	B	A	B	A	A	B	A	B	A	B	A	B	A	B	A	B	A	B	A	B	A	A	A	A	B	
Conover loam	90	100	90	100	80	100	90	100	80	90	100	90	100	70	90	80	100	70	90	50	70	60	...	70	90	60	70	100	2	1	Good to excellent crop land
Coloma loamy sand	40	60	40	60	30	50	40	50	50	40	60	30	50	30	50	40	60	...	...	50	70	30	...	40	70	70	60	30	7	5	Fair to poor crop land

* The indexes in columns A refer to yields obtained under the more common practices of management that prevail in this county and include the use of some commercial fertilizer and lime; those in columns B refer to yields obtained under the more intensive or best practices that include the greater use of legumes, green manures, manures, commercial fertilizers, and improved plant varieties.

landscape, especially the kinds of slopes, are an important factor affecting the use and value of land.

The possible criteria for the classification of land on an agricultural basis are (1) gross money value of agricultural products—plant and animal; (2) net income from land; (3) selling price of land; (4) values assessed for taxation; (5) measured yields of crops; (6) physical character of the land.

Taken singly or used alone, none of the criteria listed above is entirely satisfactory except for some limited objective or for small areas of land. However, classifications, more or less subjective, including only a small number of divisions, 3 to 5, may be made and may have a considerable value in the inventory of land resources if a combination of factors, such as listed, is employed.

CHAPTER XV

SOILS AND AGRICULTURE OF ARID REGIONS

No generally accepted definition for an arid region exists. A definition based solely on total annual precipitation is faulty because it does not take into consideration factors which influence the efficiency of a given amount of precipitation in the production of crops. Some of these factors are (1) distribution of the rainfall through the year, (2) temperature, (3) wind, (4) air humidity, and (5) amount of sunshine. A comparatively low rainfall, coming very largely during the growing season, may permit relatively high crop yields, while some areas, receiving as much as 70 to 80 inches of rain, with little of it coming during the growing season, require irrigation for crop production. Frequent fogs, moderate temperature, and little wind lead to low rates of evaporation of water from the soil and reduced transpiration of plants and hence reduce the moisture required for crop production.

For the purpose of this discussion the arid region is considered to include those areas with predominantly clear weather and annual precipitations as high as 15 inches, and also areas subject to fog, but with less precipitation, in which the soils have the same characteristics as those developed in the clear areas. Without irrigation, these lands are used principally for grazing, with a very small production of small grains in areas where conditions are especially favorable and under a system of one year of crop and one of fallow for moisture accumulation. Two headings under which arid soils may be studied are suggested.

Objectives:

A. Characteristics and utilization of soil in arid regions.
B. Development and management of alkali soils.

CHARACTERISTICS AND UTILIZATION OF SOIL IN ARID REGIONS

Many misconceptions in the minds of laymen have prevailed regarding soils of the arid regions. They are not merely accumulations of more or less decomposed rock fragments with no profile character-

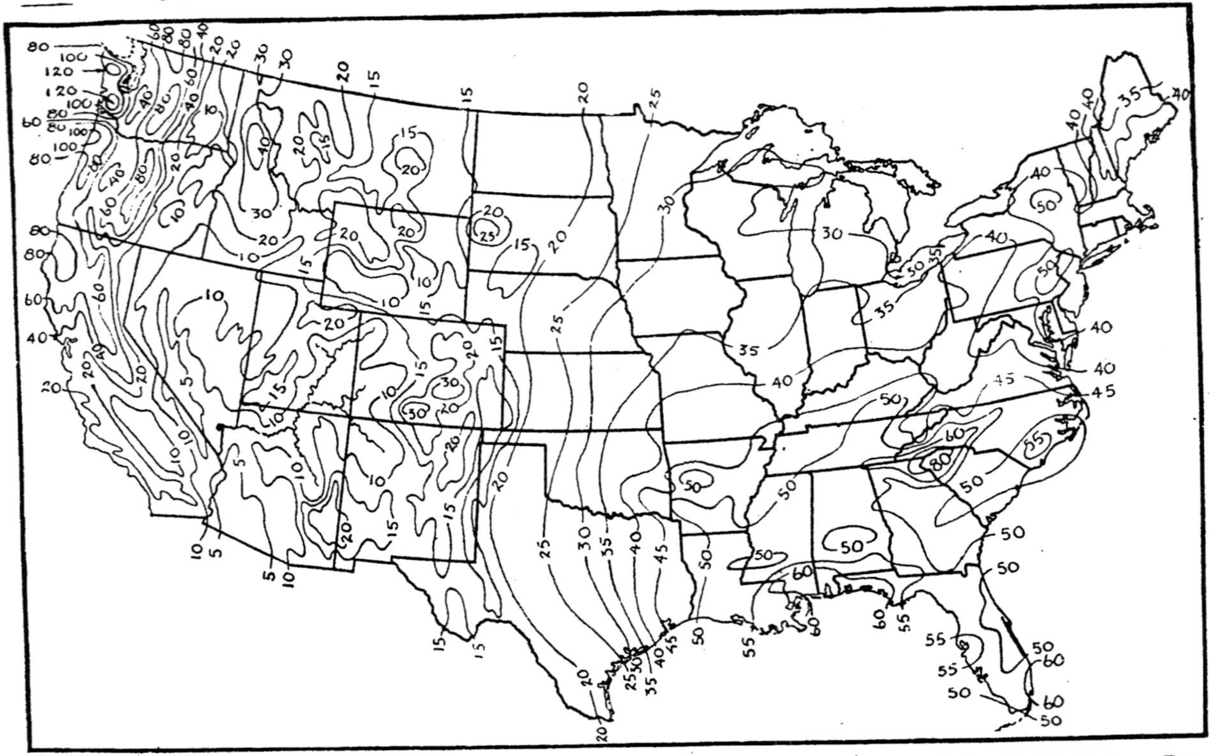

FIG. 63. A generalized map showing the total annual precipitation in different parts of the United States. Precipitation includes rain and the water equivalent of snow, sleet, and hail. Some of the smaller areas, such as those receiving 100 or 120 inches, are enlarged on the map. [Drawn from a map prepared by the Weather Bureau and presented in the "Atlas of American Agriculture."]

istics, nor are they all of a coarse texture or sandy nature. These soils exhibit much diversity in properties. There are heavy soils, medium-textured soils, and sandy soils; young soils and soils with a pronounced profile development; alkali soils and those without injurious salt accumulations in the arid regions.

The agricultural utilization of arid soils is also varied. With an abundance of water for irrigation, an intensive agriculture may be carried on, as is described in the following chapter. Without irrigation, forage for grazing ranges from the very scanty supply of the true desert regions to the thin grass cover of the arid prairies. The following questions may serve as guides in studying these soils.

Questions:

1. What physiographic locations are occupied by arid soils?
2. What were the sources and nature of the parent materials for arid soils?
3. Which groups of soils are found in the arid regions?
4. What are the characteristics of red desert soils?
5. In what respects do gray desert soils differ from red desert soils?
6. Are the deserts occupied by characteristic vegetation?
7. Do the deserts make suitable grazing lands?
8. Are the deserts subject to erosion?
9. What are the characteristics of brown soils and how are they utilized?
10. What are adobe soils?

Locations of Arid Soils. Arid soils of western United States occupy, in the main, broad valleys and basins between mountain ranges and extend up the slopes until increased precipitation or lower temperatures give rise to a different type of vegetation and to soils of different characteristics. In some cases, however, arid conditions prevail over the entire surface of low mountain ranges, which are either bare of vegetation or are occupied by a thin cover of brush. The mountains serve as barriers to clouds and to cooling winds and so produce areas of low rainfall and relatively high-mean annual temperatures. With few exceptions, the relative air humidity is very low. Similar conditions prevail to a certain extent on the high plains lying directly east of the Rocky Mountains and give rise to areas classed in the brown soil group which are included among the arid soils in this discussion. In some locations reasons other than the occurrence of mountain barriers must be sought to explain the arid conditions, but a discussion of these more unusual situations must be left to a textbook on climatology.

Sources and Nature of Parent Material. The basins in which the major portion of the arid soils has developed were occupied at some remote period by lakes or estuaries of the ocean. Evaporation of these waters has left deposits of salts or soil materials impregnated with salt. To a large extent, these deposits have been covered with sediments carried by streams and ancient glaciers from the mountains.

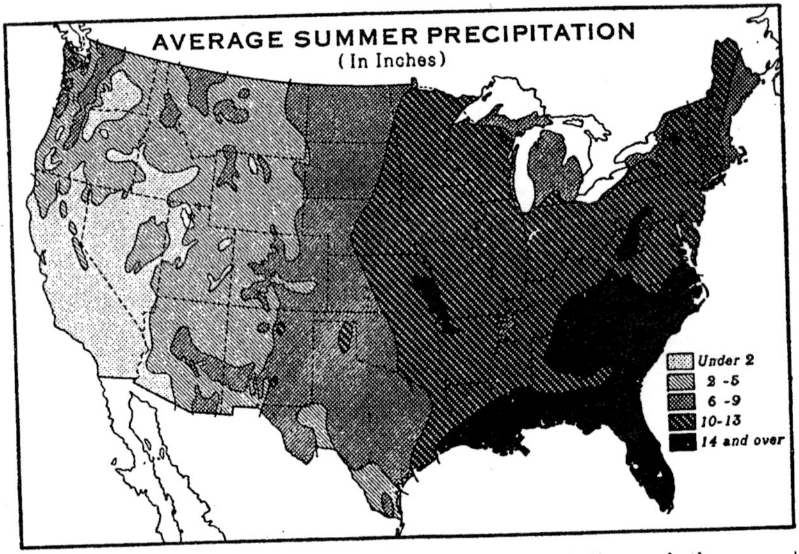

FIG. 64. Total annual precipitation does not always indicate whether or not crops can be grown without additional water. The amount of precipitation during the growing season is a better criterion. [From U.S.D.A. Misc. Pub. 260, by O. E. Baker. Drawn from a map prepared by the Weather Bureau and published in the "Atlas of American Agriculture."]

In fact, the true desert soils have developed largely on long alluvial fans which stretch outward from the mountains until they meet the fans from adjacent parallel ranges. A belt of flat, low land, subject to flooding and known as a playa, sometimes separates the fans. These alluvial fans provide the material for the development of deep, pervious, well-drained soils. The more recent alluvial deposits exhibit only the first stages of soil formation. Terraces of varying ages have furnished the material for soils with highly developed profiles. Some of the terrace soils have firmly cemented hardpans at varying depths, whereas dense claypans characterize the horizons of others.

The characteristics of these secondary soils are greatly influenced by the nature of the rocks furnishing the parent material. Granitic debris develops into pervious, medium-textured soils with desirable moisture relationships. Soils of heavier texture, which are sometimes less favorable for moisture infiltration, develop from the material contributed by basic igneous rocks. Sedimentary rocks may contribute to the formation of heavy clays, sandy soils, or soils of medium texture, depending on the proportions of shale, slate, sandstone, and limestone in the valley sediments.

The quantity of lime in the soil material has influenced the reaction of the horizons in the various profiles and has contributed to the formation of limestone nodules and of more or less cemented deposits of calcium carbonate known as *caliche.* The soluble salt content of the parent rocks and of the transporting streams has been influential in determining the alkalinity and the structure of the resulting soils.

More recent alluvium deposited by streams flowing through the lower portion of the valley floor may be rich in sand and silt and highly pervious where it is dropped near the channel or fine textured where water is impounded in depressions farther back.

On the foothills and mountain slopes residual soils of great variation in depth and texture are present. Their chemical and physical properties depend on the nature of the parent material and on moisture and temperature relationships.

In the following discussion no attempt is made to mention all the soil conditions occurring in the various areas considered. Attention is centered on the predominating soil-profile characteristics and on the accompanying vegetative cover and utilization practices.

Arid Soil Groups. Within the arid region, as defined in this chapter, are found soils of the gray and red deserts and the brown soils (except those approaching the dark-brown or chestnut soils) as described by Marbut.[1] The desert soils usually occupy areas having less than 7 to 8 inches of annual rainfall. The factor of the temperature resulting in different degrees of dehydration of hydrated iron oxides accounts for the division between the gray and red deserts. Roberts [2] has observed that in Nevada the mean annual temperature of the red deserts exceeds 60° F., whereas on the gray deserts it is much less, possibly between 45° F. and 50° F.

The brown soils occupy areas receiving 11 to 15 inches of rainfall, except in the fog belts of the Pacific coast where the precipitation may drop as low as 10 inches. Many soils designated as "adobe" are also

[1] "Atlas of American Agriculture, Soils of the United States," C. F. Marbut, U.S.D.A., 1935, pp. 76–82.

[2] For this section much material has been taken from an unpublished manuscript by R. C. Roberts, Div. Soil Survey, Bur. Plant Industry, U.S.D.A.

found in the arid region although arid conditions are not necessary for the development of adobe soils.

Red Desert Soils. These soils are in the hot, arid regions covering the extensive southern part of Nevada, Southern California, western Arizona south of the Grand Canyon, and sections of southern New Mexico and southwestern Texas. They vary in color from light-pinkish gray to reddish brown and red. The upper subsoil is more compact and of heavier texture than the surface and is usually reddish brown or red. The lower subsoil is pink or white and very rich in lime. Maximum concentration of lime carbonate occurs at approximately 14 inches although the zone may be many feet thick and frequently is consolidated. An intermixture of calcium sulphate often occurs below a depth of 2 feet. In many places the soil has blown or washed away sufficiently to leave a covering of stones on the surface, thus giving rise to the so-called "desert pavement." A coating of iron oxide on these stones gives them a polished appearance and hence is known as "desert varnish." Stones buried at shallow depths are coated on the underside with calcium carbonate, deposited by upward moving water, and those at greater depths may be entirely covered with lime.

Gray Desert Soils. These soils lack the red color found in the southern desert. They occupy the region north of the Grand Canyon and Little Colorado River in Arizona and of extreme southern Nevada, and extend northward to the Columbia River in Washington. Possibly 10 per cent of the area of the United States is included in the region.

The surface soil is gray to light-grayish brown in color and of small thickness. The soil below, to a depth of several inches, is of a more pronounced brown and grades into a more compact sometimes granular B horizon of heavier texture. The lime content increases with depth, forming a distinct zone at approximately 18 to 20 inches, which may take the form of a hardpan at this or greater depth. In both the red and gray deserts the lime accumulations in the lower horizons may develop cemented formations known as "caliche," which sometimes extend to a depth of many feet. Desert pavements develop where the soil is derived from material containing much coarse gravel and stone, but the desert varnish infrequently occurs.

Vegetation of Red and Gray Deserts. In the northern areas of red desert, creosote bush and bur sage occur in relatively large numbers with many other plants, some of which are wolfberry, Mormon tea, yucca, and prickly pear. In Southern California and Arizona the giant cactus, ocotillo, and ironwood are found. When the ground-water table is nearer the surface, tamarack, mesquite, and cat's-claw occur. In areas with higher concentrations of salts are found iodine bush, quail bush, and salt grass.

On the gray desert shadscale, bud sage, rabbitbrush, and horsebrush are prominent in the most arid parts, with big sage brush predominating where rainfall is generally 7 to 9 inches. Among plants of importance for grazing purposes, white sage, winter fat, galleta, Indian rice

FIG. 65. In the desert, grass and other short succulent vegetation grow only around the base of larger shrubs.

grass and, in the northern areas, blue grama and needle and thread grass should be mentioned. Alkali areas are characterized by the same plants found in similar locations, as on the red desert.

Vegetation on the deserts occupies only a small portion of the surface, seldom exceeding 20 per cent even in the most favored location and frequently almost disappearing. Possibly from 3 to 8 per cent coverage would be a reasonable average. It is noteworthy that grass and other small succulent vegetation is found mostly around the base of the larger shrubs. This phenomenon is explained on the basis of the presence of a very thin crust underlaid by a layer of soil containing innumerable small round pores filled with air (vesicular structure) which resists the entrance of water, except within the spread of the branches, and hence enough rain is absorbed under the shrub to supply the needs of both shrubs and smaller plants. Another explanation has been suggested by Brezeale, whose observations have led to the conclusion that the roots of the grass withdraw water from the roots of the shrubs, which they in turn have obtained from the deeper subsoil.

Utilization of Desert Soils. When supplied with water many of the desert soils are highly productive, as will be pointed out in the section dealing with irrigation. Without irrigation the desert soils are utilized almost entirely for grazing. The amount of vegetation produced which will be eaten by cattle, sheep, and horses is extremely meager, and hence carrying capacity on the year-round basis varies from virtually nothing to about 5 head of cattle per section on the red desert soils and possibly up to 10 head on the better-grassed northern portion of the gray desert. In general, the cattle are in better condition on the gray desert and graze in somewhat larger groups of about 10 or less. Sheep are more common on the gray desert, and as many as 1,000 may be herded together.

There is a strong tendency for desert soils to form a thin hard crust on the surface, which retards the infiltration of rain. Under this crust a layer with vesicular structure sometimes to a thickness of 2 or 3 inches further reduces penetration of water. The development of these layers is more pronounced in the soils of heavier texture and, as previously mentioned, between the clumps of vegetation. As a result, the vegetative cover is more dense, especially for grass, on the soils of coarser texture such as loamy sand and sands. These soils not only absorb a greater proportion of the rainfall but also give up a larger percentage of their moisture to plants. Furthermore, root penetration is deeper in such soils as the formation of hardpans of *caliche* is less frequent and, when present, is at a greater depth. In the non-irrigated areas deep, coarser-textured soils without a gravelly subsoil or substratum are accordingly the most highly prized on the desert.

Erosion on the Desert. Storms on the desert are frequently of torrential nature, which circumstance, coupled with the formation of crusts and layers of soil which resists infiltration of water, leads to severe erosion. Stream beds and dry washes, which through most of the year are entirely dry, carry torrents of water during storms and may wash out highways or cover them with streams of such depth and velocity that they are impassable for many hours. Gully formation is common, and the washes are often of great depth. Sheet erosion removes the surface soil from the more gentle slopes. Water erosion is much more pronounced on the sandy loams and on soils of heavier texture than on the sandy soils which are more receptive of rainfall.

The effects of wind erosion are evident everywhere. Each bush or clump of vegetation occupies a little mound, the soil of intervening spaces having been carried away to a greater or less extent. The formation of the desert pavement is an evidence of soil blowing as is the accumulation of wind-drifted sediment on the windward slopes of the hills and mountains. The giant sand dunes northwest of Yuma and extending some 50 to 60 miles in a northeasterly, southwesterly direction through Southern California and northern Mexico are further proof of the erosive power of desert winds.

The Brown Soil Area. In the fog areas of the Pacific coast where about 80 per cent of the rainfall occurs from December to April, brown soils develop with as low a precipitation as 10 inches. As the effect of fog and humidity diminishes, the soils are formed under higher precipitation until on the open plains and in the arid valleys they occupy areas receiving 11 to 15 inches. Although mean annual temperatures may reach 60° F. in the coastal zones, the soils, as a whole, are formed under temperatures of 45° to 50° F. Areas of brown soil are found in both eastern and western Colorado, in New Mexico, Wyoming, and Montana. Small areas are also scattered throughout the arid region where climatic conditions are somewhat too humid to permit of desert soil development.

The surface soil is grayish brown to brown and 6 to 8 inches in depth. The gray shade predominates over the brown of the humus to such an extent that sometimes it is difficult to distinguish the humus horizon from the underlying material. A thin, laminated layer has developed in the immediate surface. The B horizon is lighter in color than the surface soil, which is a distinguishing feature from the desert soils, as is also the occurrence of cracks extending from the surface to the bottom of the B horizon. The horizon is friable and generally alkaline in reaction although not calcareous. Below the B horizon the soil is calcareous but is easily penetrated by roots and has a yellowish-gray color with sometimes a brownish tinge. Maximum lime accumulation occurs at a depth of 2 or 3 feet, depending on moisture relations.

The dominant vegetation on the brown soils is composed of various short grasses of which blue grama, niggerwool, bluegrass (Poa spp.), June grass (*Koeleria cristata*), buffalo grass, and needle and thread grass are most prominent. Near the Pacific coast wild oats and brome grass make their appearance. Several of the sages, prickly pear, rabbitbrush, and bitter brush, together with pinon pine and juniper on well-drained areas and cottonwood and willows near stream beds, are representative of the larger types of vegetation.

The vegetative cover is much more dense than on the desert soils and affords a grazing capacity ranging from 8 to 30 cattle per square mile. Both cattle and sheep ranches are numerous. Stock raising is the main activity, although some dry farming is carried on. Water is usually supplied by means of windmills. Erosion is not so extensive as on the desert soils, owing to the denser cover of vegetation, although much damage is evident on some of the more sloping fields that have been plowed for crop production.

Adobe Soils. The term adobe has been used with no great uniformity of meaning. The word is of Spanish or Spanish-American origin and originally referred to sun-dried bricks. More recently it has been applied to the soil from which the bricks were made. An examination by Hendry of bricks taken from old missions and other structures in southwestern United States revealed that the soils used were not of very fine texture, as often has been assumed, but varied from fine sandy loam to heavy loam. If the natural soil in the locality where the bricks were to be made was of too fine a texture, sand was added to make it more loamlike.

Replies to inquiries addressed to a considerable number of experiment stations and a study of soil survey maps and reports led Smith [3] to the following conclusions: Adobe soils occur primarily in the western states, having been mapped or described in Colorado, Wyoming, Idaho, Oregon, Utah, Nevada, California, Arizona, Texas, and New Mexico. They are derived from both transported and residual soil material, but not from wind-deposited sediment as has been sometimes suggested. The soils are developed in areas with an annual precipitation of less than 20 inches and in which there are distinct wet and dry seasons. Much adobe land is and will continue to be used for grazing, with possibly 5,000,000 acres being used for production of cultivated crops. Although a large variety of crops is grown, including citrus fruits, alfalfa, sugar beets, rice, and vegetables, the grain crops usually are grown to best advantage.

The general conclusion of both Smith and Shaw is that adobe is a term which should be applied to heavy clay soils that exhibit a specific structural condition. Shaw [4] offered the following definition, "A soil which on drying cracks and breaks into irregular but roughly cubical blocks. The cracks are usually wide and deep and the blocks are from 20 to 50 or more centimeters across."

A comparison by Chang [5] of the physical properties of a number of adobe soils and of non-adobe clay soils of similar clay content led to

[3] "Characteristics of Adobe Soils," A. Smith, Amer. Soil Survey Assoc., Bul. 14, 1933, pp. 79–82.

[4] A Definition of Terms Used in Soil Literature, C. F. Shaw, in "Proceedings and Papers," the First International Congress of Soil Science, Com. V, Vol. 4, 1927, p. 57.

[5] An Experimental Study on the Development of Adobe Structures in Soils, C. W. Chang. "Soil Science," Vol. 52, 1941, pp. 213–227.

the following conclusion. Upon drying, adobe soils shrink extensively, developing wide, straight cracks which divide the soil into relatively large blocks or primary structural units. These blocks as well as the smaller secondary units into which they break have approximately flat sides, few in number, which meet at angles approaching right angles. In other words, the structural units are roughly cubical. They were found to have a high resistance to crushing. Soils containing a large quantity of montmorilionitic clay in a dispersed condition have an adobe structure unless other factors interfere. On the other hand, clay soils containing kaolinitic clay or montmorillonitic clay in a flocculated condition are unlikely to develop such a structure. Some factors which tend to prevent the formation of adobe structure are the presence of a considerable quantity of (1) soluble salt, (2) calcium carbonate, (3) organic matter, and (4) sand.

DEVELOPMENT AND MANAGEMENT OF ALKALI SOILS

The term *alkali*, as applied to soils, may have a restricted meaning referring only to soils with a *p*H value of 8.5 or higher and which are impregnated with soluble salts including some sodium carbonate. A more general and more common use of the term includes soils containing an excess of soluble salts, usually more than 0.2 per cent, even though no sodium carbonate is present and the *p*H value is not excessively high. In this discussion the term is used with its broader interpretation.

Soils having a *p*H of 8.5 will give a pink color with a phenolphthalein indicator and almost invariably will contain some sodium carbonate. The term *black alkali* is applied to such soils because of the brown color developed by the dispersion of organic matter through the action of the alkaline soil solution. The quantity of sodium carbonate present is usually quite small, and it is accompanied generally, although not necessarily, by high concentrations of sodium chloride, sodium sulphate, and often calcium carbonate. An important characteristic of black alkali soils is the large amount of sodium adsorbed by the colloidal fraction.

Soils containing considerable quantities of soluble neutral salts but no sodium carbonate are designated as *white alkali*. The salts most commonly present are the chlorides and sulphates of sodium. Occasionally nitrates are present and also calcium carbonate. The colloidal fraction may or may not contain adsorbed sodium. The following questions suggest themselves in relation to alkali soils.

Questions:

1. What conditions give rise to alkali soils?
2. Where do the salts come from that cause alkali?
3. Why are high concentrations of soluble salts detrimental to many plants?
4. How does alkali affect soil conditions?
5. Which plants are most tolerant of alkali?
6. How may white alkali soils be reclaimed?
7. Are special treatments needed for the reclamation of black alkali soils?
8. How does removal of alkali affect soil structure?

Chart 8. A Chart Showing the Relationship of Soil Reaction (pH) and Alkalinity

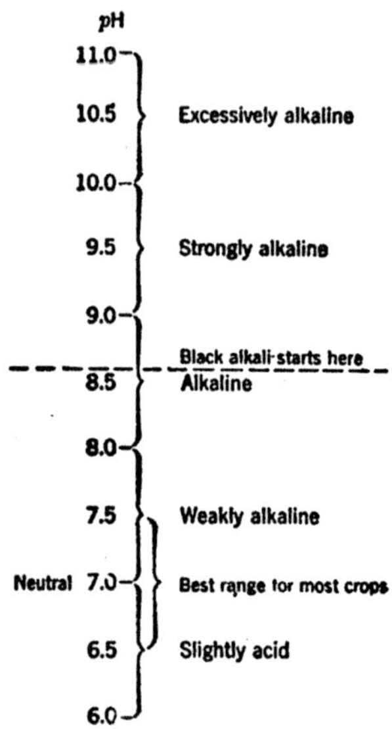

[From Ariz. Ext. Ser. Circ. 108, by W. T. McGeorge.]

Formation of Alkali Soils.[6] The ground water of arid regions usually contains considerable quantities of soluble salts. If the water table is high, large amounts of water move to the surface by capillarity and are evaporated, leaving an ever-increasing accumulation of soluble salts. Through this process the upper parts of the soil may become so highly impregnated with salts that only alkali-resistant plants will grow. In locations where the water table is too deep for appreciable capillary movement to the surface, harmful accumulations of soluble salts will not develop. The rate of salt accumulation will be deter-

[6] Many of the ideas in this section were obtained from "The Reclamation of Alkali Soils," W. P. Kelley, Calif. Agr. Exp. Sta. Bul. 617, 1937.

mined by (1) the rate of capillary movement of water to the surface, (2) the salt content of the ground water, and (3) the rapidity of evaporation. The nearer the water table is to the surface, the more rapid will be the salt accumulation, other factors being the same.

Under irrigation alkali soils have been developed by several means. (1) Excessive applications of water have raised the ground-water level sufficiently to permit of concentration of salts through evaporation. (2) Seepage from leaky canals and lateral ditches have resulted in a high water table. (3) The use of irrigation water with a high salt content has resulted in salt accumulation when (*a*) drainage is poor so that the salts cannot be leached out, (*b*) the application of water is so limited that the salts are left in the root zone in place of being washed out, and (*c*) the water contains such quantities of sodium salts that the colloids become impregnated with them and the favorable structure of the soil is destroyed.

Alkali spots may also develop in irrigated fields when for some reason a given area of soil does not absorb the water. Water that is carrying salts will move into such soil from the sides and below, and through surface evaporation salts will accumulate. Because no water enters from the surface, the situation becomes steadily worse and the spots may enlarge.

The overflow of streams by flood waters may cause a rise in the ground-water level of adjacent valleys or may also cause the seepage of flood water through the porous stream bed. Again the irrigation of nearby land of higher level may result in a higher water table. All such increases in height of ground water carry the possibility of alkali development.

Sources of Alkali Salts. Alkali salts are derived primarily from the soil and rocks through which the waters percolate that supply the water table and the irrigation streams. The salt supply of the rocks and soil may be the result of accumulation when the area was a portion of the ocean bed or of the bottom of a saline lake, as in sedimentary deposits. The chemical decomposition of minerals has also contributed some salts. In igneous rocks, chemical weathering has supplied the soluble material. Kelley has observed that alkali soils, watered with drainage from sedimentary rocks, contain a higher proportion of sulphates than do those receiving water from igneous rocks. On the other hand, more black alkali soils appear to develop when the water supply comes from an area of igneous rocks. The primary source of the sodium, calcium, magnesium, and sulphur is evident, for many minerals contain these elements. The source of the sodium carbonate is somewhat more difficult to visualize in regions of sparse vegetation where the CO_2 production from decay of organic matter and respiration of plant cells is limited. Somė sodium carbonate may

TABLE 38

THE RELATION BETWEEN SALT CONTENT OF SOIL AND RELATIVE SALINITY *

Soluble-Salt Content		Relative Salinity
Parts per million (p.p.m.)	Pounds per acre-foot (4,000,000 lb.) of soil	
0 to 700	0 to 2,800	negligible
700 to 1,500	2,800 to 6,000	weak salinity
1,500 to 3,000	6,000 to 12,000	medium-strong salinity
3,000 to 8,000	12,000 to 32,000	strong salinity
8,000 to 15,000	32,000 to 60,000	very strong salinity
Above 15,000	Above 60,000	excessive salinity

* "Interpretation of Soil Analyses," W. T. McGeorge, Ariz. Ext. Ser. Circ. 108, 1940, p. 5.

be produced through the interaction of sodium chloride or sodium sulphate with calcium carbonate. It should be remembered, however, that the quantity of sodium carbonate present even in black alkali soils is not great.

Detrimental Effects of Alkali on Plants. High concentrations of neutral salts such as sodium chloride and sodium sulphate may interfere with the absorption of water by plants through the development of a higher osmotic pressure in the soil solution than exists in the root cells. Furthermore, the wilting coefficient of soils is raised by salt accumulations and hence the quantity of water a soil will supply to plants may be reduced through the presence of alkali. Detriment to plants may result also from soluble salts when the concentration is not sufficient to influence absorption of water. The entrance of nutrient ions into root hairs is influenced by the nature and concentration of other ions present. Alkali salts may therefore result in nutritional difficulties in crops because of their inability to absorb needed nutrients from the soil.

Black alkali is much more harmful to plants than is white alkali. The highly alkaline reaction due to the presence of sodium carbonate and the large quantity of adsorbed sodium represses the availability of several nutrients, especially iron, manganese and phosphorus. Also the alkaline soil solution has a corrosive action on the bark of roots and stems. The high concentration of neutral salts which is usually present in black alkali soils affects the absorption of moisture by the plants just as it does in white alkali soils.

Detrimental Effects of Alkali on Soils. As previously mentioned, the sodium in black alkali soils results in a deflocculation of the colloids and hence in a breaking down of the soil structural units. This puddled condition renders the soil more or less impervious and retards entrance of irrigation and rain water and impedes drainage. In heavy-textured soils the penetration of roots may be restricted by the density of the deflocculated zone. Aeration is also much reduced, setting up anaerobic conditions and resulting in the formation of reduced compounds which are toxic to plants.

The presence of certain white alkali salts, largely those of calcium, tends to counteract to some extent the detrimental effects of the adsorbed sodium and the sodium carbonate. The general action of the neutral salts is to flocculate the colloids and hence preserve the normal soil structure.

Tolerance of Plants for Alkali. Plants differ markedly in their tolerance of alkali, and likewise the alkali salts differ greatly in their detrimental effects on plants. Hilgard set the limit of tolerance for the principal salts found in alkali soils as follows:

Sodium carbonate, 0.10 to 0.25 per cent
Sodium chloride, 0.20 to 0.50 per cent
Sodium sulphate, 0.5 to 1.00 per cent

Unquestionably associated conditions, such as temperature, moisture supply, organic matter content, soil texture, and supply of nutrients, influence the tolerance of plants for these salts.

Shantz[7] lists some of the most alkali-resistant native plants and the percentage of salts in soils in which they are found growing, as follows:

Greasewood, more than 0.5 per cent
Seepweed, 2.5 per cent or more
Pickleweed 1.0 to 1.5 per cent
Salt grass, somewhat less than 1.0 per cent
Alkali sacaton, about 0.5 per cent

The creosote bush of the southern deserts and sage brush of the northern deserts indicate soils containing harmless quantities of salts.

[7] "Plants as Soil Indicators," H. L. Shantz, 1938 U.S.D.A. Yearbook, p. 852.

Dense stands of desert salt bush occur on fine-textured soils containing some salt but an amount insufficient to damage crops under irrigation unless further salt accumulation is permitted.

Among cultivated crops certain members of the grass family are notably resistant to alkali damage.[8] Wheat, barley, milo maize, Bermuda grass, and Rhodes grass deserve special mention. Corn is more sensitive. Legumes as a group are quite sensitive, with the exception of numerous wild plants which make up a high percentage of the vegetation on the deserts. Alfalfa is very sensitive when young but markedly resistant when old. Root crops, particularly various members of the beet family, are comparatively tolerant. Melons and fruit trees are very susceptible to injury. Walnut and citrus trees are especially sensitive, whereas olive trees and date palms are unusually resistant.

Reclamation of White Alkali Soils. Good drainage is necessary for the reclamation of alkali soils. It is essential in the reclamation process to remove the excess salts from the root zone, and this can only be done by the application of sufficient water to wash them into the lower soil depths. Unless there is ample drainage the addition of so much water will raise the water table and hence lead to increased accumulations of salt in the surface soil rather than to a correction of the alkali condition. Sufficient drainage should be provided to reduce the ground-water level well below the zone of root penetration. Kelley states that "preferably the ground water should never be less than 8 to 10 feet below the surface and every reasonable effort should be made to prevent its rising nearer than 5 to 6 feet from the surface even for brief periods."

With ample drainage provided, one may proceed to the leaching out of the salts. In fine-textured soils the reclamation process will be slow and doubly so if the soil is underlaid by a heavy clay subsoil. In fact the presence of a dense clay layer makes difficult the removal of salts from even medium- or coarse-textured soils. It is questionable if reclamation of soils with very deep, heavy clay subsoils is feasible from an economic standpoint.

Experiments have shown that leaching is all that is needed to reclaim white alkali soils that have sufficient drainage. The addition of chemicals or plowing under of manure or green-manuring crops is unnecessary. No specific directions can be given as to the frequency of irrigation or the quantity of water to apply at each irrigation. The

[8] "Alkali Soils, Origin, Examination and Management," P. L. Hibbard. Calif. Agr. Exp. Sta. Cir. 292, 1937, p. 10.

main points to observe are (1) that the soil be kept moist so that the soil solution will not become sufficiently concentrated to damage the growing crop, (2) that sufficient water be applied at each irrigation to result in some leaching of salts into the drainage water, and (3) that the soil of each irrigation check be carefully leveled so that the water will enter the soil uniformly.

Reclamation of Black Alkali Soils. All that has been said concerning the need for drainage and the application of sufficient irrigation water to cause leaching is of as much if not more importance in the reclamation of black alkali soils than in the treatment of white alkali. Although it has been shown that the application of ample irrigation water, coupled with good farming practices, will ultimately result in the removal of black as well as white alkali, the reclamation process may be materially hastened through the application of various chemicals. The basis of the treatments is the replacement of adsorbed sodium in the colloidal fraction by calcium and the conversion of the replaced sodium and any occurring as the carbonate into neutral sodium sulphate.

The desired changes may be brought about by applications of considerable quantities of finely ground calcium sulphate (gypsum). Ground sulphur, however, will accomplish the same results somewhat more quickly and the application need not be heavy, 1,000 to 2,000 pounds per acre having sufficed in some cases. The sulphur is oxidized in the soil to sulphur trioxide, which combines with water to make sulphuric acid. Other soluble sulphates such as iron or aluminum have also proved effective. As a supply of soluble calcium is needed to complete the reactions, additions of manure or the plowing under of a green-manuring crop, in addition to the chemical treatments, are helpful. The decaying organic matter produces CO_2, which with H_2O combines with $CaCO_3$, usually present in alkali soils, to form the more soluble $Ca(HCO_3)_2$.

The growth and plowing under of sweet clover or the applying of manure is a good way to start the reclamation process whether or not chemicals are to be applied. The growing of Bermuda grass, a highly alkali-resistant plant, also has proved an advantageous method of starting alkali removal practices. The data in Table 39 show the effects of various treatments in the reclamation of a black alkali soil in California, as measured by the yields of crops. The treatments were applied in May, 1923, after which the soil was allowed to lie idle, with an occasional light irrigation and cultivation, until February, 1925. After being flooded twice to leach out salts, white sweet clover was seeded and the crop plowed under as green manure in September. The next February alfalfa was seeded.

TABLE 39

Effects of Various Soil Treatments on the Yields of Crops Grown on Black Alkali Soil (Yields in Pounds per Acre) *

Date	Crop	Sulphur 2 Tons, Gypsum 2½ Tons	Sulphur 2 Tons, Ground Limestone 2 Tons	No Treatment	Sulphur 2 Tons
1925	Sweet clover plowed under as green manure on all plots.				
1926	Alfalfa	10,554	11,129	80	11,129
1927	"	16,984	18,291	596	18,243
1928	"	18,953	20,323	1,693	19,679
1929	"	10,484	14,113	5,161	15,080
1930	"	5,000	6,450	2,095	5,565
1931	"	12,661	15,242	9,596	14,233
1932	"	14,033	15,161	14,516	18,106
1933	"	12,422	16,814	16,885	18,549
1934	"	12,097	16,289	17,823	18,385
1935	Oat hay	2,742	3,790	3,838	4,597
1936	" "	6,129	5,645	6,613	6,853

* "The Reclamation of Alkali Soils," W. P. Kelley, Calif. Agr. Exp. Sta. Bul. 617, 1937, p. 13.

It will be noted that although the yields on the untreated plot were quite low at first, they were very satisfactory after six years. So successful have been experiments in the reclamation of alkali soils that both Kelley of California and McGeorge of Arizona, the two men who have probably done more work on this problem than any other living investigators, maintain that virtually any such soil may be reclaimed if adequate drainage can be established and if the soil is not underlaid by a very difficultly pervious clay subsoil. They are of the opinion, furthermore, that any soil under irrigation may be maintained sufficiently free of alkali and in a satisfactory structural condition if recommended management practices are followed, particularly those concerning irrigation.

Effect of Alkali Removal on Soil Structure. Black alkali soils are of notoriously poor structure, becoming very hard upon drying and breaking up into clods which are very difficult to crush when plowed. When a considerable quantity of white alkali salts accompanies the black alkali, the structural condition is not so bad and a satisfactory seedbed may be prepared more easily. The leaching out of the alkali tends to leave the soil in an even worse physical condition, and hence some investigators have advised against this method of reclaiming such soils. Experiments have shown, however, that applications of calcium, iron, or aluminum sulphates, and also sulphur, followed by proper irrigation, tend to flocculate the colloids and cause the rapid development of a granular structure. Soils from which the alkali is leached without the application of any of the chemicals mentioned are much slower in developing a granular structure. The growing of a crop such as Bermuda grass or the plowing under of a green-manuring crop hastens the granulation process.

Although the structure of white alkali soils is impaired by leaching, the detriment is not so great as in black alkali. The growth of Bermuda grass or of alfalfa for a few years or the plowing under of green-manuring crops will rapidly improve the soil structure.

CHAPTER XVI

IRRIGATION

The presence of remnants of canals in the southwestern states indicates that irrigation was practiced there by the prehistoric peoples who once inhabited that region. So accurately were some of these canals located with respect to grade and convenience in water use that ditches of modern irrigating systems either follow or parallel them.

According to the census there were 21,003,739 acres of land in 91,697 projects under irrigation in 1939, in 19 western states, and when the present projects are completed they will furnish water to 31,305,949 acres. In that portion of the country lying east of the 100th meridian, which passes through the center of North Dakota and Nebraska, there were 344,922 acres receiving irrigation water in 1930, through sprinklers, porous hose, or by subirrigation, flooding, or other methods. An appreciable part of this was made up of small areas devoted largely to the production of truck crops, fruit, and nursery stock, although the total includes 11,600 acres of blueberries and a large acreage of rice. By 1939, Florida alone had 132,362 acres of irrigated land. The subject of irrigation will be considered under two general headings.

Objectives:

A. Water supply and land for irrigation.
B. Irrigation practice.

WATER SUPPLY AND LAND FOR IRRIGATION

Irrigation diminishes one of the greatest hazards in crop production, namely, inadequate water supply. In few instances is this factor entirely eliminated, however, as a shortage of water during some part of the growing season is not an uncommon occurrence on many irrigation projects. Also breaks in canals as a result of floods and destruction of water-control structures by fire may require the turning off of the water with resultant crop damage. The right to obtain water from streams and canals for irrigation purposes and the cost of the water are points to be considered. Furthermore, all land is not well adapted to irrigation and considerable cost for preparation may be involved. Answers to the following questions are pertinent in selecting land for farming under irrigation.

Questions:

1. Do farmers have a right to take water from streams crossing their lands?
2. How does one procure a water right?
3. What precautions should be taken in buying land for irrigation and in bargaining for water?
4. What kind of land is suitable for irrigation?
5. Is the preparation of land for irrigation costly?
6. Why is drainage essential in irrigation?

Obtaining Water from Streams and Canals. Water from streams is usually appropriated to individuals and canal companies according to priority rights, determined by the date the arrangements are made for water and the supply available. When two or more companies or individuals obtain water from a stream which is subject to wide variations in flow, the last to obtain water rights is the first to suffer when a scarcity arises. Although in some states the doctrine of riparian rights is recognized—that is, the owning of land along a stream entitles one to withdraw water from the stream for irrigation purposes—this right is modified by the doctrine of appropriation. In most of the western states the appropriation, diversion, and distribution of water from streams are under the direction of a state engineer or irrigation board.

Procurement of Water Rights. The present-day settler or purchaser of irrigated land usually bargains with a canal company or water users' association for his water supply. In the early days of irrigation a perpetual water right was purchased for each tract of land, and an additional annual charge was made for operation and maintenance of the canal system. At present the more common procedure is the sale of rights, along with the land, which carry an interest in the water-supplying system. Under this plan the system becomes the property of the landowners when a certain proportion of the rights has been paid for. Another method is the organization of an irrigation district which includes all land to be watered from a given system. Under this arrangement no water right is purchased, as ownership of land carries the right to water. The cost of the irrigation system, however, is levied against the land in the form of taxes.

When the government develops an irrigation project and supplies water to privately owned land, it operates through a water users' association. All the landowners become shareholders in the association through the purchase of water rights, the cost of which varies from $25 to $90 per acre. This cost is payable in 20 annual instalments. The operation and management of the irrigation system, exclusive of storage reservoirs, usually passes into the hands of the landowners when payment of water rights for the major portion of the land has been made. The maintenance and operation of the system then becomes the responsibility of the landowners, and the cost is prorated among them.

Precautions in Buying Land and in Bargaining for Water. Although it is assumed that no private company or water users' association will contract to irrigate more land than the supply of water is adequate for, it is well to investigate the following points before closing a deal. (1) What "right" does the land have to water? Was it among the first or the last to obtain rights to water from the present source? (2) What right or priority of right to water does the company with which you are bargaining have? (3) Is the source of water adequate to supply crop needs during critical periods? In some cases there is an abundance of water during flood seasons but a very limited supply at other times. Is there water for the last irrigation of late-growing crops, as alfalfa or potatoes? (4) Is the cost of water stabilized or may there be a decided increase in cost in years to come?

Selecting Land for Irrigation. In choosing land for irrigation, a careful examination should be made of the soil to determine (1) texture of soil to depth of several feet; (2) presence of a difficultly pervious stratum or of gravel within a depth of 5 or 6 feet; (3) accumulation of soluble salts in injurious quantities; (4) slope and evenness of soil surface; and (5) behavior of the soil under irrigation. A desirable soil is readily pervious to water and yet is moisture retentive. It is well if the soil will absorb sufficient moisture in 24 hours to wet it to a depth of 2 or 3 feet. Some soils are so slowly pervious that they become wet to a depth of only a foot or less in 24 hours. On the other hand, some soils are so coarse textured that water passes rapidly below the reach of plant roots, and little available moisture is retained.

A difficultly pervious stratum within the root zone will limit root penetration and may result in the waterlogging of the overlying soil. On the contrary, a stratum of gravel or coarse sand in the root zone may result in the loss of considerable irrigation water and in the storage of a limited supply for crop use.

Usually it is possible to judge the behavior of the soil under irrigation by observing similar soil in nearby irrigated fields. If this cannot be done, water may be applied to a small area for a trial. In general if the soil is too heavy, it will absorb water slowly, be sticky and hard to cultivate when damp, and will crack and bake on drying, although some very heavy soils, particularly those with an adobe-like structure, absorb water readily and are not difficult to manage. Some sandy soils, on the other hand, form plow soles[1] very readily. Soil of too coarse a texture will absorb water very rapidly and show little coherence when dry.

In the absence of chemical tests, some knowledge of the presence or absence of alkali salts may be obtained by observing the natural vegetation. This point is discussed in the section dealing with alkali soils.

The land surface should be comparatively smooth, for the cost of leveling land is high. A uniform slope of 10 to 20 feet to the mile is desirable, although much steeper slopes are in use. Land cut up by ravines, gullies, or buffalo wallows, or covered with sand dunes and hummocks should be avoided if possible.

Cost of Land Preparation for Irrigation. The cost of preparing land for irrigation varies widely. If the surface is very smooth and the slope uniform and of desirable grade, the cost is very little. On the other hand, if there is much leveling to be done, boulders to be removed, ditches to be built over rough topography, and similar improvements to be made, the cost may run to $50 or $60 an acre.

Providing Drainage. It is essential to have good drainage if irrigated land is to remain permanently productive. If the land is not naturally well drained, then drainage ditches must be provided before the land is in use many years. It is difficult to apply sufficient irrigation water for maximum crop growth without getting some excess. In fact, some excess water frequently is desirable in order to wash out alkali salts. If the drainage water is not removed, it may raise the water level, thus increasing the danger of alkali accumulation. Furthermore, the subsoil may become waterlogged with a resultant restriction of root development and crop damage.

Irrigation Ditches. The main canals of an irrigation system are constructed and maintained by the company or association. The ditches carrying the water to and over individual farms, however, are constructed and maintained by the farmers themselves.

[1] A plow sole is a layer of soil in which the clay has become deflocculated and the soil packed into a dense layer as a result of repeated plowing to the same depth.

The capacity, location, and design of farm ditches depend on several factors, among which the following deserve especial consideration: (1) depth and permeability of the soil; (2) whether the water is delivered in a continuous stream or for short periods with long intervals between; (3) the method to be used in applying the water to the land; (4) the acreage to be irrigated; and (5) the water requirements of the crops to be grown.

If the soil is very permeable there is much loss of water by seepage unless it contains sufficient silt and clay to form a seal on the sides and bottom of the ditch. In extreme cases pipes are used in place of ditches. These pipes usually are of concrete or of wood staves held together by means of bands applied by machinery. If redwood is not used the staves should be treated with preservatives to delay decay.

If water flows continuously, smaller ditches will suffice than if all the water needed must be obtained during a limited time at infrequent intervals. Likewise, large acreages and crops which require much water necessitate the construction of large ditches. In general, ditch capacities range from less than 1 to more than 10 cubic feet per second. Methods used to supply water to the land will be discussed in another section.

Water is carried across ravines, gullies, or depressions by one of three methods: (1) Wood or metal flumes are built on grade across the obstruction; (2) a pipe may be laid in the depression in the form of an inverted siphon; (3) if the depression is only a few feet deep, levees may be built on each side.

IRRIGATION PRACTICE

Artificial watering adds certain complexities to farming practice. Judgment and experience are necessary to know when to apply water and how much to apply for different crops. Irrigation offers the opportunity for distribution of weed seed over the fields. Likewise, improper watering and irrigation without adequate drainage may result in alkali accumulation. Water costs and the maintenance of ditches are also important considerations in the expense of crop production. Several questions are offered to serve as guides in studying the problems involved in irrigation farming.

Questions:

1. How may irrigation water be applied?
2. What quantity of water is needed for adequate irrigation?
3. How can a suitable irrigation system be selected?

4. By what systems is irrigation water measured?
5. What provisions are in contracts for water?
6. Is all water suitable for irrigation?
7. May water be supplied for irrigation by pumping?

Methods of Applying Irrigation Water. Choice of the various methods of applying irrigation water, of which there are about a dozen, is influenced by a consideration of (1) seasonal rainfall; (2) slope and general character of the soil surface; (3) supply of water and how it is delivered; (4) crop rotation; and (5) permeability to water of the soil and subsoil. A brief discussion of a few of the chief methods is given.

The *furrow method* frequently is used for the irrigation of fruit trees and row crops, such as potatoes, sugar beets, and corn or the grain sorghums. Furrows are made across the field, leading down the slope. Water is let into the upper end of the furrow from a "head ditch" or pipe line running across the high end of the field.

The *corrugation system* is a modification of the furrow method and is used largely in the northwestern states for the irrigation of small grain and hay crops because of the uneven topography, use of small streams of water, and prevailing methods of planting and harvesting. The irrigation furrows are placed at such distances apart and are of such length that will provide for the wetting of the soil between them by horizontal seepage. The permeability of the soil, slope of the land, and volume of water available are determining factors in spacing the furrows.

Irrigation by the *border method* is suitable on land having a slope in one direction of 1 inch to 2 feet to the 100 feet, although the system has been used on land with much steeper slopes. The field is laid out in strips varying in width from 6 to 60 feet and in length according to the head of water available, the nature of the soil, and the degree of slope. If the slope of the field is too great, the strips may be laid out diagonally across it; otherwise they run down the slope. The strips must be of such width and length that they may be entirely covered with water without letting the water in at a rate that would cause erosion of the soil or without permitting too much or too little water to soak into the soil at one end before it reaches the other end.

A simple formula has been proposed for figuring the proper area of the strips, based on two factors of (1) the head of water in cubic feet per second and (2) the nature of the soil. For sandy soils the ratio of irrigating head to acres of land is 20 to 1 and for clay soils 2 to 1. Thus with a flow of water of 10 cubic feet per second, the size of the strips would be 0.5 acre on sandy soil and 5.0 acres on clay soil.

FIG. 66. A "well-built wooden flume." [From Farmers' Bul. 1348, U.S.D.A., by James C. Marr.]

FIG. 67. Irrigation by the corrugation method. Four corrugations are being supplied from one outlet. [From Farmers' Bul. 1348, U.S.D.A., by James C. Marr.]

Fig. 68. Showing a method of distributing water from a flume into irrigation furrows. [From Calif. Agr. Ext. Service Circ. 16, by M. R. Huberty and J. B. Brown.]

Fig. 69. A method of preparing land for the border method of irrigation. [From Farmers' Bul. 1243, U.S.D.A., by Samuel Fortier.]

The strips are set off by low, broad, flat-topped mounds of earth from 5 to 12 inches high, known as levees or borders. The borders are made in this shape so that farm implements may cross them readily and so that grazing stock will not destroy them when the field is in pasture. Needless to say, the soil surface in each strip must be leveled so that water will spread across it readily when let in through the two or more headgates (depending on the width of the strip) provided in the

FIG. 70. The irrigation of a peach orchard by the contour-check method. [From Circ. 73, Calif. Agr. Ext. Ser., by J. B. Brown.]

head ditch. It is not necessary that the strips in a field be at the same level.

The earliest and simplest method of irrigation is by *flooding*. This method is applicable principally to rather smooth land with a regular and moderate slope and when there is an abundant supply of water. When the fall is slight, the water is carried in shallow ditches from 50 to 100 feet apart in the direction of the slope. With a greater slope the ditches are made to wind across the field approximately on the contour. By making little earth dams in the ditches, the water is made to overflow and cover the whole field.

In fields with irregular surface or with a considerable fall, ridges called checks are thrown up on the contour, thus dividing the land into plots. Irrigation is generally started with the plots highest up the slope. When one plot has been irrigated, water is let into the next lower one from the ditch. It is not considered good practice to let the water from one plot into the next except in irrigating rice.

Another modification of the flooding system of irrigation is known as the *basin method.* The procedure consists in forming small basins by throwing up small earth dams or levees which retain the water while it soaks into the soil. A ditch is provided between each two rows of basins and the water is run from the ditch into first one basin

FIG. 71. The basin method of irrigation. [From Farmers' Bul. 1518, U.S.D.A., by Samuel Fortier.]

and then another as the irrigation proceeds. This system frequently is used in orchards and vineyards.

Quantity of Water To Apply. An ideal application of water would be a sufficient quantity to bring the soil, to the depth of the root zone of the crop, up to its field capacity. More water may result in the waterlogging of a portion of the subsoil or in the loss of water by drainage. If much water percolates below the root zone, it may accumulate under low areas, thus raising the water table unless suitable drainage facilities are provided. On the other hand, unless enough water is applied to result in appreciable drainage, it is difficult to remove excess salts from soils in which there is a tendency for alkali to accumulate.

The quantity of water available to plants which may be stored in soils ranges from ½ to 2 acre-inches per foot of depth. If the water-holding capacity of the soil per foot in depth is known, the approximate quantity of water to apply may be obtained by multiplying this figure by the depth in feet of root penetration of the crop. The readiness with which soils absorb water influences the quantity and rate of application. Most dry soils absorb water rapidly. Little change in the rate of water intake occurs in very permeable soils as the depth of wetting increases. Some clay soils, however, absorb water very slowly after the first foot or so has become saturated. Because of this situation, it is advisable to hold the water for a long time on heavy soils and for only a short time on coarse-textured soils.

Planning the Irrigation System. Before deciding on the location, capacity, and design of irrigation and waste ditches and on the size of strips if this method of application is to be used, it is well to study the lay of the land carefully. This may well be done by dividing the land into 100-foot squares and taking the elevation of each corner of the squares. Physical features, such as the location of streams, ditches, buildings, and roads, should also be shown on the map. By use of this map various arrangements of ditches and methods of applying water may be tried out on paper, thus avoiding the expense of staking them out to determine if they are feasible. When a layout is finally selected, it is well to use temporary ditches and earth or canvas dams for the first year, to be sure the arrangement is satisfactory, before putting in permanent structures.

Measurement of Irrigation Water. The two kinds of units used in measuring irrigation water are (1) units of volume and (2) units of flow. Flow refers to the volume of water passing a given point in a

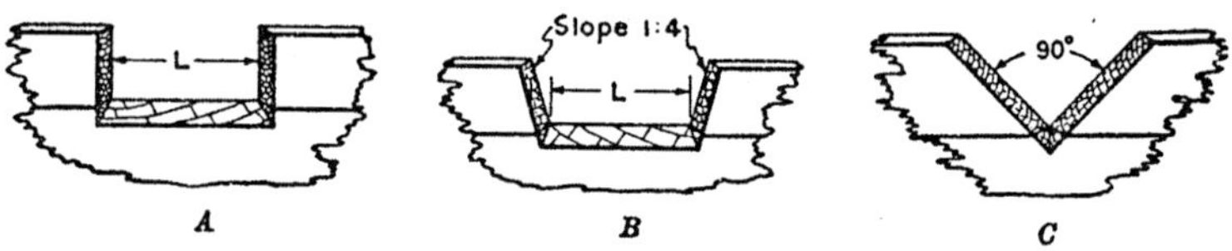

FIG. 72. Types of contracted weir notches: *A*, rectangular; *B*, trapezoidal; *C*, V-notch or 90-degree triangular. [From Calif. Agr. Exp. Sta. Bul. 588, by J. E. Christiansen.]

unit of time. For example, the cross-sectional area in square feet of a stream or ditch, multiplied by the velocity of flow in feet per second, gives the cubic feet per second. The commonly used units of flow are cubic feet per second, gallons per minute, and miner's inch. The miner's inch is the flow through an orifice of 1 square inch under a specified "head," which usually varies from 4 to 6 inches. The head is the depth of water on the upstream side to the center of the orifice. It requires 38.5 to 50 miner's inches to equal a cubic foot per second, according to the regulations of different states which specify the height of the head of water.

Units of volume in common use are cubic foot, gallon, acre-foot, and acre-inch. The volume of water required to cover an acre to a depth of 1 inch is designated as an acre-inch.

For measuring the flow of small streams and ditches, various devices such as weirs, submerged orifices, Parshall measuring flumes,[2] and current meters are used. Many types of submerged orifices are in use,

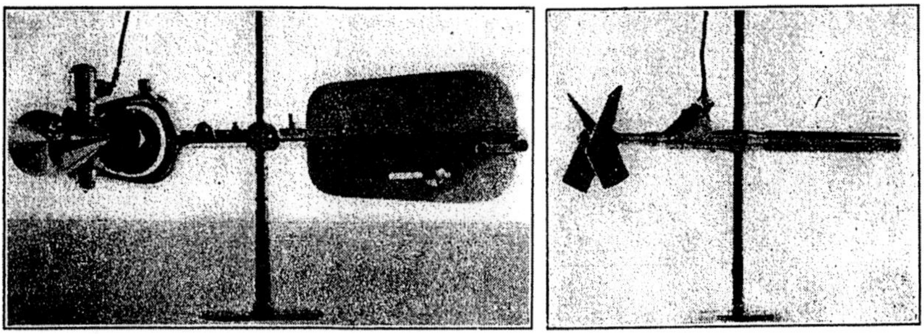

Fig. 73. Types of current meters with rod mountings. Left, Price type; right, Hoff type. Rod mountings are exchangeable for cable suspension. [From Calif. Agr. Exp. Sta. Bul. 588, by J. E. Christiansen.]

including the miner's-inch boxes. Two types of current meters are illustrated in Fig. 73. A wall or bulkhead across a ditch, with an opening cut in the top of it to permit water to flow through, is known as a weir. The opening may be rectangular, trapezoidal, or triangular and is called the weir notch. The height of the water above the bottom of the notch and the size and shape of the notch determine the quantity of water which will flow through it in a given time. The weir is one of the simplest and most commonly used means of measuring water.

Contracts for Delivery of Water. Agreements vary greatly concerning the quantity of water which is to be delivered to the land. A common method is to designate a stream of a given size, as 1 cubic foot per second for each 80 acres, with the provision that the purchaser will turn off the water when it is not needed. Sufficient water to cover

[2] The Parshall measuring flume was developed through cooperative research by the U. S. Department of Agriculture and the Colorado Agricultural Experiment Station. It is named after R. L. Parshall, the engineer who conducted the experimental work. Water is measured by this flume much as it is by a weir. A full description of the device cannot be given here.

the land to a given depth as 2 feet during the season, or, in other words, 24 acre-inches, is frequently contracted for. Most contracts carry the provision that in case of water shortage the available supply will be prorated among the purchasers.

Quality of Water. Careful attention should be given to the nature of the water to be used before an irrigation system is constructed. Sometimes the available water carries so high a concentration of soluble salts that its use for irrigation is not advisable, particularly on land which already contains a considerable concentration of alkali. Sodium salts in the water are much more objectionable than are salts of calcium and magnesium because of the tendency for sodium to cause deflocculation of the colloidal fraction of the soil and so develop an undesirable structure. Some waters also contain sufficient boron to prove toxic to plants with continued use. The following four characteristics have come to be recognized as furnishing sufficient informa-

TABLE 40

SUGGESTED STANDARDS FOR JUDGING THE QUALITY OF IRRIGATION WATER *

Rating of Water	$K \times 10^5$ at 25° C. †	Percentage Sodium $\left(\frac{Na \times 100}{Ca + Mg + Na}\right)$ ‡	Chloride in P.p.m.	Boron in P.p.m.
Suitable under most conditions	less than 75	less than 60	less than 75	less than 0.5
Suitability dependent upon soil, crop, climatic, and other factors	75–300	60–70	75–245	0.5–2.0
Unsuitable under most conditions	300 plus	70 plus	245 plus	2.0 plus

* Agr. Ext. Ser., Univ. of Calif., Mimeograph Release.

† Specific electrical conductance in reciprocal ohms at 25° C., multiplied by 10^5. This is a fairly accurate index of the total concentration of salts in a water; multiplied by 7, it gives the approximate total of dissolved solids in parts per million (p.p.m.).

‡ Concentrations expressed as milligram equivalents (m.e.) per liter are used in this computation.

tion to determine the suitability of water for irrigation purposes: (1) total concentration of salts, (2) chloride concentration, (3) content of boron, and (4) the proportion of sodium in the total bases. Owing to modifying circumstances such as soil characteristics, temperature, humidity, rainfall, and variations in the salt tolerance of plants, it is impossible to set up rigid limits for any of these four characteristics of water. As a result of experience, however, waters may be classified into three broad groups as follows: (1) those suitable under most conditions, (2) those whose suitability is determined by the modifying conditions listed above, and (3) water which is unsuitable under most conditions encountered in irrigation practice. Although the limits set forth in Table 40 should be considered approximate, they may serve as a general guide in judging of the quality of waters.

It is evident that there is no necessary correlation among the four characteristics of any water. It may be entirely satisfactory in respect to soluble salt content, for example, but may contain so much boron that its use is not advisable. If there is any question as to the quality of the water, it is well to have the water analyzed by the chemist of the state agricultural experiment station or by some other competent authority before making arrangements for its use.

Irrigation by Pumping. During the decade of 1919 to 1929, irrigation by pumping increased 62.4 per cent. The system is used in arid regions in conjunction with irrigation from streams and reservoirs, in semi-arid areas to supplement the limited rainfall, and in humid climates for the production of crops of high value per acre, especially fruits, vegetables, and nursery stock.

After years of irrigation the water table under some of the western reclamation projects became so high that it was necessary to install pumps at frequent intervals to lower it. The water so removed from the soil was used again for irrigation, thus supplementing the original supply and providing for the watering of more land. Wells of great depth and about 24 inches in diameter have also been bored at the borders of irrigated valleys where the elevation is too high to permit obtaining water from canals.

In the Great Plains states, large acreages are irrigated by pumping water from the underground flow of streams which largely sink below the surface of the pervious soil except during flood periods.

The depth from which water can be lifted economically for irrigation depends largely on the quantity of water which will be needed and on the value of the crop to be grown as well as on the cost of the fuel consumed in operating the pumps.

Before installing a pumping system, careful investigation and tests should be made to determine the constancy of the water supply, particularly in dry seasons, the quantity of water the well will deliver under constant pumping at the required rate to irrigate the desired acreage, and the depth to which the water is lowered in the well by pumping.

The water pumped from streams, ponds, and wells may be applied to the soil by any of the methods previously described. In addition, overhead sprinkling systems are frequently used on small acreages. These sprinklers may be similar to those used on lawns and golf courses but larger. More commonly, lines of pipe, carrying small nozzles every few feet, are mounted on posts at a convenient height, and are so arranged that the pipes rotate through an angle somewhat

FIG. 74. Irrigating sugar beets in Ohio with well water. The whole system with the exception of the well is portable. [From Farmers' Bul. 1846, U.S.D.A., by F. E. Staebner.]

less than 180°. The lines of pipe are placed at intervals so that they provide for the watering of the entire soil surface. At times the water is pumped into lengths of porous canvas hose which is laid on the ground at desired intervals. The pressure causes the water to seep through the pores of the hose.

Experiments are being conducted with a powerful centrifugal pump, mounted on an old automobile chassis for portability. The intake pipe is dropped into a conveniently located stream or pond, and a stream of water is thrown from a rotating nozzle over an area some 300 to 400 feet in diameter. The practicability of the system remains to be demonstrated.

CHAPTER XVII

FRUIT SOILS

One of the most important factors affecting the income of fruit farmers is the nature and the characteristics of the soil on which their fruit enterprise is located. The importance of soil is not only a common observation but is also shown by many studies on costs and returns of fruit farms on soils of varying quality. Obviously, when production costs are high, the superiority of good soils is more pronounced than when production costs are low. Costs per unit are usually low where yields are high. Good locations, with respect to soil, should be sought for all fruit enterprises. An orchard is a long-time investment, and mistakes made in planting trees on poorly adapted soils may not be realized for several years; therefore, a thorough study of the soil should be made before an orchard is set. No doubt there are many more acres of desirable soil than will ever be required for horticultural development in the United States; yet many fruit plantations have been made on marginal or submarginal land. Many of these low-yielding, badly located, marginal or submarginal orchards have been and are being forced out of production by economic pressure. Two objectives have been kept in view in discussing the soil factors in fruit growing.

Objectives:

A. Soil factors to consider in selecting a site for an orchard or small fruit plantation.

B. Soil requirements of specific fruit plants.

SELECTING A SITE FOR A FRUIT ENTERPRISE

Small fruits, exclusive of grapes, make up only about 10 per cent[1] of the total production of fruit in the United States and of this amount strawberries contribute roughly 55 per cent; consequently most of this discussion is devoted to the tree fruits. The accompanying map (Fig. 75) shows the general fruit-producing areas of the United States.

[1] "Modern Fruit Production," J. H. Gourley and F. S. Howlett, The Macmillan Co., New York, 1941, p. 13.

Fruit, to some extent, is grown throughout the United States, but, as indicated in Fig. 75, intensive production is more or less concentrated in rather well defined areas. These areas of concentration are not due to chance but have developed from the interrelationship of certain conditions. In this connection it may be well to mention some of the more important factors concerned in determining the location [2] of

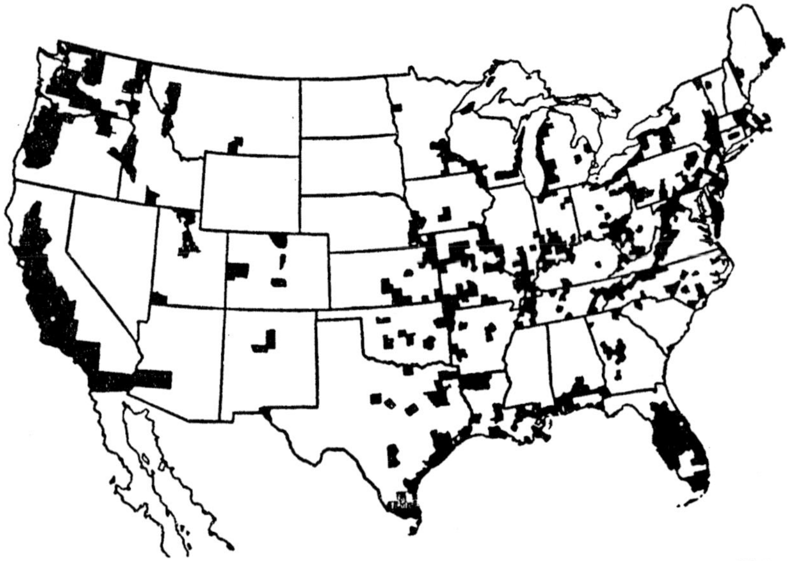

FIG. 75. Generalized map showing the fruit-producing areas of the United States. (All fruits included.) [From "Modern Fruit Production," by Gourley and Howlett, The Macmillan Company.]

the main fruit-growing regions and in choosing the site [3] for specific plantings.

The selection of the proper variety is perhaps the foundation of the fruit-growing enterprise, but, aside from that, no other factors seem so important as the site and the soil. The chief factors involved in choosing a desirable location and site may be listed as follows: (1) climate, which includes temperature conditions, rainfall, wind, sunshine, spring frosts, and hail; (2) topography, as related to erosion frost, freezing hazards, and orchard operations; (3) transportation and marketing facilities; (4) distance from market; (5) varieties; (6) supply of and demand for produce; and (7) the soil. Only those

[2] "Location" relative to fruit growing is a term generally used in relation to some town, city, county, river, etc.

[3] "Site" refers to the immediate place with respect to elevation, proximity to bodies of water, etc.

factors directly concerned with the soil are discussed here; these factors are indicated by the following questions.

Questions:

1. Do topography and elevation of the land influence fruit production?
2. Is the available moisture supply an important factor to consider?
3. Is there any relationship between depth to the ground-water table and yield?
4. How deep should a soil be that is selected for fruit growing?
5. What is the relationship between soil aeration and drainage and growth of fruit plants?
6. Is there a relationship between texture of a soil and its use for fruit growing?
7. What structural conditions of the soil are hazardous to fruit production?
8. Why should consideration be given to the content of soil organic matter?
9. Are the most fertile soils the most desirable fruit soils?

Topography and Elevation. The topography of the site is of special interest from the standpoint of erosion, frost, freezing hazards, and orchard operations. The more desirable sites are those on upland plains or broad ridges bordering lower-lying areas. Sites in the immediate vicinity of large, deep bodies of water are also favorable. When both characteristics are combined, the sites are exceptionally favorable.

The direction of the slope is usually not important, although trees on a southern exposure sometimes blossom and mature their crop earlier, which may be of significance when any variety is set close to the northern limits of its range or when the site tends to be frosty.

The slope gradient is of much concern in relation to erosion, cultivation, moving equipment, and hauling the fruit crop. The main reason for directing attention to the degree of slope is that, with all other factors constant, water-erosion losses vary directly with the steepness of the slope. In general, it can be said that all soils in humid regions lose fertility through erosion resulting from clean cultivation on slopes. Erosion is not only destructive on slopes where soil is removed but also may prove detrimental at the foot of the slope because of the deposition of coarse material over good soil. In many areas erosion is responsible for larger losses of fertility from fruit soils than any other factor of soil deterioration.

The redevelopment of eroded slopes is an expensive and time-consuming operation. Terraces may be constructed and the orchard planted on the contour, but these practices will not replace fertile top soil that has been removed, although they may be necessary to preserve what is left. In many instances where terraces appear necessary, so much of the surface soil has been removed that an orchard set on such a site will often be unproductive. The first and most important step in erosion control in orchards is selecting sites where excessive erosion has not removed the surface soil.

Increased costs of orchard operations on steep slopes may become a considerable item. This applies particularly to cultivating, spraying, and hauling. As a rule, the greater the degree of slope the less desirable the site for fruit growing.

The elevation of the land is not to be confused with altitude, height above sea level. In selecting a site for a fruit enterprise it is important to consider the elevation of the proposed area in relation to the immediately surrounding country. A good site should be high enough to prevent late spring frosts and to escape the lowest winter temperatures, but slope gradients should be gentle enough to avoid severe soil erosion when cultivated. The higher elevations permit good air drainage and thus may escape frost injury because cold air is heavier than warm air and the cold air will settle in the valleys and lower-lying areas during clear, still nights, while the warmer air will be forced upward. "In this country the effect of these inversions of temperature are experienced at practically all elevations at which fruit is grown, except in certain canyons in the western states where 'draws' of air offset the effects."[4] It is a frequent observation that fruit blossoms are injured by frost in lower areas, while those at higher elevations escape. In regions where frost damage is likely to occur, it is always well to avoid valleys, coves, swales, or other low topographic positions where air drainage is poor. Blueberries, however, are less subject to injury by spring frosts than most fruits and may be grown on frosty sites with less hazard.

Difficulty may be encountered even with what is normally the most desirable site with respect to air drainage. For example, trees located on a hilltop where no cover crop is grown may be subjected to serious winter injury in the northern areas of the United States, owing to the lack of a protective snow covering.

Available Moisture Supply. A good fruit soil is one that must meet increasing demands for moisture as the tree grows larger. The subsurface horizons should be sufficiently clayey or silty to hold a reserve supply of water close to the surface or within reach of the roots. The amount and availability of water in the soil within the region exploited

[4] Gourley, *op. cit.*, p. 87.

by the roots, particularly during droughts, appear to be of greater importance than any other single soil factor in many of the fruit-growing areas of the United States.

The more important soil factors influencing the quantity of moisture in the soil and its availability to the plant are texture, structure, organic matter content, aeration, position of water table, temperature, and soluble-salt content. These factors may affect not only the amount of moisture in the soil but also the root development and the ability of the roots to absorb water. Since the acre requirements of different fruits for water vary considerably, it is possible to select those best adapted to the special conditions of a specific site. Fruit plants with deep and extensive rooting systems, reaching abundant soil and moisture resources, give the greatest and most consistent yields. The factors affecting the available moisture supply in soils are discussed in considerable detail elsewhere.

Ground-Water Table. It is important to determine the position of the ground-water table in selecting an orchard site. The height of the water table and its persistence is of particular concern in early spring during the blossoming and fruit-setting period. This appears to be the most critical period for the tree from the standpoint of injury due to waterlogging. It is not so much a matter of the height the water table attains but rather the rapidity with which it is able to recede after a rain. The more desirable fruit soils should contain little or no free water in the top 3 or 4 feet except for a short time (2 or 3 days) after a rain.

Relation of Soil Depth to Root Penetration. As has been mentioned previously, the extent and depth of the root system of a tree have an important influence upon its productiveness and longevity. This distribution is determined essentially by the character of the soil. If shallow rooting occurs, usually the trees are more subject to drought injury, have less mechanical support, and often a less abundant supply of plant nutrients. The depth to which roots penetrate is the effective depth of a soil so far as that particular plant is concerned. From the standpoint of tree growth and production, the distribution of tree roots is relatively unimportant provided sufficient supplies of moisture and nutrients are reached.

Shallow soils should not be set to orchards. Investigations reveal that in many instances tree roots may extend 10 or 12 feet or more in depth. For most orchard plantings it is desirable to have a soil at least 6 or 8 feet deep; solid rock, a hardpan, a tight claypan, or a spring-time water table within 3 or 4 feet of the surface should be considered prohibitive. On the other hand, the small fruits such as raspberries and strawberries are shallow rooted, and the soil material below 3 or 4 feet is usually of little significance, except as it may influence soil moisture supply.

Aeration and Drainage. Soil aeration and drainage are very closely interrelated. The well-drained soils usually are well-aerated soils. One of the main factors influencing the suitability of a soil for fruit growing is its natural internal drainage. A compact, impervious soil is very unfavorable, and any practice that increases compaction should be avoided. In compact soils the pore spaces are small and are usually filled with water to the exclusion of air. Under such conditions water moves through the soil very slowly. When water enters the soil faster than it can percolate through the underlying strata, the air spaces may be completely filled with water, producing what is commonly called a waterlogged condition. Roots of fruit trees require large amounts of air in the soil and even partial waterlogging may greatly interfere with their normal respiration, growth, and ability to absorb water.

Soil color is an excellent indication of drainage conditions. Poor subsurface drainage results in inadequate aeration, which causes mottling in the subsoil. The depth at which mottling occurs indicates the depth of good subsurface drainage, and where mottling occurs near the surface the area is usually too poorly drained for a good orchard site. Poorly drained and poorly aerated profiles are not restricted to lowland positions; they may be found on uplands and even on hillsides. The time to solve or rather the time to avoid the problem of aeration and drainage is when the orchard site is selected.

It is not to be inferred, however, that a soil cannot be too well drained and aerated. The deep, dry, sandy soils represent such a condition; such soils because of their low content of moisture and of the distance to water table may prevent the penetration of roots to the underlying moist soil and are unable to supply moisture in sufficient quantity to the fruit plants during extended periods of drought. The deep, dry, sandy soils not only have a low moisture-holding capacity but are usually low in organic matter and mineral nutrients. Because of excessive aeration it is difficult to increase or even to maintain the organic matter level in these soils.

Texture. In selecting soils for fruit growing it is frequently more important to consider soil texture than it is to consider fertility. It is not the direct effects of texture that are of chief concern but rather the relationship of soil texture to drainage, aeration, ease of root penetration, moisture supply, and so forth.

The most favorable soils for the majority of the fruit plants are intermediate in texture, such as the loams which contain sand or gravel in the subsoil sufficient to permit good internal drainage and free penetration of roots but which at the same time contain sufficient clay to provide ample moisture during periods of drought. It is advisable to avoid the extremes in soil texture in selecting fruit soils.

Structure. Consideration should be given to the structural condition not simply of the surface soil but particularly to the penetrability of the subsoil. Soils which permit only shallow rooting generally produce small, weak trees; this does not necessarily imply that all soils which permit extensive rooting are good soils. Roots of trees may range farther and penetrate more deeply in deep, loose, incoherent sand but reach smaller quantities of moisture and nutrients than the roots of productive trees which may be restricted somewhat in the heavier and more fertile soils. Subsoil conditions, known as hardpans or claypans, tend to prevent extensive and deep root penetration. There is usually a marked correlation between weak, unfruitful growth and subsoils that are not penetrated by tree roots. Another common condition preventing the extensive development of deep root systems is waterlogging, owing either directly or indirectly to an unfavorable soil structure. The diagrams (Fig. 76) indicate how root development may be modified by soil structure and soil texture.

Soil Organic Matter Content. The importance of organic matter in soils is well recognized. The presence of organic matter and its decomposition in the soil brings about many desirable physical, chemical, and biological changes, which are discussed in considerable detail on p. 223. Fruit production may be influenced either directly or indirectly by organic matter. But it is of greatest concern in relation to increasing the supply of available moisture and nitrogen in the soil throughout the growing season. In most fruit-growing areas differences in the quantity of available nitrogen and moisture are usually the chief causes of differences in yields of fruit. Both of these (supply of nitrogen and moisture) are closely related to the content of soil organic matter.

Since nitrogen is the most important nutrient element to consider in orchard fertilization in most orchards and it is largely held in organic form in the soil; since organic matter performs many important beneficial functions in the soil; since the process of increasing

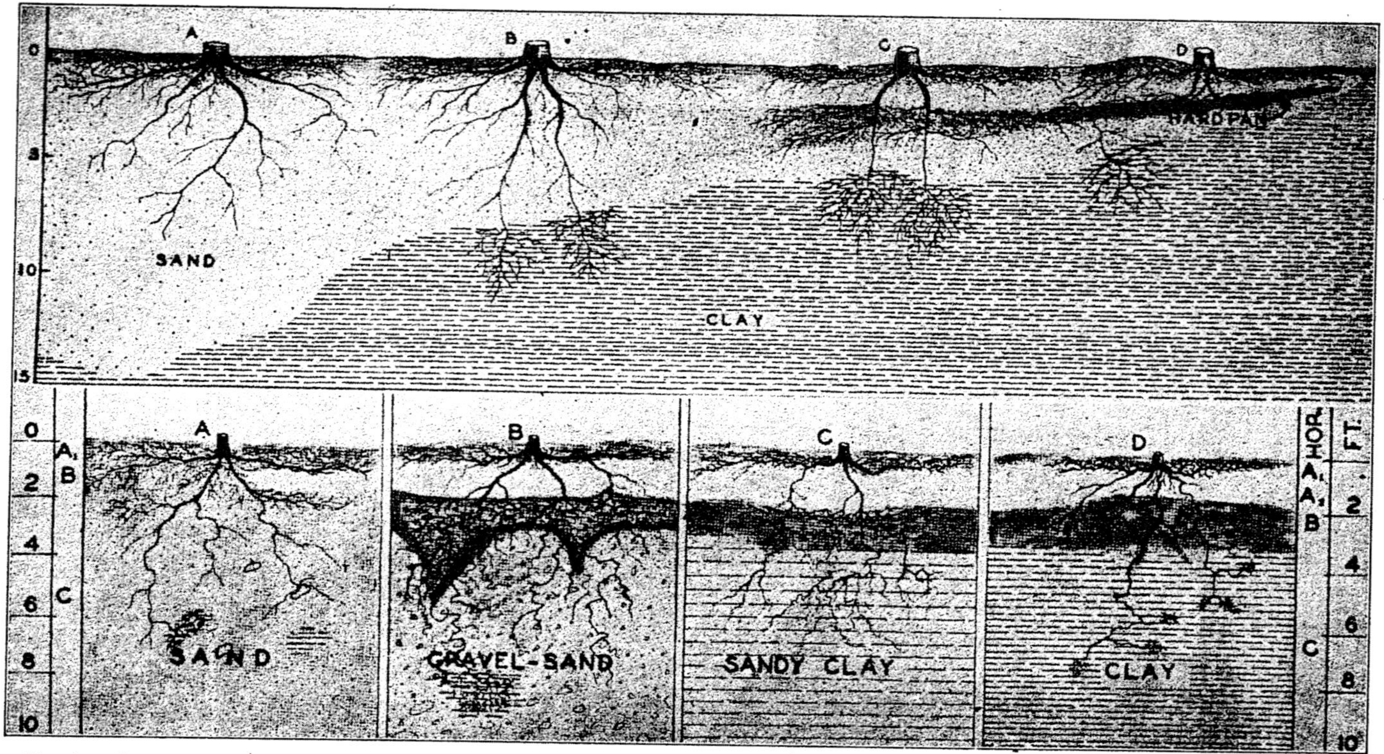

FIG. 76. Upper: Sketch showing root development in relation to soils, given a variable thickness of sand over impervious clay. Note (*D*) that a hardpan prevents deep root penetration. Lower: Sketch showing root development in relation to representative soil profiles and to the textural composition and structure of the underlying glacial deposits. It is observed that the roots range farther and penetrate deeper in deep, open sand (*A*) and are more limited in heavier soils (*D*). [From Mich. Agr. Exp. Sta. Bul. 155, by N. L. Partridge and J. O. Veatch.]

the organic matter content of the soil is a rather slow and often an expensive process; and since it is difficult to grow large yields of soil-improving crops in mature orchards, it is evident that when a choice can be made it is wise to select those soils that have a good supply of organic matter if all other conditions are favorable. These are usually the relatively uneroded, more gentle slopes of the uplands.

Supply of Plant Nutrients. Fruit trees and small fruits thrive over a wide range of soil types with respect to plant-nutrient content, and the limits are not so exacting as with some other crops, provided proper amounts of nitrogen are available. In order to understand why fruit trees are not so exacting in their nutrient requirements as the general field crops, it is necessary to point out some of the fundamental differences between these two types of crops.

Most farm crops complete their life cycle in a few months, whereas trees go through a period of development that lasts for several years. The seasonal growth period of field crops is usually shorter than that of trees. Some of the nutrients stored up in new growth of fruit trees may be used the following season in the development of fruit. Owing to the extended feeding season of fruit trees the plant nutrients in the soil can be in smaller quantities and perhaps in less available forms than in the soil of field crops and yet meet the demands of the trees. Tree roots have a greater feeding zone than most field crops because of their deeper and more extensive root system. On the other hand, orchard crops represent essentially a single-crop system which calls for the same plant nutrients in about the same ratio year after year.

Of the three elements (N, P, and K) nitrogen as a rule demands first consideration from the standpoint of the soil-nutrient supply for fruit production. The other elements are usually present in soils in sufficient quantities for normal tree growth; yet applications of phosphorus, potassium, and even calcium sometimes may affect fruit production directly or, through their effect on cover crop growth, indirectly. Where soils are too poor to grow a good cover or green-manure crop it is often necessary to use a complete fertilizer and even lime if a cultivation-cover-crop-system of orchard culture is used.

It is difficult to make generalizations regarding the fertilizer needs of orchards. The kind and amount of fertilizer needed depends primarily upon the natural fertility of the soil, the cultural methods used, the age of the tree, and the kind of fruit. As indicated above, nitrogen occupies a more important place in orchard fertilization than any of the other nutrient elements. Potassium and occasionally phosphorus may be deficient in some soils and may give excellent response when applied. In some cases deficiency symptoms of zinc, copper, manganese, and particularly boron have been observed.

Fruit trees are tolerant of a wide range of soil reaction and have responded directly to applications of lime only when the *p*H is very low. As indicated above, however, leguminous cover crops frequently respond well to liming, and thus indirectly the trees may be benefited.

In concluding this section dealing with factors to consider in selecting a soil for fruit production, it is important to emphasize that nutritional deficiencies may be remedied rather easily but that fundamental soil characteristics, such as aeration, drainage, organic content, texture, and structure, are modified with difficulty.

SOIL REQUIREMENTS OF SPECIFIC FRUIT PLANTS

It has been emphasized that the commercial orchard represents a long-time investment and that due consideration should be given to all factors involved in selecting a desirable site. Furthermore, the importance of soil factors has been stressed; and, so far as these factors are concerned, the growth of the tree is a function of the entire soil profile, rather than a function of the plow soil alone. Although it is not always possible to lay out orchards or small-fruit plantations to conform to all the soil variations, yet different kinds of fruit vary in their adaptability to soils and, when it is possible to make selections, those fruits should be grown on the soils for which they are better adapted. It is therefore necessary that consideration be given to specific soil requirements of the different fruit plants. The subject matter may be divided into two parts, as indicated by the questions which follow.

Questions:

1. What are the more important soil requirements of the different tree fruits?
2. In what respects do the soil requirements for small fruits differ from those of the tree fruits?

Tree Fruits. In this group of fruit plants, apple, pear, peach, plum, and cherry will be considered. Although there appears to be a varietal difference relative to the soil requirements in some of these fruits, only the general soil requirements will be considered here.

Apple. Apple trees require a large available water supply and are therefore well adapted to the well-drained, porous loam and silt loam soils. Sandy soils and sandy loams with clay at a depth of 6 to 8 feet are usually satisfactory. Deep, dry sandy soils and wet or poorly drained soils are to be avoided.

Pears. They are similar to apples in water requirements, but can withstand greater extremes with respect to moisture supply. They are better adapted to somewhat heavier soils than other fruit trees and grow well on well-drained clay loam soils. It is important that the soil be of suitable depth for extensive root development. Good drainage and aeration are important. Excessive fertility, particularly with respect to nitrogen, should be avoided from the standpoint of increased susceptibility to fire blight and winter injury.

Peach. Peach trees do not require so much moisture as apple and pear trees and can therefore be grown on lighter soils, although sandy loams and even silt loams, where the surface horizons are deep, are also suitable for peaches, within their climatic limitations. The soils must be well-drained; the heavy compact clay soils are least desirable. Peaches are usually most successful on soils of moderate fertility since they are more subject to winter injury and some diseases when the trees are extremely vegetative.

Plum. The soil requirements for plums are about the same as those for pears. They appear somewhat better adapted to heavier soils than other trees with the exception of pears.

Cherry. The water requirement of cherry trees is less than for apples, and consequently they are adapted to soils somewhat lighter in texture. Deep, dry sandy soils, however, are objectionable. The soil requirements of cherries are similar to those of peaches in that they are very susceptible to defective drainage.

Small Fruits. In this group of fruits, grapes, blackberries, raspberries, currants, gooseberries, dewberries, blueberries, and strawberries are considered. These crops possess many characteristics which resemble those of orchard fruits; yet the small fruits come into bearing earlier and their period of bearing is usually much shorter. Grapevines are very long lived; currants and gooseberries live for 10 to 15 years; brambles, 6 to 12 years, but strawberry plants do not bear well for more than two or three years. The small fruits may be grown on a variety of soils that are moderate in fertility, well drained, easily tilled, and have good moisture relations. In general these crops are grown in widely distributed areas in small acreages on soils well adapted to the orchard fruits. They are responsive to the presence of abundant supplies of soil organic matter.

Grape. Grapevines have a rather extensive root system and a relatively small amount of foliage. They are thus better adapted to drier soils than are apples, pears, or plums. In general they grow best on soils well suited to the stone fruits. Grapevines are able to survive on practically all soils except those that are poorly drained. The best grape soils are usually deep, well drained, of medium texture with gravelly subsoil, moderately to highly fertile, well supplied with bases, and have favorable water and air relations. Soils that are high in nitrogen may induce overgrowth and cause unproductiveness.

Raspberry and Blackberry. They are shallow rooted and require moist surface soils. They are perhaps most successful on the fertile, well-drained loams that are high in soil organic matter with clayey subsoils, having a good moisture supply. They are frequently grown on acid soils without the use of lime. Black raspberries are well suited to sandy loam soils or even to loamy sands which are well supplied with organic matter.

Dewberry. Dewberries have low water requirements and are quite successful on deep, dry sandy soils.

Strawberry. These plants are shallow rooted and require a moist surface soil for best results. It has sometimes been stated that a good corn soil is a good strawberry soil. As a rule, they do best on soils of medium texture that are well supplied with organic matter. Strawberries may be grown on any type of soil if water, nutrient supply, and drainage are adequate. They are not limited in their growth, owing to soil *p*H difference within the range normally found in mineral soils.

Currant and Gooseberry. These plants are best adapted to a cool climate and are very successful in partial shade. They grow well on any soil that produces good general farm crops. They succeed on some soils too poorly drained to be suitable for the brambles.

Highbush Blueberry. The blueberry plant is very exacting in its soil requirements. The soil must be very acid (*p*H 3.4 to 5.0), well supplied with moisture, and of the proper texture. If soils are not sufficiently acid, the plants make very little growth and many die. In Michigan,[5] a water table 14 to 22 inches below the surface has been found most satisfactory. However, excess water during the growing season is to be avoided. Open, porous soils like those found in sand or muck are best for blueberries.

In New Jersey it is recommended that land which is to be planted to blueberries consist of shallow muck which is overlying sand that is underlain with a claypan at 3 to 4 feet.

[5] "The Cultivation of the High Bush Blueberry," S. Johnston, Mich. Spec. Bul. 252, 1934, p. 49.

CHAPTER XVIII

LAWN SOILS

A well-kept lawn adds to the attractiveness of a home, be it small or large, more than any other single factor. An expanse of beautiful turf is also an unsurpassed setting for public buildings and is an essential feature of parks, cemeteries, boulevards, and recreation centers.

Given a reasonable amount of sunshine and ample water, the development and maintenance of a thick, velvety grass cover is not difficult, provided a few simple rules are followed. The presentation of these rules is the objective in the following discussion.

Objectives:

A. Soils and soil preparation.
B. Grass selection and seeding practices.
C. Fertilization and liming.
D. Mowing and watering.

SOILS AND SOIL PREPARATION

It is much easier to establish and maintain a good lawn on a soil that is suitable for the purpose than on one which is not well adapted for grass culture. Unfortunately, in making a lawn one seldom has a choice of soils and hence must bring the soil he has to as suitable a condition as possible. In answering the following questions, guidance will be given in the selection and conditioning of soil for lawns.

Questions:

1. What is the first step in the making of a lawn?
2. What kind of soil is best suited for lawn purposes?
3. How may unsuitable soil be improved?
4. How is the soil prepared for seeding or sodding?

Proper Grading Comes First. Two purposes are in view in establishing the grade for a lawn: first, to set off the home or other main feature of the landscape to best advantage and, second, to provide a good foundation or subsoil condition for the grass. A lawn that slopes gently away from the building in all directions usually produces the most pleasing effect and at the same time provides satisfactory drainage. The difficulty in establishing and maintaining grass on terraces and steep slopes dictates their avoidance when possible. When soil from the basement is to be used for building up the grade or when subsoil is to be drawn in for that purpose, the surface or black soil on the yard should first be scraped into a pile and saved for surfacing when the grading is completed. Needless to say the grade or foundation soil should be well settled and the surface leveled off before the top soil is put on.

Good drainage is indispensable if a satisfactory turf is to be maintained. Usually the grading can be done in such a way as to provide adequate drainage for the lawn around the home. Where there are large expanses of turf, as are found around country clubs, estates, and parks, an artificial drainage system may be necessary if the subsoil is of heavy clay or the site so level as to create a poorly drained condition. Turf which receives much tramping, as do football fields and playgrounds, must be carefully drained if the grass is to survive. Usually lines of 4-inch agricultural tile, placed about 15 feet apart and at a depth of 18 to 24 inches, will give adequate drainage. The tile should empty into main lines of larger tile (6 to 8 inches) which have a suitable outlet into a ditch or sewer. Tile lines should have a fall of at least $\frac{1}{10}$ foot per 100 feet.

Use Medium-Textured Soil High in Humus Content. Soils of medium texture, such as sandy loams, loams, and light silt loams, make the best surface for lawns. The humus content should be sufficiently high to give the soil a very dark color. The required thickness of the surface layer will be dependent on the nature of the foundation or subsoil, but never should less than 4 inches be used. Table 41 suggests the suitable thickness of surface soil to use over different types of subsoil.

TABLE 41

DEPTHS OF SURFACE SOIL TO USE OVER DIFFERENT TYPES OF SUBSOIL

Nature of Subsoil or Subgrade	*Thickness of Surface Soil Required*
Sandy loam or loam	4 to 6 inches
Silt loam	6 to 8 inches
Clay subsoil	8 to 10 inches
Sandy subsoil	10 to 12 inches

Improving Unsuitable Soil. If suitable top soil is not available for surfacing the lawn, a satisfactory soil may be prepared as follows: If the soil is a clay loam, containing little or no organic matter, mix with it equal volumes of sandy soil and of finely divided peat or muck. Twice as much sandy soil may be needed if the soil is a heavy clay. Likewise a sandy soil may be improved by mixing with it suitable proportions of clay loam or clay and of muck or peat. If the muck or peat is not available, a soil of acceptable properties may be made by mixing two to three volumes of sandy soil high in humus (can usually be obtained from a swampy area) with one volume of clay or clay loam. The mixing may be accomplished by running the ingredients through a concrete mixer or by placing them in layers, an inch or so thick, on the surface of the lawn, and then disking or spading and hoeing until a fairly uniform mixture is obtained.

Soil Preparation for Seeding or Sodding. Before seeding, the surface soil should be hoed, raked, and rolled repeatedly until it is free of lumps and stones, is thoroughly settled and firm, and is free of depressions, ridges, or bumps. If the soil is not firm, the lawn will soon develop irregularities which mar its beauty and are difficult to correct. Slighting soil preparation because of too much eagerness in getting the lawn seeded may result in disappointing results. There should be just enough loose soil on top, ¼ to ½ inch, to permit of covering the seed. The soil surface should be approximately an inch below the level of sidewalks or driveways.

If sod is to be used in establishing the lawn, the preparation of the base or subsoil is the same as though the lawn were to be seeded. The grade, however, is lowered sufficiently to allow for the thickness of the sod.

GRASS SELECTION AND SEEDING

Sowing the seed of grasses unadapted to the location or the conditions under which the lawn must be maintained or the use of inferior seed of low germination which contain many weed seeds results in many disappointments to the lawn maker. Improper methods of seeding also account for a high percentage of poor lawns. The selection of seed mixtures and seeding practices are discussed in answers to the following questions.

Questions:

1. How may one know what grass or grass mixture to use?
2. Are the bent grasses suitable for lawns?
3. What seeding practices are advisable?

Selection of Grass. The selection of a grass or a mixture of grasses well suited to the climatic conditions under which the lawn is to be grown and to the particular situation, as to shade, nature of soil, drainage, and amount of tramping which the grass must tolerate, requires a thorough knowledge of the characteristics of the different grasses. For the small lawn it is better to trust the judgment of a reputable seedsman and to buy a ready-prepared seed mixture from a responsible local dealer. When considerable quantities of seed are needed or when it is desired to make mixtures for special situations,

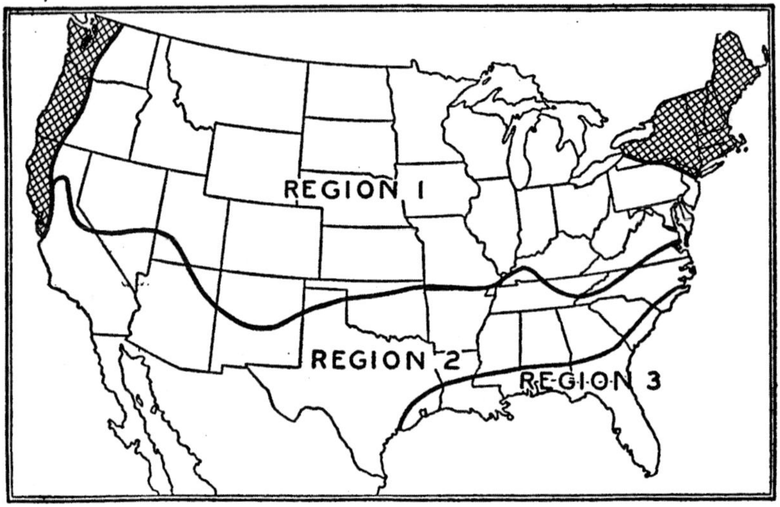

FIG. 77. Map showing regions of the United States to which various lawn grasses are adapted. Region 1: Where local conditions are not unfavorable, Kentucky bluegrass, sometimes called June grass, has no superior for general use. Region 2: Bermuda is better suited in general to this region than any other grass. Region 3: In this area carpet, Bermuda, centipede, or St. Augustine grass are used according to local conditions. [From Farmers' Bul. 1677, U.S.D.A., by H. L. Westover.]

a competent landscape architect should be consulted or information obtained from a state college or university. A few general suggestions regarding the selection of grasses for different climatic zones and soil conditions are given in the following paragraphs.

In the New England States, New York, and the coastal areas of Washington, Oregon, and northern California, soils well suited to the growth of bent grasses frequently occur. These grasses are most commonly used for lawn purposes in the cross-hatched areas in Fig. 77.

Kentucky bluegrass is adapted to a wide range of soil and climatic conditions, as is evident by the extent of Region 1 on the map. Although bluegrass will make a fine lawn when seeded alone under favorable conditions, it is customary to mix the seed with that of a more rapidly growing grass, as redtop or ryegrass. These two grasses grow quickly and help to suppress weed growth until the bluegrass is established. Very shortly they will disappear from the lawn, for they are unable to withstand the competition. For shady locations in

TABLE 42

SEED MIXTURES FOR DIFFERENT LOCATIONS IN HUMID AREAS IN REGION 1

Soil Conditions	Grass-Seed Mixtures		Rate of Seeding (Lb. per 1,000 Sq. Ft.)
Open, sunny, fertile, well-drained loam	Kentucky bluegrass	90%	1½
	redtop	5%	
	white clover	5%	
	Kentucky bluegrass	85%	
	domestic ryegrass	15%	
	Kentucky bluegrass	80%	
	domestic ryegrass	10%	
	white clover	10%	
	Kentucky bluegrass	90%	
	white clover	10%	
Shady, fertile, well-drained loam	Chewing's New Zealand fescue	50%	1½
	rough bluegrass (*Poa trivialis*)	40%	
	redtop or domestic ryegrass	10%	
Sandy soils, shady or sunny	Chewing's New Zealand fescue	85%	1½
	domestic ryegrass	15%	
	Chewing's New Zealand fescue	80%	
	domestic ryegrass	10%	
	white clover	10%	
Wet, poorly drained soils	Canadian bluegrass	50%	1
	rough bluegrass	30%	
	redtop	20%	

Region 1, Chewing's New Zealand fescue or rough-stalked bluegrass (*Poa trivialis*) is preferable to Kentucky bluegrass (Table 42). In portions of the Great Plains crested wheat grass, Western wheat grass, or smooth brome grass are sometimes used in place of bluegrass.

Over the major part of Region 2, Bermuda grass is better adapted than any other grass for lawn purposes. In higher altitudes where the climate is more mild Kentucky bluegrass may well be substituted for Bermuda. It also does reasonably well in slightly shaded locations in the northern part of the region. Further south St. Augustine grass is suitable for shady places if moisture is abundant and the lawn is carefully fertilized.

As Bermuda grass dies down in the winter, it is customary to mow it very short in the fall, loosen the soil by raking, and then seed Italian ryegrass for a winter lawn.

In the Deep South and coastal region (Region 3) carpet and centipede grasses are most commonly used. For moist soils carpet grass is preferable, whereas centipede grass does well on all except poorly drained soils. St. Augustine grass is used extensively for shady locations, especially in Florida and along the Gulf coast. Both centipede and St. Augustine grasses must be planted by using stolons, for seed is not available.

Bent Grasses. When properly cared for, most of the bent grasses make very beautiful lawns in climates to which they are adapted. When not given good care, on the other hand, bent-grass lawns are usually unsightly and the source of much annoyance to the home owner. The management of bent grass requires more knowledge of grass culture and much more work than does the care of lawns composed of other grasses. Unless one is in a position to give considerable time to the lawn or to employ a competent gardener the use of bent grass for a home lawn is not advised.

Methods of Seeding. The quantity of seed to use depends on several factors, of which the most important are (1) size of seed, (2) germination percentage, and (3) evenness of distribution. For seed of good quality, evenly distributed over the seedbed, 1 to 2 pounds per 1,000 square feet of lawn area should be sufficient. On the other hand, if the seed is of low germination, if the lawn maker is inexperienced or if other adverse conditions prevail, it is better to use more or even double the quantity of seed.

Distributing the seed evenly over the soil surface is one of the most difficult problems for inexperienced lawn makers. It is better to divide the seed into three equal portions. The first portion should be sown over the entire lawn surface while one walks back and forth (say north and south) at measured intervals. The second portion should be sown while one walks back and forth at right angles to the first traverse (east and west). Follow a diagonal course when sowing the third portion. By this method the whole lawn surface is covered with seed three times. It is well to increase the volume of each portion of the seed by mixing with it two or three times its volume of dry screened soil to make even distribution easier. If the lawn is large some type of seed distributor may well be used.

After sowing the seed, rake the lawn lightly and roll with a heavy roller to firm the soil around the seed. Covering the seed too deeply by raking heavily frequently results in a thin stand of grass, owing to the inability of the tiny shoots to push through the soil.

Late summer or early fall is the best period for seeding a bluegrass lawn. When the summer drought is over and frequent showers may be expected is a satisfactory time. The grass will grow until decidedly cold weather sets in and so will develop a well-established root system, while most weeds which start growth will be killed by a severe frost. When spring comes the grass will be all ready for an early start and will have made considerable growth before the weeds get under way. On the contrary, in spring seeding the weeds and grass start growth simultaneously and, because of their more vigorous habit of growth, the weeds may gain the advantage and crowd out much of the grass.

FERTILIZATION AND LIMING

Grass must be well nourished if it is to develop a thick, velvety turf and resist the encroachment of weeds. Young grass contains a high percentage of plant nutrients which must be supplied to it by the soil. A suitable soil reaction is also as essential for grass to make a vigorous growth as it is for any other plant. A short list of questions will call attention to the most pertinent points in grass fertilization and in the need for lime.

Questions:

1. In what soil-reaction range does grass grow well?
2. What kind and how much fertilizer should be applied in making a lawn?
3. How should the established lawn be fertilized?
4. Is there danger of damage from fertilizer?

Soil Reaction Suitable for Grass. The need for lime on lawn soils has been greatly overemphasized. Most lawn grasses will thrive in all but strongly acid soils and some in fact, such as fescue and certain bent grasses, grow well even in those soils. Unless a soil test shows a *p*H value of 5.5 or less, the money which might be spent in liming may be more advantageously used for the purchase of fertilizer. If lime is needed, approximately 100 pounds per 1,000 square feet of a finely crushed limestone or other form of calcium carbonate should be evenly applied and thoroughly mixed into the top soil as the seedbed is prepared. Hydrated lime, if used, should be applied two weeks or more before seeding, for it reacts with some fertilizer ingredients—namely, sulphate of ammonia—with liberation of ammonia gas, which is toxic to young plants.

The application of lime to established lawns is also of little value unless the soil is strongly acid. The idea that lime applications will discourage weed growth in the lawn is without foundation except in special cases.

Kind and Amount of Fertilizer. Nitrogen is the nutrient element which requires major consideration in the fertilization of lawns under most soil conditions. If the soil is very sandy, more potash will be needed than when the grass is growing on a loam or heavy sandy loam. There should always be some phosphoric acid in the fertilizer. For average conditions a 10–6–4 analysis is quite satisfactory although other grades with similar nutrient ratios may serve as well. For the new lawn 10 to 15 pounds of the fertilizer per 1,000 square feet of lawn should be thoroughly raked into the surface soil as seedbed preparation is being completed.

Fertilizing the Established Lawn. When the lawn is established, a regular schedule of fertilization should be adopted. Early in the spring, as soon or even before growth starts, about 10 pounds of the 10–6–4 or similar fertilizer should be broadcast evenly over each 1,000 square feet of lawn. A second application, consisting of an equal quantity, should be applied in about eight weeks. These two fertilizations will supply the needs of most lawns until after the heat of the summer has passed. In early fall, when moderate temperatures may be expected, and yet some six weeks before frost, a third application of the same kind and quantity of fertilizer should be made. These suggestions are for lawns which are not shaded to an appreciable extent and in which the grass is not subject to the competition of roots of trees or large shrubs for nutrients and moisture.

Shady lawns and those containing many shrubs should be fertilized more frequently. The early-spring application is the same as described above, but beginning about a month or six weeks later a light application of approximately 4 pounds of the 10–6–4 fertilizer per 1,000 square feet should be made every two weeks until early fall.

Fertilizer Damage. If fertilizer is permitted to adhere to the moist leaves of grass, there is considerable danger of the grass' turning yellow and presenting an unsightly appearance for a time, and permanent damage may be done. To avoid damage the fertilizer should be applied when the grass is dry. Only a limited area of the lawn (500 to 1,000 square feet) should be fertilized before washing the fertilizer off the grass and into the soil with water from the hose. It is advisable to use a rather coarse spray for this purpose and to wash the fertilizer off before one walks on the grass because the bruised leaves are more readily damaged. The spray should be played freely ahead of one as he advances over the fertilized area.

MOWING AND WATERING

Regularity in mowing and watering the lawn is a great aid in maintaining a beautiful turf. The grass depends for nourishment on the food manufactured in its leaves, and hence the mowing schedule must be arranged to provide an abundance of leaf surface at all times. To meet this requirement, mowing must be done frequently and at moderate height. The lawn maker may well ask himself the following questions.

Questions:

1. How frequently should the grass be mowed?
2. At what height should the grass be cut?
3. What precautions are advisable in watering the lawn?

Frequency of Mowing. No definite time schedule can be set for mowing lawn grass, for the frequency of mowing should be governed by the rate of growth. Not more than one inch of the grass leaves should be cut off at any one time, and the mowing schedule should be governed accordingly. There is no objection to mowing more often and cutting only ½ or ¾ inch off the leaves, in fact this is a far better practice than that of allowing the grass to get tall so that 1½ or more inches are cut with the mower.

Height of Mowing. The usual lawn grasses should be mowed at a height of 1½ to 2 inches when they are growing in open sunny lawns and somewhat higher when growing in shady places. The height at which bent grasses should be mowed will depend on the care they receive. Usually bent-grass lawns are mowed a little shorter than lawns composed of other grass, 1 inch to 1½ inches is a suitable height under most conditions.

The grass clippings should be allowed to fall on the lawn if the grass is mowed regularly, so that the clippings are short. They form a light mulch around the crown of the grass plants and readily decay, liberating considerable quantities of nutrients. Should the grass get especially long before mowing, it will be advisable to remove the clippings.

Watering Practices. Very few lawns receive enough water to keep them in the most luxuriant state of growth. The quantity of water transpired by rapidly growing grass is very large, much larger indeed than is realized by lawn makers in general. As grass absorbs moisture through its root system, the method of watering followed should be one that will encourage an extended root development, so that the plants will be equipped to absorb enough moisture to supply their needs during very hot periods. Frequent light applications of water wet the soil to only a shallow depth, and hence the grass roots develop largely in this thin surface layer. If the lawn is watered less frequently but each time with sufficient water to moisten the soil to a depth of 4 or 5 inches, the roots will distribute themselves throughout this deeper soil layer, and sturdier grass, better able to resist adverse conditions, will be the result.

CHAPTER XIX

SOIL RESOURCES

As food, textiles, and housing materials are largely products of the soil, either directly or indirectly, an ample acreage of productive land is one of the most essential and stabilizing resources a people can have. A nation with soil resources so limited that it must depend largely on imported food and fiber is always in a precarious economic position because not only war but also changes in trade conditions may curtail the supply of imports. Furthermore, manufacturers of many commodities must depend on agricultural workers to purchase a considerable portion of the manufactured products, even though the workers live in different countries. Unless the people who live on the land have reasonable incomes, their purchasing power is limited. It is highly important, therefore, to the city producer that there be ample good land for farming purposes and that the land be maintained in a productive state in order to afford a reasonable income to farmers and to provide an adequate food supply.

In considering the soil resources of the United States and other countries, and the adequacy of these resources to supply the needs of the peoples concerned, the following major topics should be investigated.

Objectives:

A. Acreage of producing and potential farm land in the United States.
B. Acreages of arable land in different countries, and land requirement per capita to supply food.
C. Land policies of the United States.

ACREAGE OF FARM LAND IN THE UNITED STATES

An invoice of land resources in continental United States was recently made by federal and state authorities. Liberal use was made of data contained in the 1935 census of agriculture. Objectives of the survey were, first, to determine the acreage of land suitable for cultivation without erosion damage under prevailing farming methods and with price levels equal to those prevailing during the period 1921 to

1936; secondly, to know the additional acreage that could be profitably farmed under the same price levels if the best farming methods and erosion control practices were employed. Other studies have been made to determine the acreage of first-, second-, and third-class land in our country and to indicate the proportion of the various classes of land in the different states. Answers to the following questions may well be sought in considering this topic.

Questions:

1. How much land is suitable for cultivation without special provisions to control erosion?
2. To what uses is this land being put at present?
3. Where is this good land located?
4. How much land can be farmed if soil-conserving methods are used?
5. What is the total acreage of present and potential crop land?
6. What is the distribution of land of different quality among the states?

Land Suitable for Cultivation. The 1935 census of agriculture listed 415,334,931 acres as the crop-land area of continental United States. This figure includes the acreage of (1) crops harvested, (2) land on which crops failed, and (3) land suitable for cropping but temporarily idle or fallow. The best estimates indicate that only 39.1 per cent of this crop land, amounting to 160,948,703 acres, is of a quality that can be cultivated by prevailing methods, without serious erosion, and yield well enough to give profitable returns if prices are as they were between 1921 and 1936. This means either that 60.9 per cent of the present crop land is of such poor quality that it cannot be profitably farmed or that it is subject to continued erosion under the methods by which it is being cultivated.

Present Use of Good Land. All the land which may be profitably farmed without serious erosion is not now listed in the United States census as crop land. There are some 21,611,680 acres, designated as plowable land, which are now pasture. Brush or timber occupy 19,856,239 acres. Also, 6,429,565 acres need drainage, and 2,919,500 acres could be farmed if irrigated. Summing up these figures, we find that 50,816,984 acres can be added to the present area of crop land suitable for cultivation under present practices and at the price levels mentioned, provided the needed improvements of clearing, drainage, irrigation, etc., are made. This gives a total of 211,765,687 acres of what may be called good land.

Location of Potential Crop Land. It is of interest to note where this good land, which may be brought into production if needed, is located. The location and the type of improvement needed are shown by the data in Table 43.

Plowable pasture offers the greatest opportunity for expanding our present acreage of good land for producing crops; clearing off brush and timber, drainage, and irrigation, in the order named, are the other possibilities. The West South Central states have the greatest acreage

TABLE 43

POTENTIAL CROP LAND SUITABLE FOR CULTIVATION UNDER PREVAILING FARMING METHODS AND AT PRICE LEVELS SIMILAR TO THOSE OF 1921 TO 1936 *

Group of States	Present Use or Improvement Needed				Total
	In plowable pasture	In brush or timber	Needs drainage	Needs irrigation	
	acres	*acres*	*acres*	*acres*	*acres*
New England	103,143	....	....	....	103,143
Middle Atlantic	369,129	22,843	1,984	....	393,956
South Atlantic	1,015,910	4,003,802	1,873,609	600	6,893,921
East North Central	6,568,996	3,127,172	678,496	....	10,374,664
West North Central	7,094,306	285,923	1,191,600	138,100	8,709,929
East South Central	1,502,272	3,790,762	222,021	....	5,515,055
West South Central	3,347,079	8,238,937	1,994,323	790,706	14,371,045
Pacific	417,900	274,600	205,000	236,000	1,133,500
Mountain	1,192,945	112,200	262,532	1,754,094	3,321,771
Total	21,611,680	19,856,239	6,429,565	2,919,500	50,816,984

* Compiled from data in "The Problem: The Nation As a Whole," E. J. Utz, C. E. Kellogg, E. H. Reed, J. H. Stallings, and E. N. Munns, 1938 Yearbook of Agriculture, U.S.D.A., p. 95.

which may be brought into production. More than half of this potential crop land is in timber or brush. The East North Central states have the second largest acreage of good land which is not under cultivation. Considerably more than half of this land is in pasture. The West North Central and South Atlantic states follow in order in area of land suitable for crop production if needed improvements are made.

Land Suitable for Cultivation by Use of Soil-Conserving Methods. To the 39.1 per cent of the "crop land" which is considered suitable for cultivation under present methods may be added 42.9 per cent, or 178,178,685 acres, which may be farmed profitably if soil-conserving methods are used. Soil-conserving methods of farming are discussed in Chapters VI and XIV. This leaves 18 per cent of the present land listed in the census as crop land which is considered to be either too poor to yield profitable returns at the prices used or else of such rough topography that its fertility cannot be maintained even by the use of good soil-conserving practices.

In addition to the 82 per cent (39.1 per cent + 42.9 per cent), 339,-079,482 acres of the present crop land which may be farmed by present or improved methods, there can be added a considerable area which needs clearing, drainage, irrigation, or plowing before it can be farmed. Soil-conserving methods of farming would then be needed if the land were to be cropped profitably for any length of time. The location of these lands and the improvement needed are shown in Table 44.

TABLE 44

LAND SUITABLE FOR CULTIVATION IF SOIL-CONSERVING METHODS ARE USED AND THE INDICATED IMPROVEMENTS ARE MADE *

Group of States	Present Use or Improvement Needed						
	In plowable pasture	In brush or timber	Needs drainage	Needs irrigation	Total	In cultivation	Total
	acres	*acres*	*acres*	*acres*	*acres*	*acres*	*acres*
New England	336,127	....	....	....	336,127	3,579,142	3,915,269
Middle Atlantic	1,620,920	73,276	1,984	....	1,696,180	12,910,992	14,607,172
South Atlantic	3,419,817	11,552,831	2,977,376	600	17,950,624	28,209,286	46,159,910
East North Central	12,313,011	5,878,757	690,611	....	18,882,379	57,346,837	76,229,216
West North Central	20,248,118	599,393	1,213,900	349,800	22,411,211	128,410,635	150,821,846
East South Central	5,138,369	7,203,114	222,021	....	12,563,504	19,727,870	32,291,374
West South Central	5,986,678	15,704,103	2,025,803	872,671	24,589,255	51,886,792	76,476,047
Pacific	1,162,500	958,400	239,700	2,087,000	4,447,600	15,349,922	19,797,522
Mountain	2,497,136	181,670	292,426	2,538,658	5,509,890	21,658,006	27,167,896
Total	52,722,676	42,151,544	7,663,821	5,848,729	108,386,770	339,079,482	447,466,252

* Compiled from data in "The Problem: The Nation As a Whole," E. J. Utz, C. E. Kellogg, E. H. Reed, J. H. Stallings, and E. N. Munns, 1938 Yearbook of Agriculture, U.S.D.A., p. 95.

The West North Central states have by far the largest area of land which may be cropped profitably by use of soil-conserving methods, and incidentally about five-eighths of this land is now being cultivated regardless of the extent to which soil-conserving methods are being used. The East North Central states have the next largest acreage of land that falls into the class under discussion, with slightly less than three-fourths of the land being cultivated. The West South Central states and the South Atlantic states offer about equal opportunities for enlarging the acreage of crop land by use of soil-conserving farming methods. Approximately 43 per cent of the land which should be cultivated only by special methods to maintain its productivity is already under cultivation. This situation is deserving of careful consideration, for the question immediately arises as to how long these soils will remain in the productive class under present methods of management.

Plowable pasture again occupies the greatest acreage which may be brought into cultivation, if need be, and again the clearing of brush and timber offers the second largest opportunity to increase crop acreage.

Total Acreage of Present and Potential Crop Land. Summing up the data presented in the preceding paragraphs, it is found that there are some 160,948,703 acres of land in cultivation in the United States which may be profitably farmed with present methods without serious deterioration through erosion. To the above, 50,816,986 acres may be added by plowing up pasture, clearing, draining, and irrigating. There is a total, then, of 211,765,689 acres suitable for profitable farming with present methods. If soil-conserving methods were universally adopted, this area could be enlarged by 286,565,455 acres, of which a considerable area is now being cultivated regardless of the damage to the soil. This gives a grand total of 498,331,144 acres which may be used with reasonable profit to the farmer for the production of food, hay, and fiber crops if conditions demand and if soil-conserving methods are used.

Distribution of Land of Different Quality among the States. A study has been made, with the purpose of dividing the land into five groups, on the basis of the land's capacity to produce grains, grasses, and legumes without irrigation, drainage, or fertilization. The group names used are excellent, good, fair, poor, and unfit for crop production. In Table 45 is presented the total percentage of land in the different groups and the percentage of land in each class found in each of the 42 states having excellent or good land. The distribution of the first three classes of land is also shown graphically in Fig. 78.

The high percentage of excellent and good land in the corn belt states and the states immediately adjoining them is worthy of consideration. The question may well be asked as to what conditions gave rise to the development of so much land of superior quality in that location. It should be remembered, furthermore, that some of the

TABLE 45

Approximate Acreage of Land; Excellent, Good, Fair, Poor, or Unfit for General Crops without Irrigation, Drainage, or Fertilization in the Several States and in Continental United States *

Quality Rating	Excellent	Good	Fair	Poor	Unfit
% of total land	5.3%	11.1%	18.1%	19.1%	46.4%
United States	101,038,000	210,935,000	345,872,000	362,559,000	881,735,000
Iowa	25,983,000	6,906,000	1,393,000	1,007,000	345,000
Illinois	14,777,000	6,847,000	6,223,000	6,622,000	961,000
Minnesota	12,022,000	12,139,000	7,511,000	6,899,000	13,145,000
Missouri	8,675,000	13,833,000	12,304,000	4,259,000	4,915,000
Nebraska	8,121,000	9,690,000	9,640,000	10,908,000	10,758,000
Indiana	5,262,000	6,438,000	7,744,000	2,437,000	1,153,000
Ohio	4,214,000	6,234,000	10,440,000	3,810,000	1,375,000
Kansas	3,765,000	15,172,000	15,965,000	11,207,000	6,095,000
South Dakota	3,052,000	8,775,000	8,960,000	15,552,000	12,821,000
Wisconsin	2,820,000	14,530,000	8,397,000	4,564,000	5,271,000
Michigan	2,251,000	8,961,000	5,387,000	7,229,000	13,223,000
Oklahoma	1,701,000	12,795,000	15,269,000	7,826,000	6,856,000
Texas	1,591,000	19,461,000	54,974,000	35,950,000	56,098,000
Arkansas	1,453,000	7,952,000	9,231,000	8,300,000	6,707,000
Louisiana	1,289,000	2,973,000	7,347,000	12,443,000	5,024,000
Tennessee	902,000	4,770,000	9,305,000	8,240,000	3,412,000
Kentucky	864,000	5,020,000	8,956,000	7,906,000	2,877,000
Mississippi	844,000	5,336,000	8,615,000	10,929,000	4,091,000
West Virginia	389,000	1,350,000	2,987,000	7,647,000	2,828,000
Oregon	343,000	2,059,000	4,694,000	5,465,000	48,568,000
Pennsylvania	177,000	6,268,000	9,497,000	5,909,000	6,909,000
Maryland	144,000	1,875,000	1,446,000	1,859,000	973,000
New York	94,000	6,864,000	9,249,000	9,172,000	4,926,000
Washington	84,000	542,000	6,537,000	6,760,000	28,836,000
Florida	71,000	933,000	3,928,000	12,893,000	17,078,000
Connecticut	48,000	339,000	902,000	962,000	851,000
California	47,000	478,000	3,365,000	13,380,000	82,347,000
Vermont	33,000	901,000	1,268,000	2,185,000	1,452,000
Massachusetts	12,000	760,000	1,275,000	1,644,000	1,367,000
New Jersey	6,000	943,000	964,000	1,379,000	1,471,000
New Hampshire	2,000	237,000	370,000	1,672,000	3,478,000
North Dakota	....	7,366,000	17,716,000	14,152,000	5,680,000
Virginia	....	3,653,000	9,280,000	7,312,000	5,400,000
Alabama	....	3,002,000	9,528,000	10,730,000	9,491,000
Georgia	....	1,852,000	15,614,000	9,750,000	10,369,000
Maine	....	1,615,000	3,789,000	5,791,000	7,978,000
North Carolina	....	1,167,000	11,360,000	9,382,000	9,285,000
South Carolina	....	296,000	7,070,000	7,128,000	5,023,000
Colorado	....	286,000	7,041,000	13,744,000	45,270,000
Delaware	....	222,000	351,000	417,000	273,000
Idaho	....	87,000	949,000	3,729,000	48,581,000
Rhode Island	....	5,000	228,000	71,000	378,000
Distribution	¾ in corn belt area	⅓ between Texas and North Dakota	½ south of Potomac and Ohio Rivers, including Arkansas, Louisiana, Oklahoma and Texas	43% in southern states	75% of all U. S. Non-tillable land is in the 11 far western states owing to lack of rain

* "A Graffic Summary of Physical Features and Land Utilization in the United States," O. E. Baker, U.S.D.A. Misc. Pub. 260, 1937, pp. 12–14.

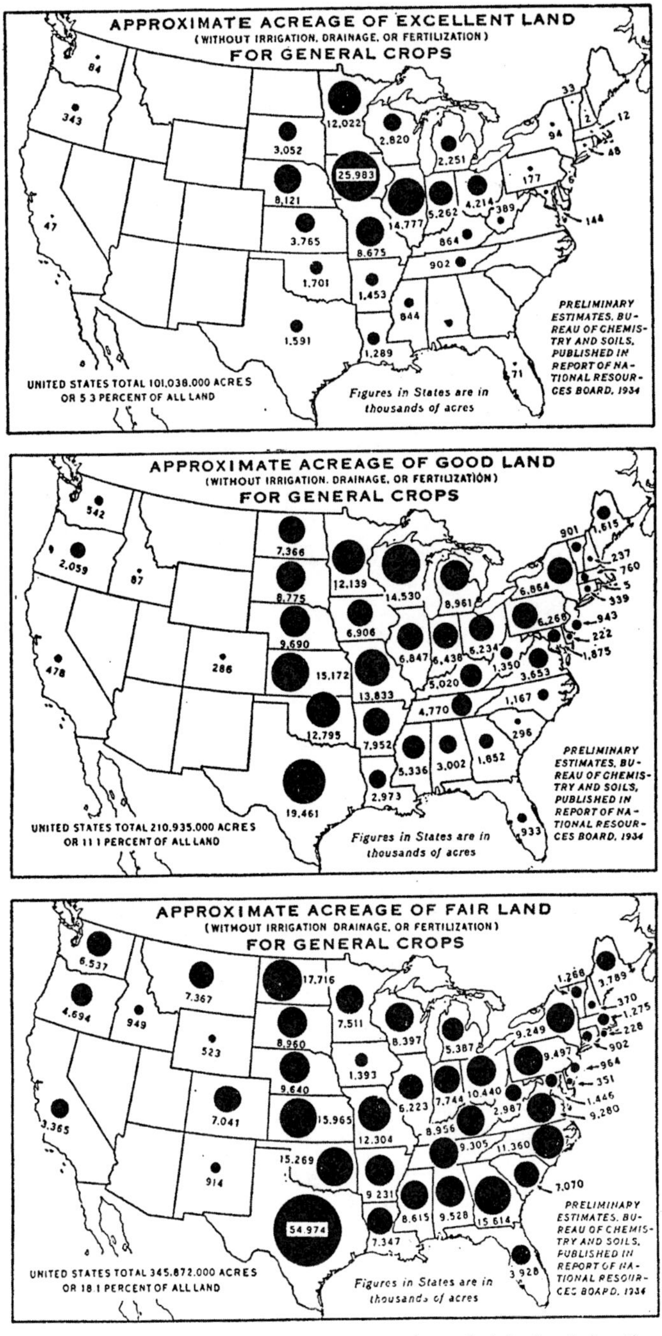

FIG. 78. The distribution of excellent, good, and fair land in the various states. [From Misc. Pub. 260, U.S.D.A.]

land designated "fair" in this classification produces superior yields when fertilized. In fact, some of the states which have very little if any "excellent" land have higher average acre yields of several of the grain crops than do the corn belt states. Likewise, the production on large acreages of irrigated land is much greater than on high-quality land without irrigation.

ACREAGES OF ARABLE LAND AND LAND REQUIREMENTS

A knowledge of the acreage of farm land in different countries in relation to population leads to a better understanding of some of the problems confronting the great nations. The interdependence of nations is brought out also by such a study. The more one learns about good farm land and the part it plays in the welfare of people, the more respect one has for the soil. Answers to the following questions will aid in an understanding of the land problem in different countries.

Questions:

1. What is meant by arable land?
2. How much arable land per capita is there in different countries?
3. What is the land requirement per capita to supply food?

Meaning of Arable. A broad and variable interpretation of the term arable is possible, depending on experience and viewpoint. Strictly speaking, arable land is land which is suitable for plowing or cultivation. No mention is made, however, of whether such land may be cultivated profitably at any set price level for agricultural products or whether the land will deteriorate rapidly under cultivation unless special tillage practices and cropping systems are used. Probably there is a tendency to apply the term arable from a purely physical viewpoint and to consider any cleared land arable which is not too stony, of too rough topography, too badly in need of drainage to grow crops, or in an area of very limited rainfall during the growing season. It is evident that land which one person or group of persons would consider arable might not be so classed by another group of persons. In view of this fact, it is difficult to compare accurately the acreages of good land in different countries. In the following discussion the comparisons are based on the best data available.

Acreage of Arable Land per Capita. The census designation of crop land as representing the arable land in continental United States gives us 415,334,931 acres. Assuming a population of 130,000,000 persons, this allows approximately 3.2 acres per capita. The preceding discussion, however, shows that much of the crop land is being farmed without profit or at the expense of serious deterioration through erosion. It has been brought out that, by improving the good land not now in crops, a total of 211,765,689 acres could be made available which may be profitably farmed by present methods without serious erosion damage. There are about 1.63 acres of such land per person. By bringing into cultivation all land which may be profitably farmed, without severe erosion, by use of soil-conserving methods there would be 498,331,142 acres, or 3.83 acres per capita.

By making use of available data, Table 46 has been prepared which gives a picture of the land per capita available for crop production in

TABLE 46

POPULATION, ARABLE LAND, AND ARABLE LAND PER CAPITA IN SEVERAL COUNTRIES *

Country	Population	Arable Land	Acres of Arable Land per Capita
Canada	11,209,000	59,517,120	5.31
China	429,494,138	189,270,800 †	0.44
Denmark	3,777,000	6,609,720	1.75
England, Wales, Scotland	46,200,000	11,505,260	0.25
Finland	3,834,662	6,439,290	1.68
France	41,980,000	51,205,570	1.22
Germany, Austria	74,827,000	54,910,570	0.73
Netherlands	8,727,321	2,556,450	0.29
Italy	43,430,000	32,117,410	0.74
Japan	72,222,700	14,889,160	0.21
U.S.S.R.	170,467,186	553,072,520	3.24
United States ‡	130,000,000	415,334,931	3.19

* Sources of data: The China Yearbook, 1935–1936, pp. 123 and 839; International Yearbook of Agricultural Statistics, 1939–1940, Introduction, p. x and pp. 2–9.

† Cultivated land. Total acreage is somewhat in excess of this figure because surveys are not complete.

‡ See p. 416.

several countries. These data show in how favorable a position the United States finds itself with regard to land resources. The precarious position of several countries with respect to soils on which to produce food and clothing for their people is very evident. Perhaps this scarcity of soil resources is a contributing factor to the social unrest in certain countries and to activities which continually disturb world peace.

Land Required To Produce Food. Studies [1] have been made of the foods included in diets of American families at various dietary levels and of the acreage of land needed to produce the food included in the diets. Four dietary levels have been considered, as shown in Table 47.

TABLE 47

CALORIES SUPPLIED AND VITAMIN CONTENT OF FOUR DIETS AND LAND REQUIRED TO PRODUCE THE FOOD

Diets	Per Capita per Day			Acres Required
	Calories	Units of vitamin A	Units of vitamin C	
1. Restricted diets for emergency use	2,675	2,746	86	1.2
2. Adequate diets at minimum cost	2,980	5,067	118	1.5
3. Adequate diets at moderate cost	2,985	5,692	168	1.8
4. Liberal diets	2,930	6,495	206	2.1

The variation in the proportion of different groups of foodstuffs in the different diets, as indicated by the calories supplied by each, is shown in Table 48, as is the cost of each diet at retail price levels prevailing during 1931–1932.

It will be noted that the low-cost diets contain much less meat, fish, eggs, fruits, vegetables, and milk and a proportionally higher amount of cereal foods than the "liberal" and "adequate at moderate cost" diets. This diet "has a lower retail value than have diets budgeted by many if not most organized social agencies for dependent families in times of national prosperity. It has a higher retail value, however, than does the food which can be procured with the relief allowances furnished thousands of families in the period of widespread unemployment during 1931–33." [2]

[1] "Diets at Four Levels of Nutritive Content and Cost," Hazel K. Stiebeling and Medora M. Ward, U.S.D.A. Circ. 296. 1933, p. 3.

[2] *Ibid.*, pp. 3 and 4.

TABLE 48

DISTRIBUTION OF CALORIES AMONG DIFFERENT FOOD GROUPS AND COST OF DIETS AT 1931–1932 PRICES

Diet	Percent of total calories from different food groups						Cost per Year in Dollars 1931–1932 Retail Prices
	Bread, flour, cereals	Milk	Fruits, vegetables	Fats	Sugars	Lean meat, eggs, fish	
1	43	12	14	17	9	5	61
2	35	18	15	17	7	8	85
3	24	19	18	18	9	12	140
4	15	19	18	18	9	21	165

The adequate diets at minimum cost are similar to those of lower cost except for a somewhat greater supply of meat, eggs, fish, and milk, with less cereals. There is little difference in diets 3 and 4 except for a decidedly less quantity of cereal foods and more meat, eggs, and fish.

The writers of Circular 296 made the following statement: "As far as can be judged from the available data, the retail money value of the food of the majority of the families in the United States in the years 1922 to 1929 was between the values of the minimum- and moderate-cost adequate diets." Studies of the costs of diets of many farm families in 11 states and of the families of wage earners in New York City and Detroit support this statement. The cost of the liberal diet corresponds to expenditures by families of skilled wage earners and of business and professional men.

The acreages of land required to produce the diets presented were calculated by O. E. Baker, Senior Agricultural Economist, Bureau of Agricultural Economics, and are based on the average yields per acre of different crops during the years 1917–1926. The acreages given are exclusive of grazing land. Food-consumption data for the period 1927–1931 indicate that 1.9 acres per capita would be needed to supply the food requirement.

By referring to the preceding data relative to the acreage of land per capita in continental United States, it is seen that there is scarcely enough land which may be farmed profitably by commonly used methods *without serious erosion damage* to supply a satisfactory diet for our people. On the other hand, by farming a large acreage by soil-depleting methods, as is being done, there are over 3 acres per person. By using soil-conserving methods where needed, by bringing into cultivation land now in need of drainage, irrigation, or clearing, and by utilizing plowable pasture, approximately 3.83 acres per person may be made available. The necessity of using soil-conserving methods in order to maintain sufficient crop land in a productive state for the future well-being of the people is evident.

LAND POLICIES OF THE UNITED STATES

In the early history of our country the people lived to a large extent directly on the products of the soil they tilled. In those days the importance of the farm in the production or creation of wealth, both for the individual and for the political units through taxation, was appreciated to a much greater extent than at present because industry now provides employment for so many people and has become a fertile source of income to tax-levying agencies. Problems having to do with land received accordingly considerable attention.

In the United States some 30,000,000 people were living on farms in 1930, and another 23,000,000 were classed as rural but non-farm residents. The urban population was composed of 70,000,000 persons. Likewise in many other countries an appreciable proportion of the population lives on the land. In addition many persons are engaged in businesses which involve the processing and marketing of farm products and the manufacture and sale of materials used exclusively in connection with farming. Furthermore, there were some 400,000,000 acres of publicly owned land in 1930. As land is so closely connected with the lives and incomes of so many people, it would seem that there should be some definite national policy concerning the ownership, management, and use of land. The question of land policies may well be considered under the following headings.

Questions:

1. Why was the government so anxious to have land pass into private ownership?
2. By what means was ownership of land transferred to individuals and organizations?
3. What attention was paid to quality of land in passing it from public to private ownership?
4. What benefits have resulted from getting land into private ownership?
5. How has misfortune come to people through improper land use?
6. What problems has the land situation brought to public agencies?
7. Are efforts being made to rectify past mistakes?

Government-Owned Land. When our government was organized, all land not in private ownership became federal property. The states and territories held title to no land when they were organized. State ownership of land was a later development by government grant, through purchase, and as a result of tax delinquency. The government was in need of income, and federally owned property brought in little if any returns, whereas privately owned land could be taxed. Farm products were also needed for export so that manufactured goods might be purchased and trade developed with European countries. It was desired also to develop the resources of the country so that the nation might become stronger economically and greater in population. As a result, a great effort was made to get land into private ownership. As the states developed and came into possession of land, they followed the same policy in order to get the land on the tax roll.

Government Sales and Grants of Land. Large tracts of land were opened to homesteaders after the Homestead Act in 1862, the areas made available by this method being pushed farther and farther westward as the country became settled. In fact it was not until the passage of the Taylor Act in 1934 (amended in 1936) that public land ceased to be available to homesteaders. In order to induce railroads to extend their lines westward and to help them finance the construction, the government made railroad companies large grants of land. These lands were then sold to pioneer farmers. A real estate department which would dispose of these lands was an important unit in most railroad organizations.

Grants were made to help finance and encourage the development of schools and thus hastened the movement of property into private hands. Large acreages in timbered areas were sold to lumber companies and, where mineral resources were thought to be present, mining companies purchased extensive tracts.

Land Was Land. In the early days there was so much land that little thought was given to the question of soil fertility except by the owners of the older farms near the eastern seaboard, such as Washington and Jefferson. Most land was fairly productive because of its virgin fertility. Anyway, if soil gave out, there was an unlimited supply and so there was no need for concern. Land was opened for homesteading or was sold at extremely low prices without thought of its suitability for farming. The consuming idea was to get people on the land so that it could be placed on the tax roll.

Advantages of Private Ownership of Land. The past land policy of promoting private ownership of land has resulted in individual benefits to many people and in the rapid production of wealth. The opportunity to obtain land by homesteading or by purchase at low cost has greatly hastened the development of our country. Furthermore, the possibility of obtaining ownership of land and of passing on that ownership to their children has encouraged industry and initiative and fostered self-reliance, independence, and pride of citizenship in our people. The right to own property is no small part of our democratic liberties.

Settlers on Poor Land. The policy of getting land into private ownership with little regard to the quality of the land or its fitness for the proposed use could be expected to result in much misuse of land and consequent human grief and suffering. Thousands of families seeking farm homes and an opportunity to make a living from the soil have found themselves located where successful farming was impossible under conditions which have prevailed. Some of the circumstances which have contributed materially to the unsatisfactory farming conditions are (1) soils of low natural fertility because of limited plant-food content or low moisture-holding capacity or both; (2) a geographical location such that marketing of farm products either was too expensive or required so much time that the products deteriorated in quality; (3) climatic conditions such as a short growing season or low temperature during the growing period of crops which limited yields or selection of crops to too great an extent; and (4) rainfall that was too limited or not properly distributed during the growing period for profitable crop production.

In far too many instances large tracts of unproductive land have fallen into the hands of unprincipled operators who have sold farm-sized units to settlers who were not capable of selecting land suitable for farming. The family savings were used as a part payment for the land, with a mortgage on the entire acreage given as security for payment of the balance. When the mortgage was foreclosed, the family was left penniless and the land was resold to another unfortunate settler.

An equally discouraging fate has been experienced by settlers who took up homesteads in areas of low rainfall. A succession of drought years, sometimes accompanied by soil blowing, has necessitated abandonment of the farm. At times the acreage allotment was too small for extensive grazing for which the land was fitted, and the settler was forced to engage in a type of farming unsuitable to his location. The result has been disastrous to both the farmer and the soil.

Problems Resulting from Poor Land Use. The inability of farmers, heedlessly located on inferior soils, to make a living has resulted in a great welfare load for local and state agencies dealing with these problems. Schools had to be provided for the children of these families and roads maintained for their convenience. On the other hand, little or no taxes could be collected from them. In some states large acreages of such lands have reverted to state ownership because of tax delinquency and the prospects are for a much further increase in state-owned land through this channel. When the land reverts to public ownership the state is confronted with the problem of the best use for it. This problem is causing much concern and the question of what adjustment can be made to put these families on a self-supporting basis is yet more urgent. The early policy of getting land into private ownership regardless of the quality of the soil hastened the economic development of the public domain but has ultimately resulted in the return of very large acreages to public ownership, often entailing much human grief, and has left some very serious social and economic problems to be solved.

Plans for Better Land Use. Past land policies have resulted in individual benefits and rapid production of wealth, but they have resulted also in waste of soil resources and in a great deal of human distress. As a result of the various and more recent conservation movements coupled with the economic depression, governmental administrators have become aware of the necessity of correcting the mistakes of the past and of preventing the further serious misuse of land. Various measures for better land use and soil conservation have been devised and are being put into effect.

In order to achieve better land use, basic information about the present extent and distribution of our land resources must be obtained. Land classification and land inventory work are being carried on by such agencies as the National Resources Committee, the United States Bureau of Agricultural Economics, the Division of Soil Survey of the United States Bureau of Plant Industry, the United States Soil Conservation Service, and several state experiment stations connected with the land-grant colleges.

Land-planning movements also have been initiated for the purpose of achieving better land use. The ideal objective of rural land planning is to achieve for separate natural and economic land divisions the uses for which they are best adapted under particular local conditions, and then to adjust local plans to state and national plans. Plans which have the sanction of public opinion may result in better land use without legal compulsion, or may be preliminary to laws regulating use.

Rural-zoning ordinances, analogous to city-zoning ordinances, have been adopted in a few states. One of the ostensible purposes of rural zoning is to prevent further agricultural use of submarginal land, or to restrict use to that land which is most suitable, thereby preventing waste in human effort and at the same time protecting land dedicated to non-agricultural uses.

Soil-conservation districts may be established by law in some states. The primary purpose of this measure is to restore lands which have deteriorated because of soil erosion and to maintain soil productivity by good management practices. Districts may be any size, and by agreement of landowners or farm operators regulations for land use may be imposed by law. Acquisition, by purchase by national and state agencies, of privately owned poor land and of exploited and so-called submarginal land is another measure proposed for better land use. Lands so acquired may be used for forest, parks, game refuges, public hunting grounds, and other non-agricultural purposes.

Educational programs are also regarded as an essential part of movements for better land use. Extension divisions of land-grant colleges and the United States Department of Agriculture establish demonstration projects and arrange meetings and conferences to promote the better use of land.

GLOSSARY

ACID SOIL. A soil giving an acid reaction (below *p*H 7.0). A soil having a preponderance of H ions over OH ions in the soil solution.

ADHESION. The attraction or union of unlike materials. The particles or substances are attached more or less firmly together by their adjacent surfaces.

ADSORPTION. The concentration of material at the contact zone (interface) of two substances. Usually applies to the molecular or ionic state of division; for example, NH_3 gas concentrated on the surface of charcoal. Potassium ions concentrated on the surface of soil colloids.

AGGREGATE (*soil*). A single mass or cluster of soil particles, such as a clod, crumb, or granule.

ALKALI SOIL. A soil containing alkali salts, usually sodium carbonate or in which the colloidal complex contains a large amount of sodium, with a *p*H value of 8.5 and higher.

ALKALINE SOIL. Any soil that is alkaline in reaction; i.e. above *p*H 7.0.

ALLUVIAL SOILS. Azonal group of soils, developed from transported and relatively recently deposited material (by streams) characterized by a weak modification (or none) of the original material by soil-forming processes.

ALLUVIUM. Fine material, such as sand, silt, clay or other sediments deposited on land by streams.

AMMONIFICATION. Formation of ammonium compounds or ammonia, as in soils, by soil organisms.

ANION. An ion carrying a negative charge of electricity.

AZONAL SOILS. Any group of soils without well-developed profile characteristics, owing to their youth or conditions of parent material or relief, that prevent the development of normal soil-profile characteristics.

BASE (*ionic*) EXCHANGE. Denotes an exchange of bases. Refers to the reaction which certain insoluble constituents of soils undergo when they are brought in contact with a salt solution. The cations of the salt replace the bases from the soil.

BASE-EXCHANGE CAPACITY. Denotes the maximum quantity of base that the soil is capable of adsorbing from a neutral solution of a monovalent cation. Expressed in terms of milliequivalents per 100 grams of soil.

BOG SOILS. An intrazonal group of soils with a muck or peaty surface soil underlain by peat, developed under swamp or marsh types of vegetation; mostly in a humid or subhumid climate.

BROWN SOILS. A zonal group of soils having a brown surface horizon which grades into lighter-colored soil and finally into a layer of carbonate accumulation; developed under short grasses, bunch grasses, and shrubs in a temperate to cool semi-arid climate.

BUFFERING. The resistance of a substance to an abrupt change in acidity or alkalinity.

CALCAREOUS SOIL. Soil containing sufficient calcium carbonate (often with magnesium carbonate) to effervesce visibly to the naked eye when treated with hydrochloric acid. Soil alkaline in reaction, owing to the presence of free calcium carbonate.

CALICHE. A more or less cemented deposit of calcium carbonate or of mixed calcium and magnesium carbonates, characteristic of soils of warm or hot desert and semi-arid regions.

CAPILLARY CAPACITY (*field capacity*). The amount of water held in the soil after the excess or gravitational water has drained away.

CAPILLARY WATER. Water held by adhesion and surface-tension forces as a film around particles and in the capillary spaces. Moves in any direction in which capillary tension is greatest.

CARBON-NITROGEN RATIO. The relative proportion, by weight, of organic carbon to nitrogen in a soil. The number obtained by dividing the percentage of organic carbon in a soil by the percentage of nitrogen.

CARBONATE ACCUMULATION, SOIL HORIZON OF. A developed soil horizon, beneath the surface, containing more calcium (or magnesium and calcium) carbonate than the soil above it or the soil material below it.

CATCH CROP. A crop seeded with one of the regular crops in a rotation or between the growing periods of two regular crops, for the purpose of adding organic matter and, in some cases, nitrogen also to the soil.

CATEGORY (*soil classification*). Any one of the subdivisions of the system of classification in which soils are arranged on the basis of their characteristics. Beginning with the lowest category, soils are classified on the basis of progressively fewer characteristics into groups of progressively higher or more inclusive categories.

CATION. An ion carrying a positive charge of electricity.

CHERNOZEM SOILS. A zonal group of soils having a deep, dark-colored to nearly black surface horizon, rich in organic matter, which grades below into lighter-colored soil and finally into a layer of lime accumulation; developed under tall and mixed grasses in a temperate to cool subhumid climate. From the Russian for black earth. Sometimes spelled TSCHERNOSEM.

CHESTNUT SOILS. A zonal group of soils having a dark-brown surface horizon which grades below into lighter-colored soil and finally into a horizon of lime accumulation; developed under mixed tall and short grasses in a temperate to cool and subhumid to semi-arid climate. They occur on the arid side of chernozem soils, into which they grade.

CLASS, SOIL. Classes of soil based on the relative proportion of soil separates.

CLAY. The small mineral soil grains or particles, less than 0.002 mm. in diameter.

CLAYPAN. A dense and heavy soil horizon underlying the upper part of the soil; hard when dry and plastic or stiff when wet; presumably formed in part by the accumulation of clay brought in from the horizons above by percolating water.

COHESION. The union or attraction of substances of like characteristics, as the attraction of one water molecule for another.

COLLOID, SOIL. The term colloid is used in reference to matter, both inorganic and organic, having very small particle size and a correspondingly high surface area per unit of mass. Many mineral colloids exhibit crystalline structure. Colloid comes from the Greek words for glue and like.

COLLUVIUM. Heterogeneous deposits of rock fragments and soil material accumulated at the base of comparatively steep slopes through the influence of gravity.

COVER CROP. A crop used to cover the soil surface; to decrease erosion and leaching, shade the ground, and offer protection to the ground from excessive freezing and heaving.

DEFLOCCULATE. To separate or break down soil aggregates into their component particles. Usually refers to particles of colloidal dimensions.

DENITRIFICATION. The reduction of nitrates to nitrites, ammonia, and free nitrogen, as in soil by soil organisms, particularly certain anaerobic organisms (those living or active in the absence of air or free oxygen).

DISPERSION. The destroying of soil structure (breaking up the granules) so that each individual soil particle behaves as a unit.

DRIFT. Material of any sort deposited in one place after having been moved from another. For example, glacial drift includes glacial deposits, unstratified (till) and stratified glacial outwash materials.

DRUMLIN. An oval hill of glacial drift, normally compact and unstratified, usually with its longer axis parallel to the movement of the ice responsible for its deposition.

ELUVIATION. The movement of soil material from one place to another within the soil, in solution or in suspension, where there is an excess of rainfall over evaporation. Horizons that have lost material through eluviation are referred to as eluvial and those that have received material as illuvial.

EROSION, SOIL. Removal of soil material from the solum by wind or running water.

FAMILY, SOIL. A category in soil classification between series and great soil group; a taxonomic group of soils having similar profiles, composed of one or more distinct soil series.

FERTILIZER. A material supplying one or more of the plant nutrients in a condition suitable for application to the soil.

FERTILITY (*of soil*). The quality that enables a soil to provide the proper compounds, in the proper amounts and in the proper balance for the growth of specified plants when other factors, such as light, temperature, and the physical condition of the soil, are favorable.

FLOCCULATE. To aggregate individual particles into small groups or clusters. Usually refers to particles of colloidal dimensions.

FLOOD PLAIN. The nearly flat surface subject to overflow along streams.

FRIABLE. Easily crumbled in the fingers; non-plastic.

GENESIS, SOIL. Mode of origin of the soil; refers particularly to the processes responsible for the development of the solum from the unconsolidated parent material.

GRANULATION. The cementation of particles into masses as grains, aggregates, or clumps, essentially a result of flocculation and aggregation.

GRAVITATIONAL WATER. The water that moves under the force of gravity; it is not retained by the soil.

Gray-Brown Podsolic Soils. A zonal group of soils having a comparatively thin organic covering and organic-mineral layers over a grayish-brown leached layer which rests upon an illuvial brown horizon; developed under deciduous forest in a temperate moist climate.

Great Soil Groups (*soil classification*). A group of soils having common internal soil characteristics; includes one or more families of soils.

Green-Manure Crop. Any crop grown and plowed under for the purpose of improving the soil, especially by the addition of organic matter.

Hardpan. A hardened or cemented soil horizon. The soil may have any texture and is compacted or cemented by iron oxide, organic material, silica, calcium carbonate, or other substances.

Horizon, Soil. A layer of soil approximately parallel to the land surface with more or less well-defined characteristics that have been produced through the operation of soil-building processes. Each layer differs from the one above or below in some characteristic.

Humus. The well-decomposed, more or less stable part of the organic matter of the soil.

Hydrolysis. A double decomposition reaction in which water is a reactant.

Hygroscopic Coefficient. The maximum amount of water (percentage based on weight of dry soil) adsorbed on the surface of soil particles from an atmosphere slightly below 100 per cent relative humidity.

Hygroscopic Water. Water which is adsorbed from an atmosphere of water vapor and held on the surface of particles by forces of adhesion.

Igneous Rock. A rock produced through the cooling of melted mineral material.

Immature Soil. A young or imperfectly developed soil.

Intrazonal Soil. Any of the great groups of soils with more or less well-developed soil characteristics that reflect the dominating influence of some local factor of relief, parent material, or age over the normal effect of the climate and vegetation.

Ion. An electrically charged element or group of elements in an electrolyte. An electrically charged particle.

Lacustrine Deposits. Materials deposited by lake waters.

Land, Arable. Land which, in its present condition, is physically capable, without further substantial improvement, of producing crops requiring tillage.

Land Reclamation. Making land capable of more intensive use by changing its character, environment, or both through operations requiring collective effort. Does not include the clearing of stumps, brush, and stones from land, or simple techniques of erosion control that can be effected by the individual.

Land Type. Land uniformly possessed of particular distinguishing characteristics. A natural land type is land having a particular set of defined natural characteristics, principally of soil, climate, relief, stoniness, and native vegetation.

Landscape (*as used in soil geography*). The sum total of the characteristics that distinguish a certain area on the earth's surface from other areas. These characteristics are the result not only of natural forces but of human occupancy and use of the land.

LAND-USE PLANNING. A development of plans for the uses of land that will, over a long period, best serve the general welfare, together with the formulation of ways and means of achieving such uses.

LATERITE SOILS. The zonal group of soils having very thin organic and organic-mineral layers over reddish leached soil that rests upon highly weathered material, relatively rich in hydrous alumina, iron oxide, or both, and poor in silica; usually deep red in color. Laterite soils are developed under the tropical forest in a hot, moist, or wet-dry climate with moderate to high rainfall.

LATERIZATION. The characteristic process which tends toward the production of laterites and lateritic soils. Essentially it is the process of the silica removal with consequent increase in the alumina and iron oxide content and decrease in base-exchange capacity of the soil.

LEACHING. Removal of materials in solution.

LIME, AGRICULTURAL. Any compound of calcium and/or magnesium used to correct the harmful effects of acid soils on plant growth.

LIMESTONE. A general name for rocks composed essentially of calcium carbonate. There are several kinds of limestone varying in physical and chemical composition.

LOESS. Wind-transported material of fine texture.

MARINE MATERIAL. Material deposited in the waters of oceans and seas and exposed by elevation of the land or the lowering of the water level.

MARL. A soft, earthy deposit consisting chiefly of calcium carbonate mixed with sand, clay, organic matter, and other impurities in varying proportions. Frequently used as a liming material.

MATURE SOIL. A soil with well-developed characteristics, produced by the natural processes of soil formation, and in equilibrium with its environment.

MECHANICAL ANALYSIS. The separation by mechanical means of the different size groups (separates) and determining the percentage of each group in a given soil sample.

METAMORPHIC ROCK. A rock the constitution of which has undergone pronounced alteration. Such changes are generally effected by the combined action of pressure, heat, and water, frequently resulting in a more compact and more highly crystalline condition of the rock. Gneiss, schist, and marble are common examples.

MINERAL. A naturally occurring combination of inorganic elements and ions in the form of salts or other compounds either in crystalline or amorphous condition.

MINERAL SOIL. A general term used in reference to any soil composed chiefly of mineral matter. The mineral material is dominant over the organic matter in determining the characteristics of the soil.

MOISTURE EQUIVALENT. The amount of water retained (percentage) by a soil against a force of 1,000 times that of gravity.

MORPHOLOGY, SOIL. The physical constitution of the soil, including the texture, structure, porosity, consistence, and color of the various soil horizons, their thickness, and their arrangement in the soil profile.

MOTTLED (*mottling*). Irregularly marked with spots of different colors.

MUCK SOIL. An organic soil, composed of highly decomposed organic material, with a considerable amount of mineral soil material, finely divided and with few fibrous remains of the original plants.

NEUTRAL SOIL. A soil that is not significantly acid or alkaline; strictly one having a pH of 7.0.

NITRIFICATION. Formation of nitrates from ammonia as in soils by soil organisms.

NITROGEN FIXATION. In soils, the assimilation of free nitrogen from the air by soil organisms, making the nitrogen eventually available to plants.

NORMAL SOIL. A soil having a profile in equilibrium with the two principal forces of the environment—native vegetation and climate—usually developed on the gently undulating upland; with good drainage; from any parent material; not of extreme texture or chemical composition; that has been in place long enough for biological forces to exert their full effect.

NUTRIENTS, PLANT. The elements or groups of elements taken in by the plant, essential to its growth, and used by it in the elaboration of its food and tissue.

ORGANIC SOILS. Soils containing organic matter in sufficient quantities to dominate the soil characteristics. Frequently all soils containing 20 per cent or more organic matter by weight are arbitrarily designated as organic soils.

OXIDATION. When an element burns in oxygen, the product is a compound called an oxide and the process is called oxidation. In a broader sense, where the positive valence of an element is increased, that element is oxidized.

PARENT MATERIAL. The slightly altered or unweathered material beneath the solum; similar to that from which the soil was formed.

PARENT ROCK. The rock from which parent materials of soil are formed.

PEAT. Unconsolidated soil material consisting largely of undecomposed or slightly decomposed organic matter accumulated under conditions of excessive moisture.

PEDALFER. A soil in which there is a zone of alumina and iron oxide accumulation in the profile but with no horizon of carbonate accumulation.

PEDOCAL. A soil with a horizon of accumulated carbonates in the soil profile.

PEDOLOGY. The science of the soil in which different soils are considered as natural units and attention is given to development, physical, chemical, and biological relationships and to their dynamic nature.

pH. A notation to designate or indicate the degree of acidity or alkalinity of systems. Technically, the common logarithm of the reciprocal of the hydrogen-ion concentration (grams per liter) of a system.

PHASE, SOIL. That part of a soil type having minor variations in characteristics used in soil classification from the characteristics normal for the type, although they may be of great practical importance. The variations are chiefly in such external characteristics as relief, stoniness, or erosion.

PLASTIC. A soil which can be readily molded or deformed without rupture; pliable, puttylike.

PODSOL SOILS. A zonal group of soils having an organic mat and a very thin organic mineral layer above a gray leached layer which rests upon an illuvial dark-brown horizon; developed under the coniferous or mixed forest or under heath vegetation in a temperate to cold moist climate. Iron oxide and alumina, and sometimes organic matter, have been removed from the A and deposited in the B horizon.

PODSOLIC SOILS. Soils that have been formed wholly or partly under the influence of the podsolization process.

PODSOLIZATION. A general term referring to that process (or those processes) by which soils are depleted of bases, become acid, and have developed eluvial A horizons (surface layers of removal) and illuvial B horizons (lower horizons of accumulation). Specifically the term refers to the process by which a podsol is developed, including the more rapid removal of iron and alumina than of silica, from the surface horizons, but it is also used to include similar processes operative in the formation of certain other soils of humid regions.

POROSITY, SOIL. The degree to which the soil mass is permeated with pores or cavities. It is expressed as the percentage of the whole volume of the soil which is unoccupied by solid particles.

PRAIRIE SOILS. The zonal group of soils having a very dark brown or grayish-brown surface horizon, grading through brown soil to the lighter-colored parent material at 2 to 5 feet, developed under tall grasses, in a temperate, relatively humid climate. This term is not applied to all dark-colored soils of the treeless plains but only to those in which carbonates have not been concentrated in any part of the profile by the soil-forming processes.

PRODUCTIVITY (*of soil*). The capability of a soil for producing a specified plant or sequence of plants under a specified system of management.

PROFILE, SOIL. A vertical cross section of the soil from the surface into the underlying unweathered material.

PUDDLE. To deflocculate or to destroy the granular structure of the soil.

REACTION, SOIL. The degree of acidity or alkalinity of the soil mass expressed in pH values or in words.

REDUCTION. Whenever an element loses valence it is reduced; that is, a decrease in positive valence or an increase in negative valence. Often used to designate the loss of oxygen from a compound.

RELIEF. The elevations or inequalities of a land surface, considered collectively.

RESIDUAL SOIL. Soils formed by the weathering of rock and minerals in place; sedentary materials.

ROCK. A combination of crystals or fragments of the same or different minerals into large masses; may occur in either solid or unconsolidated condition.

SALINE SOIL. A soil containing excessive amounts of neutral or non-alkaline salts, usually chlorides and sulphates.

SALT. The product, other than water, of the reaction of a base with an acid.

SAND. Small rock or mineral fragments having diameters ranging from 1 to 0.05 mm.

SEDENTARY SOILS. Soils formed in place without the addition of transported material.

SEDIMENTARY ROCK. A rock composed of particles deposited from suspension in water.

SEPARATE, SOIL. One of several groups of soil particles having definite size limits.

SERIES, SOIL. A group of soils having genetic horizons similar in differentiating characteristics and arrangement in the soil profile, except for the texture of the surface soil, and developed from a particular type of parent material.

A series may include two or more soil types differing from one another in the texture of the surface soils.

Silt. Small mineral soil grains the particles of which range in diameter from 0.05 to 0.002 mm. (or 0.02 to 0.002 mm. in the international system).

Soil. The natural medium for the growth of land plants on the surface of the earth. A natural body on the surface of the earth in which plants grow, composed of organic and mineral materials.

Solonchak. An intrazonal group of soils having a high concentration of soluble salts; usually light colored; without characteristic structural form; developed under salt-loving grass or shrub vegetation mostly in an arid, semi-arid, or subhumid climate.

Solonetz Soils. An intrazonal group of soils having a variable surface horizon of friable soil underlain by dark hard soil, ordinarily with columnar structure; usually highly alkaline; developed under grass or shrub vegetation, mostly in a subhumid or semi-arid climate.

Solum. The upper part of the soil profile, above the parent material, in which the processes of soil formation are taking place. In mature soils this includes the A and B horizons, and the character of the material may be and usually is greatly unlike that of the parent material beneath. Living roots and life processes are largely confined to the solum.

Specific Gravity. The ratio of the weight of dry soil to that of water which will occupy only the volume of the soil particles alone (pore space excluded). It is the ratio between the weight of an object and the weight of water it will displace.

Stratified. Composed of, or arranged in, strata or layers, as stratified alluvium. The term is applied to geological materials. Those layers in soils that are produced by the processes of soil formation are called horizons, while those inherited from the parent material are called strata.

Strip Cropping. Strip cropping is a practice of growing ordinary farm crops in long strips of variable widths, across the line of slope, approximately on the contour, in which dense-growing crops are seeded in alternate strips with clean-tilled crops.

Structure, Soil. The morphological aggregates in which the individual soil particles are arranged.

Subsoil. Roughly, that part of the solum below plow depth.

Surface Soil. That part of the upper soil of arable soils commonly stirred by tillage implements or an equivalent (5 to 8 inches) in non-arable soils.

Terrace (*for control of runoff, soil erosion, or both*). A broad surface channel or embankment constructed across the sloping lands, on or approximately on contour lines, at specific intervals. The terrace intercepts surplus runoff, to retard it for infiltration or to direct the flow to an outlet at non-erosive velocity.

Texture, Soil. The relative proportion of the various size groups of individual soil grains. The coarseness or fineness of the soil.

Till, Glacial. A deposit of earth, sand, gravel, and boulders transported by glaciers. Till is unstratified.

Till Plain. A level or undulating land surface covered by glacial till.

Tilth, Soil. The physical condition of the soil in relation to plant growth; a term indicating the conditions of soil structure produced by tillage or cultivation.

Top Soil. A general term applied to the surface portion of the soil, including the average plow depth (surface soil) or the A horizon, where this is deeper than plow depth.

Tundra Soils. A zonal group of soils having dark-brown highly organic layers over grayish horizons which rest on an ever-frozen substratum; developed under shrubs and mosses in cold, semi-arid to humid climates, that is, in Arctic regions.

Type, Soil. A group of soils having genetic horizons similar in differentiating characteristics, including texture and arrangement in the soil profile, and developed from a particular type of parent material.

Varnish, Desert. A glossy coating of dark-colored compounds, probably composed largely of iron oxides, covering pebbles, stones, and large rock surfaces exposed in hot deserts.

Vesicular Structure. Soil structure characterized by round or egg-shaped cavities or vesicles.

Volume Weight. The weight of a given volume of dry soil, in its natural structural conditions, in comparison to the weight of an equal volume of water. It is sometimes referred to as the *apparent specific gravity*.

Water Requirement of Plants (*transpiration ratio*). The pounds of water transpired by a plant per pound of dry matter produced above ground.

Water Table. The upper limit of the part of the soil or underlying material wholly saturated with water.

Weathering. The physical and chemical disintegration or decomposition of rocks and minerals under natural conditions.

Wilting Coefficient (*wilting point*). The percentage of water in the soil (based on dry weight of the soil) when permanent wilting of plants occurs. It refers to that moisture content at which soil cannot supply water at a rate sufficient to maintain the turgor of a plant and it permanently wilts.

Zonal Soil. Any one of the great groups of soils having well-developed soil characteristics that reflect the influence of the active factors of soil genesis—climate and living organisms, chiefly vegetation.

In the compilation of the Glossary extensive use was made of the glossary in "Soils and Men," U.S.D.A. Yearbook, 1938.

INDEX